WITHDRAWN
WRIGHT STATE UNIVERSITY LIBRARIES

D1710787

Ovarian Cancer

State of the Art and Future Directions
in Translational Research

ADVANCES IN EXPERIMENTAL MEDICINE AND BIOLOGY

Editorial Board:

NATHAN BACK, *State University of New York at Buffalo*
IRUN R. COHEN, *The Weizmann Institute of Science*
ABEL LAJTHA, *N.S. Kline Institute for Psychiatric Research*
JOHN D. LAMBRIS, *University of Pennsylvania*
RODOLFO PAOLETTI, *University of Milan*

Recent Volumes in this Series

Volume 615
PROGRAMMED CELL DEATH IN CANCER PROGRESSION AND THERAPY
Edited by Roya Khosravi-Far, and Eileen White

Volume 616
TRANSGENIC MICROALGAE AS GREEN CELL FACTORIES
Edited by Rosa León, Aurora Gaván, and Emilio Fernández

Volume 617
HORMONAL CARCINOGENESIS V
Edited by Jonathan J. Li

Volume 618
HYPOXIA AND THE CIRCULATION
Edited by Robert H. Roach, Peter Hackett, and Peter D. Wagner

Volume 619
PROCEEDINGS OF THE INTERAGENCY, INTERNATIONAL SYMPOSIUM ON CYANOBACTERIAL HARMFUL ALGAL BLOOMS (ISOC-HAB)
Edited by H. Kenneth Hudnell

Volume 620
BIO-APPLICATIONS OF NANOPARTICLES
Edited by Warren C.W. Chan

Volume 621
AXON GROWTH AND GUIDANCE
Edited by Dominique Bagnard

Volume 622
OVARIAN CANCER
Edited by George Coukos, Andrew Berchuck, and Robert Ozols

A Continuation Order Plan is available for this series. A continuation order will bring delivery of each new volume immediately upon publication. Volumes are billed only upon actual shipment. For further information please contact the publisher.

George Coukos • Andrew Berchuck • Robert Ozols
Editors

Ovarian Cancer

State of the Art and Future Directions
in Translational Research

Editors
George Coukos
University of Pennsylvania Medical Center
Philadelphia, PA
USA
gcks@mail.med.upenn.edu

Robert Ozols
Fox Chase Cancer Center
Philadelphia, PA
USA
robert.ozols@fccc.edu

Andrew Berchuck
Duke University Medical Center
Durham, NC
USA
berch001@mc.duke.edu

ISBN: 978-0-387-68966-1   e-ISBN: 978-0-387-68969-2
DOI: 10.1007/978-0-387-68969-2

Library of Congress Control Number: 2008920270

© 2008 Springer Science+Business Media, LLC
All rights reserved. This work may not be translated or copied in whole or in part without the written permission of the publisher (Springer Science+Business Media, LLC, 233 Spring Street, New York, NY-10013, USA), except for brief excerpts in connection with reviews or scholarly analysis. Use in connection with any form of information storage and retrieval, electronic adaptation, computer software, or by similar or dissimilar methodology now known or hereafter developed is forbidden.
The use in this publication of trade names, trademarks, service marks, and similar terms, even if they are not identified as such, is not to be taken as an expression of opinion as to whether or not they are subject to proprietary rights.

Printed on acid-free paper

9 8 7 6 5 4 3 2 1

springer.com

Contents

Part I Ovarian Cancer Detection and Pathogenesis

Potential and Limitations in Early Diagnosis of Ovarian Cancer 3
Nicole Urban and Charles Drescher

SMRP and HE4 as Biomarkers for Ovarian Carcinoma When Used
Alone and in Combination with CA125 and/or Each Other 15
Ingegerd Hellstrom and Karl Erik Hellstrom

Classification of Ovarian Cancer: A Genomic Analysis 23
Michael P. Stany, Tomas Bonome, Fred Wamunyokoli, Kristen Zorn,
Laurent Ozbun, Dong-Choon Park, Ke Hao, Jeff Boyd, Anil K. Sood,
David M. Gershenson, Ross S. Berkowitz, Samuel C. Mok,
and Michael J. Birrer

Epigenetic Markers of Ovarian Cancer 35
Caroline A. Barton, Susan J. Clark, Neville F. Hacker,
and Philippa M. O'Brien

Role of Genetic Polymorphisms in Ovarian Cancer Susceptibility:
Development of an International Ovarian Cancer
Association Consortium 53
Andrew Berchuck, Joellen M. Schildkraut, C. Leigh Pearce,
Georgia Chenevix-Trench, and Paul D. Pharoah

MicroRNA in Human Cancer: One Step Forward in Diagnosis
and Treatment 69
Lin Zhang, Nuo Yang, and George Coukos

Ovarian Carcinogenesis: An Alternative Hypothesis 79
Jurgen M.J. Piek, Paul J. van Diest, and René H.M. Verheijen

BRCA1-Induced Ovarian Oncogenesis 89
Louis Dubeau

Role of p53 and Rb in Ovarian Cancer 99
David C. Corney, Andrea Flesken-Nikitin, Jinhyang Choi,
and Alexander Yu. Nikitin

Ovulatory Factor in Ovarian Carcinogenesis 119
William J. Murdoch

Part II Ovarian Cancer Therapeutics

Gynecologic Oncology Group (GOG-USA) Trials in Ovarian Cancer 131
Robert F. Ozols

Intraperitoneal Chemotherapy for Ovarian Cancer 145
Mark A. Morgan

Ovarian Cancer: Can We Reverse Drug Resistance? 153
David S.P. Tan, Joo Ern Ang, and Stan B. Kaye

**Syngeneic Mouse Model of Epithelial Ovarian Cancer:
Effects of Nanoparticulate Paclitaxel, Nanotax®** 169
Katherine F. Roby, Fenghui Niu, Roger A. Rajewski, Charles Decedue,
Bala Subramaniam, and Paul F. Terranova

**Individualized Molecular Medicine: Linking Functional
Proteomics to Select Therapeutics Targeting the PI3K
Pathway for Specific Patients** 183
Mandi M. Murph, Debra L. Smith, Bryan Hennessy, Yiling Lu,
Corwin Joy, Kevin R. Coombes, and Gordon B. Mills

Defective Apoptosis Underlies Chemoresistance in Ovarian Cancer 197
Karen M. Hajra, Lijun Tan, and J. Rebecca Liu

**Nanoparticle Delivery of Suicide DNA for Epithelial Ovarian
Cancer Therapy** 209
Janet A. Sawicki, Daniel G. Anderson, and Robert Langer

Biological Therapy with Oncolytic Herpesvirus 221
Fabian Benencia and George Coukos

Cancer Immunotherapy: Perspectives and Prospects 235
Sonia A. Perez and Michael Papamichail

Regulatory T Cells: A New Frontier in Cancer Immunotherapy 255
Brian G. Barnett, Jens Rüter, Ilona Kryczek, Michael J. Brumlik,
Pui Joan Cheng, Benjamin J. Daniel, George Coukos, Weiping Zou,
and Tyler J. Curiel

Inhibitory B7 Family Members in Human Ovarian Carcinoma 261
Shuang Wei, Tyler Curiel, George Coukos, Rebecca Liu,
and Weiping Zou

Role of Vascular Leukocytes in Ovarian Cancer Neovascularization 273
Klara Balint, Jose R. Conejo-Garcia, Ron Buckanovich, and George Coukos

**Heparin-Binding Epidermal Growth Factor-Like Growth Factor
as a New Target Molecule for Cancer Therapy** 281
Shingo Miyamoto, Hiroshi Yagi, Fusanori Yotsumoto,
Tatsuhiko Kawarabayashi, and Eisuke Mekada

Index 297

Contributors

Joo Ern Ang
Section of Medicine, Institute of Cancer Research, Royal Marsden Hospital,
Sutton, Surrey, UK

Caroline Barton
Ovarian Cancer Center, Garvan Institute, 384 Victoria Street Darlinghurst Sydney
NSW 2010 Austrailia, c.barton@garvan.org.au

Fabian Benencia, PhD
Department of Biomedical Sciences, Ohio University, Athens, Ohio 45701,
benencia@oucom.ohiou.edu

Andrew Berchuck, MD
Director, Division of Gynecologic Oncology, Barbara Thomason Ovarian
Cancer Professor, Duke University Medical Center, Box 3079,
Durham, NC 27710

David C. Corney, BS
Department of Biomedical Sciences, Cornell University, Ithaca, New York 14853
Andrea Flesken-Nikitin, BS, Department of Biomedical Sciences, Cornell
University, Ithaca, New York 14853

George Coukos, MD, PhD
Center for Research on the Early Detection and Cure of Ovarian Cancer,
University of Pennsylvania, Philadelphia, PA 19104, gcks@mail.med.upenn.edu

Jinhyang Choi, DVM, MS
Department of Biomedical Sciences, Cornell University, Ithaca, New York 14853

Georgia Chenevix-Trench, PhD
NHMRC Senior Principal Research Fellow, The Queensland Institute of
Medical Research, 300 Herston Road, Herston, QLD 4006, Australia

Tyler J. Curiel, MD, MPH
Department of Internal Medicine, Tulane University School of Medicine,
New Orleans, Louisiana 70112-2699, tcuriel@tulane.edu

Charles Decedue
Higuchi Biosciences Center, University of Kansas, Lawrence, KS, USA,
decedue@ku.edu

Ingegerd Hellstrom, MD, PhD
Department of Pathology, Harborview Medical Center, University of
Washington, Seattle, WA 98104, ihellstr@u.washington.edu

Karl Erik Hellstrom, MD, PhD
Department of Pathology, Harborview Medical Center, University of
Washington, Seattle, WA 98104

Stan B. Kaye
Section of Medicine, Institute of Cancer Research, Royal Marsden
Hospital, Sutton, Surrey, UK

Alexander Yu. Nikitin, MD, PhD
Department of Biomedical Sciences, Cornell University, Ithaca, New York 14853

Fenghui Niu
Higuchi Biosciences Center, University of Kansas, Lawrence, KS, USA,
niu@ku.edu

William Murdoch, PhD
Department of Animal Science, University of Wyoming, Laramie, WY 82071,
wmurdoch@uwyo.edu

C. Leigh Pearce, PhD
Assistant Professor, Department of Preventive Medicine,
University of Southern California, Los Angeles, CA

Jurgen M.J. Piek
Department of Obstetrics and Gynaecology, VU University Medical Center,
Amsterdam, The Netherlands

Michael Papamichail, MD, PhD
Cancer Immunology and Immunotherapy Center, Saint Savas Cancer Hospital,
Athens, Greece

Paul Pharoah PhD, FRCP
Cancer Research UK Senior Clinical Research Fellow, Dept of Oncology,
University of Cambridge, Cambridge, United Kingdom

Sonia A. Perez, PhD
Cancer Immunology and Immunotherapy Center, Saint Savas Cancer Hospital,
Athens, Greece

Katherine F. Roby
Department of Anatomy & Cell Biology, University of Kansas Medical Center,
Kansas City, KS, USA, kroby@kumc.edu

Roger A. Rajewski
Higuchi Biosciences Center, University of Kansas, Lawrence, KS, USA
rajewski@ku.edu

Bala Subramaniam
Department of Chemical and Petroleum Engineering, University of Kansas
Lawrence, KS, USA, bsubramaniam@ku.edu

Joellen M. Schildkraut, PhD
Professor, Duke University Medical Center, Division of Prevention Research, Department of Community and Family Medicine, and The Duke Comprehensive Cancer Center, Box 2949, Durham, NC 27710

David S.P. Tan
Section of Medicine, Institute of Cancer Research, Royal Marsden Hospital, Sutton, Surrey, UK

Paul F. Terranova
Department of Molecular & Integrative Physiology, University of Kansas Medical Center, Kansas City, KS, USA, pterrano@kumc.edu

Paul J. van Diest
Department of Pathology, University Medical Center Utrecht, Utrecht, The Netherlands

René H.M. Verheijen
Department of Obstetrics and Gynaecology, University Medical Center Utrecht, Utrecht, The Netherlands

Part I
Ovarian Cancer Detection and Pathogenesis

Potential and Limitations in Early Diagnosis of Ovarian Cancer

Nicole Urban and Charles Drescher

1 Ovarian Cancer Screening May Reduce Mortality in the Future but Many Challenges Remain

Five-year survival rates for invasive epithelial ovarian cancer have changed little in recent decades, remaining constant at about 30% when cancer has spread outside the ovaries, and about 90% when disease is confined to the ovaries. Ten-year survival for ovarian carcinoma varies greatly according to the stage at diagnosis (1) and survival is best when cancer is confined to the ovary at the time of diagnosis (Fig. 1); even patients with high-grade serous tumors do well if they are diagnosed while the tumors are confined to the ovary (Fig. 2).

The goal of screening is to reduce mortality by detecting cancer early. The potential reduction in mortality is great, because currently fewer than 25% of cases are confined to the ovary at diagnosis. Interest in diagnostic markers that can be measured in blood products is particularly high, as several promising marker panels have been reported in the last decade (2, 3). However, using these markers to detect ovarian cancer early enough to reduce mortality remains challenging because screening needs to identify cancer before symptoms occur, early enough that the disease is still curable. It is well established that the best screening tests detect cancer *before* it becomes invasive, by identifying precursor lesions and enabling prevention of invasive cancer through early intervention.

In considering the challenges inherent in ovarian cancer screening, it is helpful to distinguish among diagnostic, early detection, and risk markers. Figure 3 depicts the behavior of three hypothetical markers as cancer progresses through a precursor lesion stage, an early invasive stage, metastasis, and death. Markers A, B, and C are equally elevated at the time of diagnosis, but they are not equally good early detection markers because their behavior prior to diagnosis varies. Marker A performs well as a diagnostic marker because it is highly elevated in women with cancer who present clinically with symptoms, but it does not provide signal until the disease is well advanced. Marker B is a better early detection marker because it elevates while the disease is still potentially curable, signaling preinvasive as well as invasive disease. Marker C elevates even earlier; hence, it might be useful as a risk marker to predict disease in the future especially if precursor conditions are unknown or

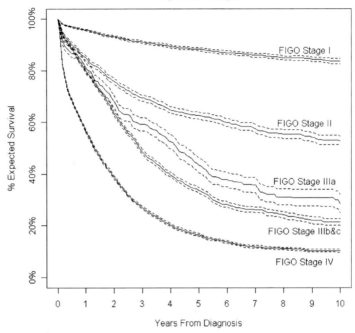

Fig. 1 Ten-year survival for ovarian cancer varies greatly according to FIGO stage at diagnosis, only when the cancer is confined to the ovary is long-term survival above 80%

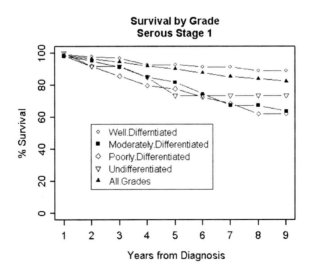

Fig. 2 Ten-year survival is over 60% when the cancer is confined to the ovary at the time of diagnosis even for serous ovarian cancers that are poorly differentiated

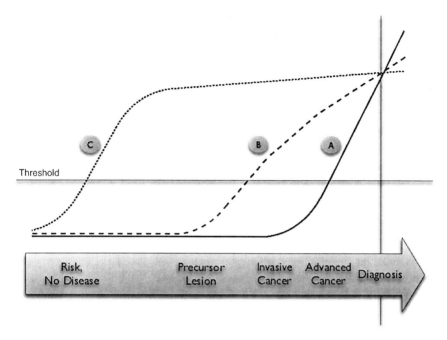

Fig. 3 Conceptual framework for determining the clinical utility of a serum marker. The signal provided by a screening test prior to symptoms and clinical diagnosis determines its utility as a diagnostic (a), early detection (b), or risk (c) marker. Reproduced from (4) with permission from Future Medicine Ltd

undetectable. Screening for elevated risk can reduce disease incidence if preventive treatment is available; for example, screening for and treating high cholesterol/triglycerides and high blood pressure effectively reduces the adverse events associated with cardiac disease. A similar use of screening for ovarian cancer risk markers may be important to explore because of the many challenges to early detection of curable invasive lesions.

2 Good Early Detection Serum Marker Candidates Complement CA125 and Show Stability Over Time

The potential for reducing ovarian cancer mortality through earlier diagnosis and treatment is great, but available screening approaches such as CA125 and transvaginal sonography (TVS) often fail to detect early, asymptomatic disease; in addition they can lead to unnecessary surgery. The hope for early detection remains high, however, because emerging technologies are facilitating identification of novel markers that complement CA125. Many serum biomarkers have been identified for ovarian cancer, including CA125 (5), prolactin (6), mesothelin (7), HK11 (8), osteopontin (9), HE4 (10), B7-H4 (11), and SPINT2 (12).

To date, only CA125 has been shown to detect ovarian cancer prior to symptoms. CA125 above 30 U mL^{-1} was used to select postmenopausal women for ultrasound screening in a pilot trial in the UK. Prevalence screening (22,000 women) yielded sensitivity of 85% and 58% at 1-year and 2-year follow-up, respectively, and specificity of 99.6%. Results of the 2-arm RCT (11,000 per arm) suggest that survival was better in the screened group (72.9 vs. 41.8) and that the positive predictive value was acceptable at 20% (13). For a disease as rare as ovarian cancer, specificity of 99.6% is needed in a screening test with 80% sensitivity to achieve a positive predictive value (PPV) of 10%. A high PPV is important because definitive diagnosis requires major abdominal surgery. Results of the pilot trial in the UK suggest that use of a marker panel including CA125 to select women for TVS and/or surgery may be a cost-effective screening strategy. These results are consistent with predictions of a microsimulation model of ovarian cancer screening (14) that uses three interrelated components to estimate screening outcomes. Assumptions were made regarding the natural history of ovarian cancer in the absence of diagnosis and treatment; disease detection as a function of characteristics of the woman, her cancer, and detection modalities used; and survival as a function of age of the woman and the stage of her disease at the time of diagnosis. The model predicted that using *rising* CA125 to select women for TVS is a cost-effective approach to screening, and that frequent screening may be needed to realize benefits if the disease progresses quickly from a curable to an incurable condition.

On the basis of these and other observations, statistical methods have been developed for using marker history to improve screening performance. Methods such as the Risk of Ovarian Cancer algorithm (ROCA) (15) or the Parametric Empirical Bayes (PEB) decision rule (16) are particularly useful when marker levels rise (or fall) as cancer develops relative to an individual woman's usual marker levels. As illustrated in Fig. 4, a marker's levels in the absence of cancer may vary more among women than within an individual woman over time, rising (or falling) significantly relative to a woman's usual level only in the presence of cancer. This is characteristic of many potentially useful markers including CA125.

Testing for change over time in a marker can improve sensitivity without loss in specificity. In the PEB approach (16), at each screen, a woman's serum is tested for deviance from her own normal value of the marker. The threshold for positivity can be set such that a targeted percent of women are referred for further work-up at each screen, so that sensitivity is maximized within desired specificity. Women's characteristics such as age are accounted for using the PEB rule. The risk-of-ovarian cancer algorithm (ROCA) is similar but tests specifically for exponential rise in CA125 using call-backs for repeat testing (15).

Markers that are specific to malignancy are needed to avoid identification of benign ovarian conditions that are much more common than ovarian cancer. Several such markers are under evaluation. For example, the human epididymis protein 4 (HE4/WFDC2) (17) has been studied independently by several institutions and is

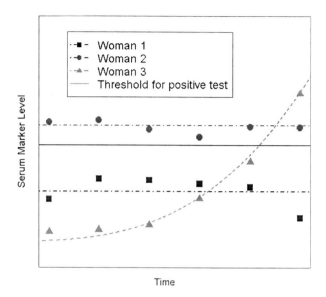

Fig. 4 Conceptual framework for developing decision rules for markers such as CA125. When marker levels vary less over time within an individual woman than among women in a screening population, change over time in a marker can signal cancer earlier than a single threshold rule

found to be promising as a marker for ovarian carcinoma (10). It is a secreted glycoprotein that is overexpressed by serous and endometrioid ovarian carcinomas (18). It is one of the several genes showing *in silico* chromosomal clustering and displaying altered expression patterns in ovarian cancer (19). Evaluation in serum suggests that HE4 is as sensitive as CA125 and more specific in that it detects fewer benign tumors; it is also stable over time in healthy women (10).

Similarly, mesothelin (MSLN) has been shown to be a soluble protein present in serum, and is potentially useful in a diagnostic panel including CA125. Mouse monoclonal antibodies were used in a sandwich ELISA to measure MSLN in serum (7). Receiver operating characteristics (ROC) curve analysis was used to evaluate the value added of MSLN to a composite marker including CA125, using 53 cases and 220 controls (20). Logistic regression was used to define a composite marker including CA125 and MSLN. The composite marker is a linear combination of the markers in the panel. Marker levels were converted to logs and standardized. Logistic regression was used to estimate the weights for each marker (21), controlling for menopausal status: CM = 1.4 × CA 125 + 1.0 × MSLN. The CM can be analyzed as if it were a single marker in ROC curves (Fig. 5) and in longitudinal algorithms such as the PEB for use in screening.

HE4, MSLN, CA125, and 15 other candidate markers were further evaluated in 200 blinded serum specimens from ovarian cancer cases and healthy women, including 41 healthy controls from a screening study (20 contributed blood two times one year apart), 47 otherwise healthy women undergoing pelvic surgery without tubal/

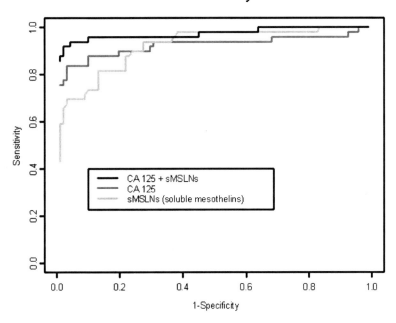

Fig. 5 Addition of MSLN to a panel that includes CA125 improves detection of ovarian cancer

ovarian pathology, 24 surgical controls with benign ovarian conditions and 68 cases including 11 stage 1, 5 stage 2, 39 stage 3, 11 stage 4, and 2 unstaged ovarian cancers. All epithelial cancer histologies were represented, including 34 serous, 7 endometrioid, 3 mucinous, 2 clear cell, 17 other and 5 undifferentiated. Several good marker candidates were identified, including 8 markers with sensitivity > 50% at 80% specificity. Three had sensitivity >50% at 85% specificity, two had sensitivity >50% at 90% specificity, and one had sensitivity >50% at 99% specificity as well as sensitivity >75% at 95% specificity.

Markers that performed well as individual diagnostic markers were further evaluated in clinical samples for their contribution to a panel including CA125, at Fred Hutchinson Cancer Research Center (FHCRC) in Seattle; some of these and other novel markers were evaluated similarly at Dana Farber Cancer Center (DFCC) in Boston. At both institutions, clinical samples were obtained from women with ovarian cancer at the time of diagnosis, prior to any treatment including surgery. Markers that showed univariate sensitivity of at least 30% at 95% specificity included seven markers from the FHCRC panel and five markers from the DFCC panel. Six markers improved the sensitivity of CA125 at 95% specificity. Eight markers showed correlation >0.5 for samples taken 1 year, apart from the same woman, suggesting stability over time within women. The screening performance of these markers can be improved using a longitudinal screening algorithm such as the PEB (22). Research to identify the markers that provide signal early, when the disease may still be curable, is currently underway but results are not yet available.

3 The Best Candidates for Use in an Early Detection Panel Provide Signal Prior to Symptoms, Early in the Disease Process

Estimates of the lead times of candidate markers are needed to accurately predict the markers' contribution to an effective early detection panel for use in a screening program. Promising serum markers have been evaluated independently in clinical samples at several institutions, but their lead times remain unknown. To address this need, a 2-year validation study has been initiated by the National Cancer Institute (NCI) in the US to evaluate candidate markers, using the serum repository of the NCI Prostate, Lung, Colorectal, and Ovarian Cancer (PLCO) screening trial, a repository that contains serial preclinical samples from over 100 women who have been diagnosed with ovarian cancer as well as serial samples from healthy comparable controls. Preclinical samples are needed to estimate the markers' lead times and the probability that cancer will be diagnosed within an arbitrary period such as 2 years, as a function of marker levels and change. The PLCO trial is a large, multicenter randomized controlled screening trial that includes collection and storage of 6 serum samples collected one year apart from 37,000 healthy women randomized to the screened arm, as well as 10 years of follow up for cancer diagnosis for all 74,000 women participating in the trial (Table 1).

It has long been recognized that collaboration is needed to identify and validate the best diagnostic and early detection panels, as promising results from single-institution studies have seldom been reproducible. The NCI collaboration is a Phase II/III Validation Study (23) of a Consensus Panel of Early Detection Serum Markers led by Dan Cramer (DFCC) and Nicole Urban (FHCRC). A 2-year study began in

Table 1 The Prostate, Lung, Colon, and Ovary (PLCO) trial provides serial serum samples from healthy post-menopausal women with follow up for cancer. Reproduced from (4) with permission from Future Medicine Ltd

Centers collaborating	10
Arms	2
Study population	Women aged 55–74
Endpoint	Cause-specific mortality
Size	74,000 total (37,000 in each of two arms)
Power	88% for 35% mortality reduction (1-sided test)
Enrollment period	3 years
Duration of screening	4 screens 1 year apart[a]
Duration of follow-up	Minimum of 10 years postrandomization
Screening protocol	Annual TVS, CA-125, bimanual pelvic exam

[a] The PLCO design was revised to continue screening for an additional 2 years using only CA125

August 2005, which includes investigators from five Ovarian Cancer Specialized Program of Research Excellence (SPORE) sites (DFCC, FHCRC, Fox Chase Cancer Center, MD Anderson Cancer Center, and University of Alabama, Birmingham), three Early Detection Research Network (EDRN) sites (FHCRC, DFCC, U Pittsburgh), and the PLCO trial at NCI. In the first year, a new set of Phase II (clinical) specimens will be used to evaluate the most promising diagnostic markers including HE4 and MSLN as well as an expanded panel of markers measured by bead-based assays. In the second year, the best diagnostic markers will be evaluated in PLCO (preclinical) specimens to predict their utility as early detection or risk markers. Using data from analysis of PLCO preclinical blood samples, diagnosis of ovarian cancer within 2 years (or another arbitrary period) of a blood draw can be predicted. Because the women who contributed blood samples were all participating in screening, cancer could have been detected by CA125, TVS or symptoms, or symptoms. Blood samples were not collected from women allocated to the control group of the PLCO trial.

The NCI collaborative study will test the hypothesis that a panel of biomarkers will have better performance characteristics than any single marker, and yield a longer lead time than CA125 alone. Over 20 putative biomarkers have been evaluated by SPORE and EDRN investigators using bead-based (Luminex®) assays as well as standard ELISA. In the first year, candidate biomarkers will be evaluated in a new set of 160 cases (80 early-stage and 80 late-stage), 160 surgical controls, 480 general population controls, and serial samples collected 1 year apart in 40 healthy controls. Samples will be provided by five ovarian cancer SPORE institutions for blinded measurement of assays at three laboratory sites: DFCC, FHCRC, and U Pittsburgh. A consensus panel will be identified including the biomarkers that are most informative on their own or most complementary when used together, within specimen volume constraints. For as many markers in the consensus panel as possible, bead-based assays will be developed and evaluated for their reproducibility, validity, and performance relative to standard ELISA. Bead-based assays, multiplexed if possible, will be used in PLCO specimens to preserve PLCO specimen volume for future studies.

In the second year, PLCO preclinical samples from approximately 100 cases and 1,000 matched controls will be used to estimate the lead time of each individual marker and establish the best marker combination. Markers that show elevation within a year prior to diagnosis will be evaluated using the entire preclinical history to estimate the lead time for each marker and the marker panel. A small amount of serum from prediagnostic specimens from the PLCO cases and controls will be made available for the study. Some will be allocated for testing bead-based (Luminex® platform) assays, and the remainder will be used for high-priority markers that can be measured only by standard ELISA. Any remaining sera from false positive and false negative cases will be used to discover additional biomarkers that complement the existing panel, using novel high throughput proteomic discovery platforms.

A research challenge is that specimen quantities are limited in the stored samples from the PLCO trial. Accordingly, to the extent possible, bead-based assays will be used to measure candidate markers in the PLCO samples to minimize specimen requirements. In preparation for this and other validation studies using preclinical specimens, bead-based assays have been developed for top marker candidates, including CA125 and HE4. For CA125, four commercially available monoclonal antibody pairs were tested on a bead-based platform to select the best pair with respect to assay feasibility (affinity) and accuracy in assessing known antigen concentrations. Two CA125 bead-based assays were optimized and evaluated in serum samples using the two best pairs of CA125 antibodies, and one HE4 assay was similarly optimized using the only available monoclonal antibody pair. These three bead-based assays were then measured blinded in a triage set of 64 cases, 55 screening controls, and 70 surgical controls, most of which had been previously characterized for CA125 and HE4, using ELISA. Each bead-based assay was evaluated for reproducibility, validity, and screening performance (24).

The best CA125 bead-based assay uses antibodies from RDI, with a correlation between replicates of 0.99 overall and 0.83 in screening controls. Its correlation with CA125II is 0.95 overall and 0.64 in screening controls. The HE4 bead-based assay showed correlation between replicates of 0.95 overall and 0.86 in healthy controls, and its correlation with ELISA was 0.95 overall and 0.86 in screening controls. A composite marker (CM) was constructed for CA125 and HE4, defined as a linear combination of the HE4 and CA125 (RDI antibody pair) bead-based assays. Using published methods (21), marker levels were converted to logs and standardized, and logistic regression was used to estimate the weights for each marker: $CM = 0.56 \times CA125 + 1.20 \times HE4$. Its diagnostic performance was measured by the area under the curve (AUC) for the ROC curve estimated using the triage set described earlier. Performance for the CM using bead-based assays for cases vs. all controls (AUC = 0.91) was better than that of the CA125II RIA assay used alone (AUC = 0.87), the bead-based CA125 assay used alone (AUC = 0.85), or the HE4 bead-based assay used alone (AUC = 0.89) (24). Interassay CVs for the bead-based assays were found high by commercial ELISA standards but have been recently improved by normalizing across plates.

These analyses suggest that bead-based assays for HE4 and CA125 combine to form a panel that performs better than either marker used alone, particularly at the very high specificities needed in screening programs. Multiplexed bead-based assays may reduce specimen requirements even further. The availability of assays that require 15 µL or less of serum, such as those described earlier, may make it possible to explore the behavior of candidate markers in stored samples from the Women's Health Initiative (WHI) (25) as well as those from the PLCO. The WHI population is restricted to women aged 50–79 at entry and represents the average-risk, postmenopausal population from which the majority of ovarian cancers arise. A total of 68,000 women were randomized in the clinical trial (CT) and 93,000 women were enrolled in the observational study (OS). The women provided self-reported demographics, reproductive, medical, and family history, and lifestyle data as well as blood samples at baseline and either 1 year (CT) or 3 years (OS) later. Table 2 reports the number of women for whom samples are currently available for biomarker validation from the OS, reported by months elapsed between the blood draw and the cancer diagnosis.

Table 2 Women's Health Initiative (WHI) samples are appropriate for estimating lead time and best clinical use of each candidate marker: Samples are available for over 250 women of whom 70 provide two samples 3 years apart, and the second sample obtained within 2 years of the cancer diagnosis

Months from draw to diagnosis	Number of cases after baseline blood draw ($n = 250$)	Number of cases after 3-year blood draw ($n = 100$)
0–6	11	20
6–12	19	14
12–18	25	16
18–24	19	20
24–30	17	5
30–36	20	7
36–42	24	7
42–48	26	4
48+	88	7

Note: This table contains information on incident ovarian cancer cases in the OS through August 2003 currently available for biomarker validation work

Although they are less well-suited to describing marker behavior over the preclinical phase of the disease in individual cases, preclinical samples from the WHI have several advantages over the PLCO samples. First, the WHI is larger: As of August 2006, 374 and 243 cases of ovarian cancer have been diagnosed in the observational study (OS) and the clinical trial (CT) components of the WHI, respectively, providing samples that could potentially be used both to develop and to validate a screening decision rule. Second, samples from the WHI allow unbiased estimation of markers' lead time relative to clinical detection and diagnosis, whereas in the PLCO many of the cases were detected by screening using CA125 or TVS or both. Third, because some of the blood samples were obtained many years prior to diagnosis, and follow up of over 10 years has been completed, the relative risk of a cancer diagnosis within 5 or 10 years can be estimated from data generated by the WHI and the behavior of each marker as cancer develops (Fig. 3) can be determined.

The availability of preclinical samples from the PLCO and WHI trials will greatly improve our understanding of the behavior of candidate markers during the preclinical phase of epithelial ovarian cancer. However, analysis of these samples cannot reveal the presence or absence of invasive disease at the times prior to diagnosis when the preclinical serum samples were obtained. Accordingly, these samples may be most useful for estimating the relative risk of subsequent ovarian cancer diagnosis on the basis of marker levels or changes in marker levels. Knowledge of the presence or absence of disease at the time a marker first provides signal requires a prospective screening study in which surgical intervention is triggered by the marker. Until such a study is initiated and completed, it may be useful to invoke a different screening paradigm focusing on markers that predict, rather than detect, disease (4). Particularly useful would be markers that could predict

5-year or 10-year risk that would provide indications for risk-reducing surgery. Women identified as high risk would not be expected to have invasive disease at the time of surgery, but some might have premalignant changes that could be confirmed using markers detectable in ovarian or tubal tissue.

4 Significant Progress Can be Expected in the Future

We can reduce ovarian cancer mortality through screening if (1) cancer detected early can be cured, (2) biomarkers in the blood can signal early cancer, (3) available technology can identify biomarkers, (4) appropriate research can be conducted to demonstrate screening efficacy, and (5) biomarkers can be used cost-effectively for cancer screening. In the last decade, methods have been developed for discovering and prioritizing candidate markers, predicting the cost-effectiveness of alternative screening strategies, combining markers for use in a panel, using marker history in a longitudinal decision rule for early detection, and evaluating specimen-efficient bead-based assays for use in validation research. Several candidate markers have been identified that perform well as diagnostic markers, and studies are underway to evaluate their potential as early detection markers. It is likely that additional markers will be needed to detect ovarian cancer early enough to reduce mortality through screening, including risk markers that detect precursor lesions or signal developing disease several years before it becomes invasive and potentially incurable. New proteomics technologies that make discovery in serum possible are likely to revolutionize the field in the near future.

Acknowledgements A team of investigators contributed to this work including *Martin McIntosh, Garnet Anderson, Nathalie Scholler, Michel Schummer, Beth Karlan,* and *Dan Cramer*. We are indebted to *Kathy O'Briant* for her able management of the ovarian cancer specimen repository at FHCRC. We are grateful for generous funding from the Pacific Ovarian Cancer Research Consortium (POCRC)/SPORE in Ovarian Cancer (P50 CA83636, N.U.), the Department of Defense/CDMRP (DAMD17-02-1-0691, N.U.), and the Canary Foundation.

References

1. Ries, L. A. G., D. Harkins, et al. (2006). SEER Cancer Statistics Review, 1975–2003. Bethesda, MD: National Cancer Institute.
2. Petricoin, III, E. F., A. M. Ardekani, et al. (2002). "Use of proteomic patterns in serum to identify ovarian cancer." *Lancet* **359**(9306): 572–577.
3. Mor, G. I., Visintin, et al. (2005). "Serum protein markers for early detection of ovarian cancer." *Proc Natl Acad Sci USA* **102**(21): 7677–7682.
4. Urban, N. and C. Drescher (2006). "Current and future developments in screening for ovarian cancer." *Women's Health* **2**(5): 733–742.
5. Bast, R. C., Jr., T. L. Klug, et al. (1984). "Monitoring human ovarian carcinoma with a combination of CA 125, CA 19–9, and carcinoembryonic antigen." *Am J Obstet Gynecol* **149**(5): 553–559.

6. Jha, P., A. Farooq, et al. (1991). "Use of serum prolactin for monitoring the therapeutic response in ovarian malignancy." *Int J Gynaecol Obstet* **36**(1): 33–38.
7. Scholler, N., N. Fu, et al. (1999). "Soluble member(s) of the mesothelin/megakaryocyte potentiating factor family are detectable in sera from patients with ovarian carcinoma." *Proc Natl Acad Sci USA* **96**(20): 11531–11536.
8. Diamandis, E. P., A. Okui, et al. (2002). "Human kallikrein 11: a new biomarker of prostate and ovarian carcinoma." *Cancer Res* **62**(1): 295–300.
9. Kim, J. H., S. J. Skates, et al. (2002). "Osteopontin as a potential diagnostic biomarker for ovarian cancer." *JAMA* **287**(13): 1671–1679.
10. Hellstrom, I., J. Raycraft, et al. (2003). "The HE4 (WFDC2) protein is a biomarker for ovarian carcinoma." *Cancer Res* **63**(13): 3695–3700.
11. Salceda, S., T. Tang, et al. (2005). "The immunomodulatory protein B7-H4 is overexpressed in breast and ovarian cancers and promotes epithelial cell transformation." *Exp Cell Res* **306**(1): 128–141.
12. Matsuzaki, H., H. Kobayashi, et al. (2005). "Plasma bikunin as a favorable prognostic factor in ovarian cancer." *J Clin Oncol* **23**(7): 1463–1472.
13. Jacobs, I. J., S. J. Skates, et al. (1999). "Screening for ovarian cancer: a pilot randomized controlled trial." *Lancet* **353**(9160): 1207–1210.
14. Urban, N., C. Drescher, et al. (1997). "Use of a stochastic simulation model to identify an efficient protocol for ovarian cancer screening." *Contr Clin Trials* **18**(3): 251–270.
15. Skates, S. J., U. Menon, et al. (2003). "Calculation of the risk of ovarian cancer from serial CA-125 values for preclinical detection in postmenopausal women." *J Clin Oncol* **21**(90100): 206s–210s.
16. McIntosh, M. W. and N. Urban (2003). "A parametric empirical Bayes method for cancer screening using longitudinal observations of a biomarker." *Biostatistics* **4**(1): 27–40.
17. Schummer, M., W. V. Ng, et al. (1999). "Comparative hybridization of an array of 21,500 ovarian cDNAs for the discovery of genes overexpressed in ovarian carcinomas." *Gene* **238**(2): 375–385.
18. Drapkin, R., H. H. von Horsten, et al. (2005). "Human epididymis protein 4 (HE4) is a secreted glycoprotein that is overexpressed by serous and endometrioid ovarian carcinomas." *Cancer Res* **65**(6): 2162–2169.
19. Israeli, O., A. Goldring-Aviram, et al. (2005). "In silico chromosomal clustering of genes displaying altered expression patterns in ovarian cancer." *Cancer Genet Cytogenet* **160**(1): 35–42.
20. McIntosh, M., C. Drescher, et al. (2004). "Mesothelin/MPF antigen(s) as a diagnostic and screening marker for ovarian carcinoma when used alone or combined with CA-125." *Gynecol Oncol* **95**: 9–15.
21. McIntosh, M. W. and M. S. Pepe (2002). "Combining several screening tests: optimality of the risk score." *Biometrics* **58**(3): 657–664.
22. McIntosh, M. W., N. Urban, et al. (2002). "Generating longitudinal screening algorithms using novel biomarkers for disease." *Cancer Epidemiol Biomarkers Prev* **11**(2): 159–166.
23. Pepe, M. S., R. Etzioni, et al. (2001). "Phases of biomarker development for early detection of cancer." *J Natl Cancer Inst* **93**(14): 1054–1061.
24. Scholler, N., M. Crawford, et al. (2006). "Bead-based ELISA for validation of ovarian cancer early detection markers." *Clin Cancer Res* **12**(7): 2117–2124.
25. (2006). Women's Health Initiative Scientific Resources Website, available at: www.whiscience.org.

SMRP and HE4 as Biomarkers for Ovarian Carcinoma When Used Alone and in Combination with CA125 and/or Each Other

Ingegerd Hellstrom and Karl Erik Hellstrom

1 There is a Need for Biomarkers to Detect Ovarian Carcinoma by Assaying Serum and/or Other Body Fluids

Assays measuring tumor antigens in serum have the advantage that they are noninvasive, quick, and relatively inexpensive. Early detection as well as monitoring of disease in treated patients requires high specificity and sensitivity and constant levels of circulating marker unless there is a change in the patient's clinical status.

CA125 is the present "gold standard" for diagnosis of ovarian carcinoma using serum samples (1–4). However, it is elevated in several nonmalignant conditions, which can lead to false-positive results (5). There is a need for additional markers to improve sensitivity with retained or better specificity, and many new biomarkers have been introduced and continue to be evaluated. Our group has focused on soluble mesothelin-related proteins (SMRP) and on HE4, a protease that is secreted into serum. In immunohistological studies of ovarian cancer samples with little or no detectable CA125 expression, mesothelin and HE4 stood out as the most promising markers, when reactivity with normal tissues was taken into account (6). Other biomarkers in this study included HK4, HK6, OPN, claudin 3, DF3, VEGF, MUC1, and CA19-9.

2 SMRP as Marker for Diagnostic Assays of Serum and Urine

With the goal to obtain monoclonal antibodies (MAbs) for therapy, our group immunized mice with human ovarian carcinoma cells in the mid-1990s. This work resulted in MAb569, which reacts with ovarian carcinomas and has low reactivity with normal tissues except for the mesothelium. N-terminal amino acid sequencing of the antigen recognized by MAb 569 showed identity with the sequence of mesothelin, a tumor marker first described by Pastan's group (7), except for the lack of a 24 bp insert. By following our standard procedures for characterizing antigens detected by MAbs (8), we found the MAb569-defined antigen in supernatants of antigen-positive tumor cells and subsequently in malignant effusions, suggesting that it may be a marker for serum-based diagnosis. This finding was surprising because studies by Pastan's group had indicated that mesothelin is stably expressed at the cell surface and not released in to tumor culture supernatants or body fluids from cancer patients (9).

To develop a double determinant (sandwich) immunoassay, as our group had done for other tumor antigens in the past (10), additional MAbs were generated by immunizing mice with Mab569 immunoaffinity-purified antigen and applied MAbs to two different epitopes to construct a "sandwich" ELISA specific for mesothelin (11). In the initial study, a SMRP variant with an 82 bp insert was also detected. In view of Pastan's finding indicating that mesothelin is not soluble, we speculated that this variant is the molecule that is measured with the original ELISA (11).

A "blinded" study was performed in collaboration with B. Robinson's group from the University of Western Australia in Perth. We demonstrated the value of our SMRP-specific ELISA for the diagnosis of patients with mesothelioma (12). Eight of 40 individuals who had been exposed to asbestos but were clinically cancer free had increased levels of circulating SMRP. Importantly, three of those individuals subsequently developed mesothelioma within 15, 26, and 69 months, dying after 3, 6, and 6 years, respectively, and one developed lung carcinoma. In contrast, none of the 32 subjects with normal SMRP levels got mesothelioma or lung cancer within 6 years of follow up, suggesting a potential predictive value of the assay.

Another "blinded" study was performed together with Dr. N. Urban and her colleagues at the Fred Hutchinson Cancer Research Center in Seattle. It showed that SMRP has similar sensitivity and specificity as CA125 for diagnosis of ovarian carcinoma and that a combination of CA125 with SMRP has higher sensitivity than either assay alone. Like CA125, SMRP has temporal stability, suggesting that repeated studies on the same high risk subjects may facilitate earlier diagnosis (13).

A third study was performed in collaboration with Dr. N. Sardesai and his colleagues at Fujirebio Diagnostics, Inc. It indicated that SMRP is released into urine of patients with ovarian carcinoma and that the measurements of SMRP in urine, using the original ELISA, offer promise for detection of ovarian carcinoma. If confirmed by ongoing studies, the ease by which urine can be obtained would facilitate frequent studies on subjects that have high genetic risks of developing ovarian cancer.

As illustrated in Fig. 1, three mesothelin variants have been identified (14): one without inserts (variant 1), one with a 24 bp insert (variant 2), and one with an 82 bp insert (variant 3). To explore which variants are released into the circulation from ovarian carcinoma cells, we created recombinant fusion proteins of the three variants, immunized the mice with them, and obtained specific MAbs. Flow cytometry on live cells was performed with MAbs to the different mesothelin variants and showed that a MAb to variant 1 identifies as many tumors as a MAb to all three variants, while variants 2 and 3 are expressed infrequently (15). The published ELISA (11) was found to recognize variants 1 and 3 and has much higher sensitivity (68% vs. 15%) and specificity than a newly constructed ELISA specific for variant 3 (15). SMRP released into ascites from a patient with ovarian carcinoma was shown to have a molecular weight of approximately 40 kDa.

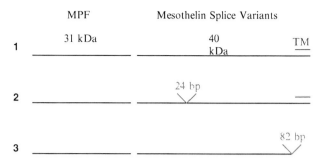

Fig. 1 Mesothelin variants

According to amino acid sequencing, variants 1 and 2 were found in the ascites, and variant 3 could not be excluded (15). A standard curve was constructed to measure SMRP with a limit of detection of 200 pg/ml, and an assay for clinical use is marketed by Fujirebio Inc, in Europe and Australia, for monitoring mesothelioma patients. Today, there is an agreement between Pastan's and our group that mesothelin is released from antigen-positive tumor cells as a useful diagnostic marker for serum assays (16).

3 Autoantibodies to Mesothelin

Autoantibodies to tumor-associated antigens have been detected in many cancer patients (17–22) and are sometimes found to correlate with the clinical state. We have started to investigate whether patients with ovarian cancer form antibodies to mesothelin, whether such antibodies can also be found in healthy subjects, and whether the presence of anti-mesothelin antibodies provides clinically useful information. Native mesothelin was purified by Mab569 immunoaffinity chromatography from urine of patients with ovarian cancer and used to coat ELISA plates. Sera were added from patients with ovarian cancer at various stages, as well as from control donors, and bound autoantibodies were detected with anti-human IgG antibody as a probe. Antibodies were detected in a fraction of sera from both patients with ovarian cancer (Fig. 2) and healthy women (data not shown). Experiments are ongoing to find out whether the presence of these antibodies provides information on diagnosis or monitoring of ovarian cancer.

There are several reasons why antibodies to mesothelin can have important functions that relate to the development and progression of ovarian cancer. Anti-mesothelin antibodies have been shown in vitro to prevent binding of mesothelin to CA125 and thereby impact cellular adhesion (23). Furthermore, antibodies can be cytotoxic in the presence of complement, mediate antibody-dependent cellular cytotoxicity in the presence of NK cells or macrophages, and the generation and expansion of T cell responses to tumor antigens may be impacted by such antibodies as

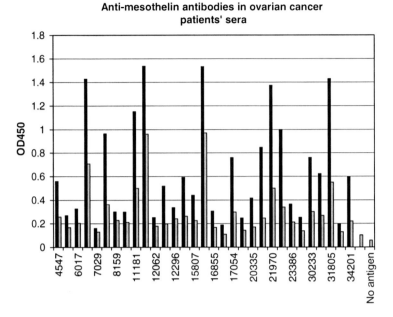

Fig. 2 Data from an ELISA detecting antibodies in the sera from patients with ovarian cancer (31 patients studied). Values above 0.4 are considered positive. Patients studied at two serum dilutions (1:20 *first*, 1:80 *second bar*)

well (24). Anti-cancer therapy is likely to influence antibody formation, not only by decreasing the number of tumor cells releasing antigen but also by acting directly on antibody forming cells, as in the case of cytotoxic drugs, and changes in antibody levels are likely to influence the amount of SMRP, which is detectable by ELISA.

4 HE4 as a Marker for Ovarian Carcinoma

The *WFDC2* (HE4) gene (25), which is a member of the disulfide-core family of secreted proteins, is amplified in ovarian carcinoma (26). On the basis of this already published information, we decided to evaluate HE4 as a biomarker for ovarian cancer and hence made fusion proteins, immunized mice, and constructed a Sandwich ELISA. "Blinded" studies on sera from postmenopausal women with ovarian carcinoma and controls were then carried out in collaboration with Dr. N. Urban and her colleagues at FHCR. They showed the sensitivity of the HE4-based ELISA to be equivalent to that of CA125, but that HE4 was found to be less frequently positive in women with nonmalignant disease, i.e., to be more specific. Therefore,

HE4 may complement CA125 for diagnosis and monitoring of ovarian cancer (27). Like CA125 and SMRP, HE4 has temporal stability, which should make longitudinal studies possible to facilitate earlier diagnosis.

According to an ongoing, collaborative study with Dr. E. Friedman's group in Israel on 329 sera from 111 patients with clinical evidence of ovarian carcinoma, 68% of the patients were positive for CA125, 57% for HE4, and 65% for SMRP. A combination of all three markers detected 85% of the patients (E. Friedman et al., unpublished data).

5 Detection of Other Tumors by Assaying for SMRP or HE4

Mesothelin is overexpressed by carcinomas of the pancreas (unpublished findings in collaboration with Dr. P. Goedegebuure), indicating that assays for SMRP in serum and other body fluids should be evaluated as possible aids to diagnose and monitor patients with that tumor. Recent immunohistological studies have demonstrated expression of HE4 in most adenocarcinomas of the lung. This suggests that it may be a biomarker for serum assays also for those tumors, a matter that needs to be studied further (28).

6 Summary

Assays measuring SMRP (mesothelin) and HE4 (a secreted protease) in serum and other body fluids (including urine for SMRP) are likely to be clinically useful for patients with ovarian cancer, as data indicate that they complement CA125 for diagnosis and monitoring of patients. Both markers have temporal stability, as does CA125, which may be utilized to facilitate earlier diagnosis by performing longitudinal studies on high risk subjects. Preliminary data show autoantibodies to native mesothelin in some patients with ovarian carcinoma and in some healthy women. We are presently studying their relationship to the patients' clinical state to learn whether measurements of antibody levels provide information that can aid diagnosis and monitoring of treated patients. Prospective studies are needed to establish the clinical relevance of our findings.

Acknowledgments Our early, published work on SMRP was supported by NIH, while our more recent studies, including work published in 2006 or still unpublished, were supported by a grant from Fujirebio Diagnostics, Inc. In addition to various collaborators on our published studies, we acknowledge collaboration with Y. Yang, E. Cutter, and J. Jaffar in our group; Drs. N. Sardesai and T. Verch at Fujirebio Diagnostics, Inc; Dr. E. Friedman and several of his colleagues at Sheba Medical Center in Israel; Drs. N. Kiviat and E. Swisher at University of Washington; and Dr. N. Urban at FHCRC. We also thank Dr. M.L. Disis at University of Washington for advice and discussions.

References

1. Bast, R. C. J., Siegal, F. P., Runowicz, C., Klug, T. L., Zurawski, V. R. J., Schonholz, D., Cohen, C. J., and Knapp, R. C. Elevation of serum CA125 prior to diagnosis of an epithelial ovarian carcinoma. Gynecol Oncol, *22:* 115–120, 1985.
2. Einhorn, N., Sjovall, K., Knapp, R. C., Hall, P., Scully, R. E., Bast, R. C. J., and Zurawski, V. R., Jr. Prospective evaluation of serum CA125 levels for early detection of ovarian cancer. Obstet Gynecol, *80:* 14–18, 1992.
3. Einhorn, N., Bast, R. C. J., Knapp, R. C., Tjernberg, B., and Zurawski, V. R. J. Preoperative evaluation of serum CA 125 levels in patients with primary epithelial ovarian cancer. Obstet Gynecol, *67:* 414–416, 1986.
4. Skates, S. J., Menon, U., MacDonald, N., Rosenthal, A. N., Oram, D. H., Knapp, R. C., and Jacobs, I. J. Calculation of the risk of ovarian cancer from serial CA-125 values for preclinical detection in postmenopausal women. J Clin Oncol, *21:* 206–210, 2003.
5. Fung, M. F., Bryson, P., Johnston, M., and Chambers, A. Cancer Care Ontario Practice Guidelines Initiave Gynecology Cancer Disease Site Group. Screening postmenopausal women for ovarian cancer: a systematic review. J Obstet Gynaecol Can, *26:* 717–728, 2004.
6. Rosen, D. G., Wang, L., Atkinson, J. N., Yu, Y., Lu, K. H., Diamandis, E. P., Hellstrom, I., Mok, S. C., Liu, J., and Bast, R. C. J. Potential markers that complement expression of CA125 in epithelial ovarian cancer. Gynecol Oncol, *99:* 267–277, 2005.
7. Chang, K. and Pastan, I. Molecular cloning of mesothelin, a differentiation antigen present on mesothelium, mesotheliomas, and ovarian cancers. Proc Natl Acad Sci USA, *93:* 136–140, 1996.
8. Linsley, P. S., Horn, D., Marquardt, H., Brown, J. P., Hellström, I., Hellström, K. E., Ochs, V., and Tolentino, E. Identification of a novel serum protein secreted by lung carcinoma cells. Biochemistry, *25:* 2978–2986, 1986.
9. Chowdhury, P. S., Viner, J. L., Beers, R., and Pastan, I. Isolation of a high-affinity stable single-chain Fv specific for mesothelin from DNA-immunized mice by phage display and construction of a recombinant immunotoxin with anti-tumor activity. Proc Natl Acad Sci USA, *95:* 669–674, 1998.
10. Brown, J. P., Hellstrom, I., and Hellstrom, K. E. Use of monoclonal antibodies for quantitative analysis of antigens in normal and neoplastic tissue. Clin Chem, *27:* 1592–1596, 1981.
11. Scholler, N., Fu, N., Yang, Y., Ye, Z., Goodman, G., Hellstrom, K., and Hellstrom, I. Soluble member(s) of the mesothelin/megakaryocyte potentiating factor family are detectable in sera from patients with ovarian carcinoma. Proc Natl Acad Sci USA, *96:* 11531–11536, 1999.
12. Robinson, B. W. S., Creaney, J., Lake, R. L., Nowak, A., Musk, A. W., deKlerk, N., Wintzell, P., Hellstrom, K. E., and Hellstrom, I. Mesothelin-family proteins and diagnosis of mesothelioma. Lancet, *362:* 1612–1616, 2003.
13. McIntosh, M. W., Drescher, C., Karlan, B., Scholler, N., Urban, N., Hellstrom, K. E., and Hellstrom, I. Combining CA125 and SMR serum markers for diagnosis and early detection of ovarian carcinoma. Gynecol Oncol, *95:* 9–15, 2004.
14. Muminova, Z. E., Strong, T. V., and Shaw, D. R. Characterization of human mesothelin transcripts in ovarian and pancreatic cancer. BMC Cancer, *4:* 19–29, 2004.
15. Hellstrom, I., Raycraft, J., Kanan, S., Sardesai, N. Y., Verch, T., Yang, Y., and Hellstrom, K. E. Mesothelin-variant 1 is released from tumor cells as a diagnostic marker. Cancer Epidemiol Biomarkers Prev, *15:* 1014–1020, 2006.
16. Hassan, R., Remaley, A., T, Sampson, M. L., Zhang, J. L., Cox, D. D., Pingpank, J., Alexander, R., Willingham, M. C., Pastan, I., and Onda, M. Detection and quantitation of serum mesothelin, a tumor marker for patients with mesothelioma and ovarian cancer. Clin Cancer Res, *12:* 447–453, 2006.
17. Klein, G., Clifford, P., Klein, E., and Stjernsward, J. Search for tumor specific immune reactins in Burkitt lymphoma patients by the membrane immunofluorescence reaction. Proc Natl Acad Sci USA, *55:* 1628–1635, 1966.

18. Morton, D. L., Malmgren, R. A., Holmes, E. C., and Ketcham, A. Demonstration of antibodies against human malignant melanomas by immunofluorescence. Surgery, *64:* 233–240, 1968.
19. Hellstrom, I., Hellstrom, K. E., Pierce, G. E., and Yang, J. P. Cellular and humoral immunity to different types of human neoplasms. Nature, *220:* 1352–1354, 1968.
20. Old, L. J. and Chen, Y. T. New paths in human cancer serology. J Exp Med, *187:* 1163–1167, 1998.
21. Cramer, D. W., Titus-Ernstoff, L., McKolanis, J. R., Welch, W. R., Vitonis, A. F., Berkowitz, R. S., and Finn, O. J. Conditions associated with antibodies against the tumor-associated antigen MUC1 and their relationship to risk for ovarian cancer. Cancer Epidemiol Biomarkers Prev, *14:* 1125–1131, 2005.
22. Goodell, V. and Disis, M. L. Human tumor cell lysates as a protein source for the detection of cancer antigen-specific humoral immunity. J Immunol Methods, *299:* 129–138, 2005.
23. Rump, A., Morikawa, Y., Tanaka, M., Minami, S., Umesaki, N., Takeuchi, M., and Miyajima, Q. Binding of ovarian cancer antigen CA125/MUC16 to mesothelin mediates cell adhesion. J Biol Chem, *279:* 9190–9198, 2004.
24. Hellstrom, K. E. and Hellstrom, I. Novel approaches to therapeutic cancer vaccines. Expert Rev Vaccines, *2:* 517–532, 2003.
25. Kirchhoff, C., Habben, I., Ivell, R., and Krull, N. A major human epididymis-specific cDNA encodes a protein with sequence homology to extracellular proteinase inhibitors. Biol Reprod, *45:* 350–357, 1991.
26. Schummer, M., Ng, W., Bumgarner, R., Nelson, P., Schummer, B., Bednarski, D., Hassell, L., Baldwin, R., Karlan, B., and Hood, L. Comparative hybridization of an array of 21500 ovarian cDNAs for the discovery of genes overexpressed in ovarian carcinomas. Gene, *238:* 375–385, 1999.
27. Hellstrom, I., Raycraft, J., Hayden-Ledbetter, M., Ledbetter, J. A., Schummer, M., McIntosh, M. W., Drescher, C., Urban, N., and Hellstrom, K. E. The HE4 (WFDC2) protein is a biomarker for ovarian carcinoma. Cancer Res, *63:* 3695–3700, 2003.
28. Bingle, L., Cross, S., High, A. S., Wallace, W. A., Rassl, D., Yan, G., Hellstrom, I., Campos, M. A., and Bingle, C. D. WFDC2 (HE4): a potential role in the innate immunity of the oral cavity and resiratory tract in the development of adenocarcinomas of the lung. Respiratory Res, *7:* 61–80, 2006.

Classification of Ovarian Cancer: A Genomic Analysis

Michael P. Stany, Tomas Bonome, Fred Wamunyokoli, Kristen Zorn, Laurent Ozbun, Dong-Choon Park, Ke Hao, Jeff Boyd, Anil K. Sood, David M. Gershenson, Ross S. Berkowitz, Samuel C. Mok, and Michael J. Birrer

1 Introduction

Ovarian cancer is the most lethal gynecologic cancer, accounting for over 16,000 deaths. In the US annually (1). The poor prognosis of this disease is due to the lack of reliable screening tools, the late stage of disease at the time of diagnosis, the high rate of recurrence of the disease, and the poor response to chemotherapy in the recurrent setting. Patients who present with stage I disease have over 90% 5-year survival, while those diagnosed with stage III disease have less than 20% 5-year survival (2). However, only 25% of patients with ovarian cancer are diagnosed with stage I disease (1). Despite new chemotherapeutic regimens and radical surgical debulking procedures, only minimal improvement in overall survival has been appreciated in the last several decades.

Epithelial ovarian cancer encompasses four major histotypes: papillary serous, endometrioid, mucinous, and clear cell. These histotypes resemble various müllerian cell types, with serous tumors resembling Fallopian tube, endometrioid tumors resembling uterine endometrium, mucinous tumors resembling the endocervix, and clear cell tumors resembling endometrial glands during pregnancy (2). Tumors are graded from 1 to 3, with grade 3 being the most poorly differentiated. Tumors of low malignant potential (LMP) display the atypical cellular features of cancer, but do not invade into the ovarian stroma (2).

The clinical characteristics of ovarian tumors vary according to histology and grade. Although clear cell ovarian cancer usually presents with earlier stage of the disease, there is a higher rate of recurrence among these patients. In fact, the 5-year survival for patients with stage I of clear cell ovarian cancer is only 60% (2). Late stage clear cell cancer also carries a worse prognosis when compared with papillary serous cancer, with an overall median survival of 12 months compared with 22 months (3). Mucinous cancers have a poorer response to chemotherapy and worse survival than the other epithelial ovarian histotypes (4). Patients with endometrioid ovarian cancers tend to have a better prognosis. Higher tumor grade correlates with poorer prognosis. Patients with LMP tumors have a 5-year survival over 95%. Among patients with early stage of disease, the survival drops down from 97% for patients with grade I tumors to 50% for those with grade III tumors (5).

These clear clinical differences among ovarian tumors likely reflect different underlying molecular mechanisms. Elvcidating how these tumors vary from a molecular biology standpoint can help us understand the pathogenesis and clinicopathologic characteristics of these tumors. Since gene expression is a critical determinate for many molecular features, gene expression profiling is a powerful approach to determine the underlying mechanism for these biologic and clinical differences.

Recent advances in molecular technology have provided the ability to perform whole genome expression profiling. This technique provides a global analysis of the transcriptional activity in ovarian tumors, which can then be correlated to pathologic and clinical determinates. Microarray determines the expression of genes by measuring mRNA levels using the ability of mRNA to hybridize to the DNA template. Over a dozen commercial platforms are currently available, each utilizing different technologies to fabricate their microarray chip. Much progress has been achieved in the microarray field since it was first introduced over a decade ago, and whole genome expression profiling, analyzing over 37,000 genes, is now possible on a single microarray chip.

2 Choice of Normal Ovarian Control

Identifying genes whose expression is altered during the transformation process relies on comparing malignant cells with their normal counterpart. Expression profiling of normal ovarian epithelium has utilized several sources of "normal" cells, including whole ovary samples (WO), ovarian surface epithelium exposed to short-term culture (NOSE), and immortalized ovarian surface epithelium cell lines (IOSE). WO has the advantage of providing a large amount of RNA; however, a large stromal component may mask true genomic expression differences within the epithelial component. Short-term cultures of ovarian surface epithelium scrapings provide a robust sample of ovarian surface epithelia, but are exposed to tissue culture conditions. Cultured media may select a subset of cells that are not representative of the original culture, altering overall gene expression. Immortalization methods of ovarian surface epithelia have utilized SV40 large T-antigen (6) and telomerase immortalization techniques (TIOSE) (7). These immortalized cells are exposed to tissue culture conditions, and they can demonstrate a large increase in chromosomal imbalance that could cause gene expression differences due to the immortalization process (8).

Another option for obtaining normal ovarian cells involves the preservation of ovarian surface epithelium brushings (OSE) without culturing (9). Brushings are obtained at the time of surgery followed by direct immersion in solution to preserve RNA quality. These cells are not immortalized and are not exposed to culture.

Zorn et al. (10) analyzed these five types of ovarian surface epithelium by comparing each group's expression profile with that of a set of 24 serous ovarian carcinomas. Hierarchical clustering and multidimensional scaling of the expression profiles of these groups demonstrated very distinct clusters (Fig. 1). In fact, the differences between "normal" samples were larger than those among the cancer specimens. When any two "normal" groups were compared with one another, the Pearson correlation coefficient for all combinations ranged from 0.04 to 0.54 (Table 1). When the individual gene lists were compared with the gene expression profile for the set of serous ovarian cancer samples, there was a majority of genes that were unique to each list. No gene appeared on all five lists. From their analysis, WO, NOSE, IOSE, TIOSE, and OSE had distinct expression profiles, and they concluded that OSE brushings seemed to be the most reasonable control as this sample was not affected by culture conditions or immortalization techniques.

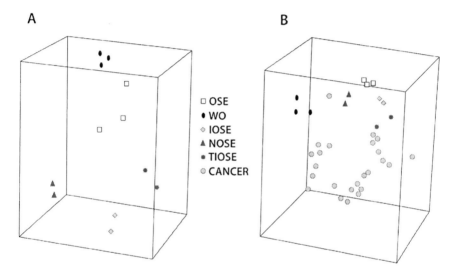

Fig. 1 Multidimensional scaling (MDS) of the expression profiles of ovarian specimens. MDS allows assessment of the likeness of the samples' expression patterns by compressing their gene expression profiles into a three-dimensional space. Samples with similar profiles cluster relatively close. (**a**) Unsupervised MDS of the normal groups. (**b**) Unsupervised MDS of the normal groups and the serous ovarian carcinoma samples

Table 1 Pearson correlation coefficient (r) of expression profiles between different normal ovarian epithelium samples in a group and between groups

Group	Within group	With OSE	With WO	With IOSE	With TIOSE	With NOSE
OSE	0.78	1	0.47	0.23	0.04	0.36
WO	0.86	0.47	1	0.10	0.22	0.39
IOSE	0.73	0.23	0.10	1	0.28	0.54
TIOSE	0.93	0.04	0.22	0.28	1	0.27
NOSE	0.90	0.36	0.39	0.54	0.27	1

3 Genes Expression Profiling of Different Tumor Grades

Serous ovarian tumors represent about 50% of all ovarian cancer. A small proportion of these tumors are classified as those of low malignant potential (LMP). LMP tumors display an atypical nuclear structure and can be metastatic, but because they lack stromal invasion, they are not characterized as "cancer." There is a debate as to how LMP tumors relate to the other frankly invasive tumors. One hypothesis states that LMP tumors are a distinct disease from invasive carcinoma, while another hypothesis argues that these tumors represent an early precursor lesion that eventually develops into malignant disease. Given that LMP tumors with micropapillary features have a lower overall survival when compared with LMPs without these features (11), it has been hypothesized that this small subset of LMP tumors with micropapillary histology can develop into low-grade invasive ovarian cancer (12).

Bonome et al. addressed this debate by evaluating the biological relationship among serous LMP, low-grade, and high-grade invasive ovarian carcinomas (13). They generated global gene expression profiles for 66 microdissected serous tumors. Unsupervised hierarchical clustering demonstrated distinct clusters that differentiated LMP and high-grade tumors. Of note, the majority of low-grade tumors clustered with LMP tumors. This strongly supports that LMP tumors are a unique disease entity from high-grade invasive cancer. Low-grade invasive cancers are essentially indistinguishable from LMP tumors.

Gene ontological analysis between high-grade tumors and LMP tumors found statistically significant differences ($p < 0.001$) between the number of genes involved in mitotic cell cycle, M phase, mitosis, G2-M transition, and cytokinesis. The majority of genes were upregulated in the high-grade specimens when compared with normal ovarian surface epithelium (Table 2). Specifically, genes linked to cell proliferation that were upregulated on the microarray analysis in high-grade tumors but not in LMP tumors included *PDC4, CCNDBP1, E2F3, CDC2, CCNB1,*

Table 2 GO categories associated with cell cycle progression in high-grade, low-grade, and LMP tumors

Gene ontology category	Late-stage high-grade		Early-stage high-grade		LMP/low-grade	
	Present	Number of genes	Present	Number of genes	Present	Number of genes ®
Mitotic cell cycle	Yes	70	Yes	58	No	0
M phase	Yes	66	Yes	55	No	0
Mitosis	Yes	51	Yes	43	No	0
G2/M transition	Yes	14	Yes	14	No	0
Cytokinesis	Yes	28	No	0	No	0

There was a statistically significant difference ($p < 0.05$) in number of genes associated with cell cycle progression that were over-expressed in high-grade tumors that were not differentially expressed in low-grade and LMP tumors

CCNB2, ASK, STMN1, CCNE1, MCM4, MCM5, MCM7, RFC4, FEN1, STK6, CENP-A, CDC20, EIF4G1, PTTG, and *PCNA.*

Major differences between *p53* and its associated genes were noted between LMP tumors and high-grade tumors. LMP tumors displayed elevated levels of p53 RNA and its principal effector CDKNIA, while high-grade tumors did not. Dysregulated genes that were unique to LMP tumors included UBE2D1 and ADNP. Both are negative regulators of p53 (14), and they were down-regulated in LMP tumors. PPM1A, which has been shown to increase the overall level of p53, was over-expressed in LMP tumors (15). LMP tumors also demonstrated an over-expression of important targets of p53. PML, which modulates apoptosis (16), and GDF15, which mediates growth arrest (17) are both over-expressed in LMP tumors but not in high-grade tumors. These results demonstrated clear differences between LMP tumors and high-grade tumors among p53-modulated genes, suggesting this pathway may play an important role in the distinct phenotypic differences between these two tumor grades.

4 Gene Expression Profiling of Different Tumor Histotypes

Gene expression profiling has also been used to characterize differences among the four main histotypes of ovarian cancer. Marquez et al. (18) compared whole genome expression profiles of serous, endometrioid, mucinous, and clear cell ovarian tumors with each other and to mucosal scrapings of normal fallopian tube, endometrium, and colon. Hierarchical clustering displayed grouping by the individual histotypes. They found a statistically significant correlation between serous tumors and normal fallopian tube, mucinous tumors with normal colonic epithelium, and endometrioid and clear cell tumors with normal endometrium. Their analysis utilized whole genome expression profiling, and when comparing the individual histotypes, mucinous cancers displayed a greater number of dysregulated genes than the other histotypes.

A comparison of tumors of similar histotypes across different organs has also been analyzed (19). Serous, endometrioid, and clear cell cancer histotypes of ovarian and endometrial origin were compared using a cDNA microarray. Although distinct expression patterns were appreciated among serous and endometrial tumors with respect to their organs of origin, clear cell tumors demonstrated a similar gene expression pattern for tumors originating in the ovary and endometrium (Fig. 2). This unique gene signature for clear cell tumors is consistent with an earlier study (20). Furthermore, in Zorn et al.'s analysis, expression profiling of renal clear cell cancers demonstrated that they were unable to be distinguished from clear cell tumors from the ovary or endometrium. This common pattern of clear cell tumor gene expression among these organs may represent a common precursor cell or similar processes of transformation.

Clear cell ovarian cancer is a rare histotype, and its clinical course presents a poorer prognosis. Patients with early stage disease have a higher recurrence rate,

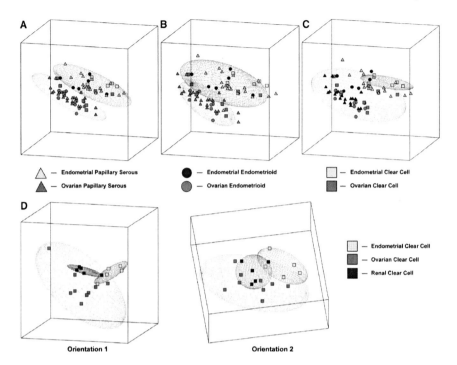

Fig. 2 Graphic depiction of the principal component analysis of ovarian and endometrial cancers categorized by histology. The ellipses represent a region where an additional sample of particular groups would fall with a 95% confidence interval. **a.** Analysis of serous tumors demonstrates nonoverlapping ellipses separating endometrial (*top*) and ovarian (*bottom*) specimens. **b.** Analysis of endometrioid tumors demonstrates nonoverlapping ellipses separating endometrial (*top*) and ovarian (*bottom*) specimens. **c.** Analysis of clear cell tumors showing overlapping of endometrial (*top*) and ovarian (*bottom*) specimens. **d.** Analysis of clear cell tumors of ovarian, endometrial, and renal origin demonstrate three overlapping elliptical regions, with two different orientations (1 and 2)

with 37% of stage IC patients recurring (21). Clear cell tumors also have a lower response rate to standard platinum and taxane chemotherapy, which is the usual first line therapy for ovarian cancer. Clear cell ovarian tumors have chemotherapy response rates as low as 15%, when compared with over 70% in serous tumors (22). In the analysis by Zorn et al. (19), genes that were common in clear cell tumors of the ovary and endometrium included *ANXA4* and *UGT1A1*. Both genes have been associated with chemoresistence, wherein *ANXA4* has been associated with paclitaxel resistance and *UGT1A1* has been shown to detoxify the active metabolite of irinotecan, SN-38 (23, 24).

The conclusions from these expression profiling studies of ovarian cancers of different histologies suggest that clear cell tumors represent a unique disease. Clear cell ovarian tumors are clinically and biologically distinct tumors from the other ovarian histotypes. As such, clinical trials addressing the optimal treatment of these tumors are needed. Research is needed that will hopefully identify molecular pathways that are unique to clear cell tumors, exposing targets for chemotherapeutic intervention.

Mucinous tumors also represent a rare ovarian cancer subtype, with the majority of tumors being benign (2). Although advanced stage disease represents the minority of mucinous tumors, this group has been found to have a worse prognosis and poorer response to chemotherapy when compared with other epithelial ovarian cancers, with chemotherapy response rates as low as 26% (4). Invasive mucinous ovarian tumors frequently have coexisting cells of varying malignancy, transitioning between benign and malignant cells on the same tumor (Fig. 3). Furthermore, identical K-*ras* mutations are frequently found in coexisting LMP and invasive epithelia within the same mucinous tumor (25). This suggests a progression model for mucinous ovarian tumors.

To evaluate the potential reasons behind the biological and clinical differences between mucinous tumors and other epithelial ovarian tumors, Wamunyokoli et al. performed global gene expression of mucinous cystadenomas, tumors of low malignant potential, and cystadenocarcinomas (26). The expression profiles of the mucinous tumors were compared with OSE and serous tumors. Unsupervised hierarchical clustering and binary tree analysis showed clustering of OSE with serous LMP tumors, and clustering of grade III serous tumors with invasive and LMP mucinous tumors (Fig. 4). Serous tumors had distinct clustering between LMP tumors and advanced stage tumors, while mucinous tumors of all grades had a high misclassification rate among the grades. Furthermore, the clustering of mucinous tumors with advanced stage serous tumors suggests the existence of a set of genes that may account for the poorer prognosis of mucinous ovarian cancer.

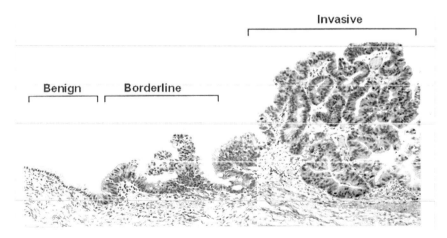

Fig. 3 This mucinous tumor specimen demonstrates close regions that display cells that are invasive as well as those with low malignant potential

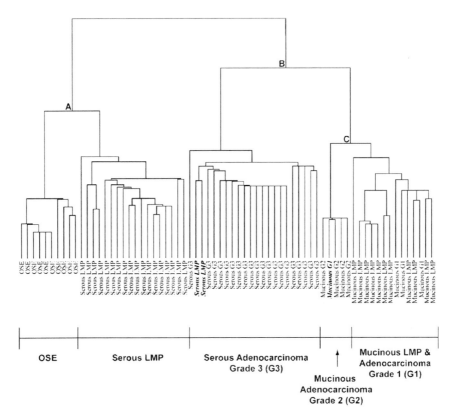

Fig. 4 Unsupervised hierarchical clustering of OSE, mucinous cystadenomas, mucinous LMP tumors, mucinous adenocarcinomas, and serous tumors. This dendogram illustrates OSE specimens grouping independently of serous LMP tumors, while high-grade serous tumors were closely associated with mucinous LMP and invasive tumors

To analyze what genes may be associated with the development of the mucinous phenotype, gene lists for mucinous cystadenomas and cystadenocarcinomas were evaluated to identify coregulated pathways. Genes common and unique to the different grades of tumors were also identified. Genes that were found to be upregulated in LMP tumors and cystadenocarimomas, but not OSE and cystadenomas, included *NET1* and *ERBB3*, suggesting the involvement of these genes in transformation. These genes have been found to increase tumorigenicity and promote invasiveness (27, 28). Genes involved in multidrug resistance, such as *ABCC3* and *ABCC6* (29) were upregulated in mucinous LMP tumors and cystadenocarimomas, but not cystadenomas. This is consistent with the known lower response rate of mucinous tumors to chemotherapy. Genes that modulate cell morpholophy, such as *CDC42*, *ECT2*, *IQGAP2*, and *Cortactin* (30, 31), were found to be upregulated in mucinous cystadenocarcinomas, but not in mucinous

LMP tumors. The differential regulation of these genes by tumor grade suggests a role of these genes in tumor progression.

5 Conclusion

Gene expression profiling can be used to evaluate ovarian cancer and identify genes and pathways important in tumor transformation and progression. Serous LMP tumors and low-grade serous cancer appear to have pathogenetic pathways that differ from high-grade serous cancer, implying these tumors are separate entities. Recent studies evaluating clear cell, serous, and mucinous epithelial ovarian cancer have found molecular differences that could explain their different clinical and biologic phenotypes. Clear cell ovarian tumors have an expression profile that is distinct from the other ovarian histologies, and similar to clear cell tumors originating in other organs. Mucinous ovarian cancer pathogenesis, unlike serous tumors, represents a continuum. Mucinous benign tumors may develop from ovarian inclusion cysts, acquire further KRAS mutations and other molecular changes to become mucinous borderline tumors, and then progress to low-grade and subsequently high-grade tumors. With gene expression profiling, novel pathways for these tumors will eventually be identified, exposing new and more specific targets for chemotherapy.

References

1. Jemal, A., T. Murray, E. Ward, A. Samuels, R. C. Tiwari, A. Ghafoor, E. J. Feuer and M. J. Thun (2005). "Cancer statistics, 2005." *CA Cancer J Clin* **55**(1): 10–30.
2. Hoskins, W. J. (2005). *Principles and practice of gynecologic oncology*. Philadelphia, Lippincott.
3. Goff, B. A., R. Sainz de la Cuesta, H. G. Muntz, D. Fleischhacker, M. Ek, L. W. Rice, N. Nikrui, H. K. Tamimi, J. M. Cain, B. E. Greer and A. F. Fuller, Jr. (1996). "Clear cell carcinoma of the ovary: a distinct histologic type with poor prognosis and resistance to platinum-based chemotherapy in stage III disease." *Gynecol Oncol* **60**(3): 412–7.
4. Hess, V., R. A'Hern, N. Nasiri, D. M. King, P. R. Blake, D. P. Barton, J. H. Shepherd, T. Ind, J. Bridges, K. Harrington, S. B. Kaye and M. E. Gore (2004). "Mucinous epithelial ovarian cancer: a separate entity requiring specific treatment." *J Clin Oncol* **22**(6): 1040–4.
5. Shimada, M., J. Kigawa, Y. Kanamori, H. Itamochi, T. Oishi, Y. Minagawa, K. Ishihara, Y. Takeuchi, M. Okada and N. Terakawa (2005). "Outcome of patients with early ovarian cancer undergoing three courses of adjuvant chemotherapy following complete surgical staging." *Int J Gynecol Cancer* **15**(4): 601–5.
6. Jazaeri, A. A., C. J. Yee, C. Sotiriou, K. R. Brantley, J. Boyd and E. T. Liu (2002). "Gene expression profiles of BRCA1-linked, BRCA2-linked, and sporadic ovarian cancers." *J Natl Cancer Inst* **94**(13): 990–1000.
7. Wernert, N., C. Locherbach, A. Wellmann, P. Behrens and A. Hugel (2001). "Presence of genetic alterations in microdissected stroma of human colon and breast cancers." *Anticancer Res* **21**(4A): 2259–64.

8. Tsao, S. W., N. Wong, X. Wang, Y. Liu, T. S. Wan, L. F. Fung, W. D. Lancaster, L. Gregoire and Y. C. Wong (2001). "Nonrandom chromosomal imbalances in human ovarian surface epithelial cells immortalized by HPV16-E6E7 viral oncogenes." *Cancer Genet Cytogenet* **130**(2): 141–9.
9. Shridhar, V., J. Lee, A. Pandita, S. Iturria, R. Avula, J. Staub, M. Morrissey, E. Calhoun, A. Sen, K. Kalli, G. Keeney, P. Roche, W. Cliby, K. Lu, R. Schmandt, G. B. Mills, R. C. Bast, Jr., C. D. James, F. J. Couch, L. C. Hartmann, J. Lillie and D. I. Smith (2001). "Genetic analysis of early- versus late-stage ovarian tumors." *Cancer Res* **61**(15): 5895–904.
10. Zorn, K. K., A. A. Jazaeri, C. S. Awtrey, G. J. Gardner, S. C. Mok, J. Boyd and M. J. Birrer (2003). "Choice of normal ovarian control influences determination of differentially expressed genes in ovarian cancer expression profiling studies." *Clin Cancer Res* **9**(13): 4811–8.
11. Longacre, T. A., J. K. McKenney, H. D. Tazelaar, R. L. Kempson and M. R. Hendrickson (2005). "Ovarian serous tumors of low malignant potential (borderline tumors): outcome-based study of 276 patients with long-term (> or = 5-year) follow-up." *Am J Surg Pathol* **29**(6): 707–23.
12. Shih Ie, M. and R. J. Kurman (2004). "Ovarian tumorigenesis: a proposed model based on morphological and molecular genetic analysis." *Am J Pathol* **164**(5): 1511–8.
13. Bonome, T., J. Y. Lee, D. C. Park, M. Radonovich, C. Pise-Masison, J. Brady, G. J. Gardner, K. Hao, W. H. Wong, J. C. Barrett, K. H. Lu, A. K. Sood, D. M. Gershenson, S. C. Mok and M. J. Birrer (2005). "Expression profiling of serous low malignant potential, low-grade, and high-grade tumors of the ovary." *Cancer Res* **65**(22): 10602–12.
14. Zamostiano, R., A. Pinhasov, E. Gelber, R. A. Steingart, E. Seroussi, E. Giladi, M. Bassan, Y. Wollman, H. J. Eyre, J. C. Mulley, D. E. Brenneman and I. Gozes (2001). "Cloning and characterization of the human activity-dependent neuroprotective protein." *J Biol Chem* **276**(1): 708–14.
15. Ofek, P., D. Ben-Meir, Z. Kariv-Inbal, M. Oren and S. Lavi (2003). "Cell cycle regulation and p53 activation by protein phosphatase 2C alpha." *J Biol Chem* **278**(16): 14299–305.
16. Pearson, M. and P. G. Pelicci (2001). "PML interaction with p53 and its role in apoptosis and replicative senescence." *Oncogene* **20**(49): 7250–6.
17. Li, P. X., J. Wong, A. Ayed, D. Ngo, A. M. Brade, C. Arrowsmith, R. C. Austin and H. J. Klamut (2000). "Placental transforming growth factor-beta is a downstream mediator of the growth arrest and apoptotic response of tumor cells to DNA damage and p53 overexpression." *J Biol Chem* **275**(26): 20127–35.
18. Marquez, R. T., K. A. Baggerly, A. P. Patterson, J. Liu, R. Broaddus, M. Frumovitz, E. N. Atkinson, D. I. Smith, L. Hartmann, D. Fishman, A. Berchuck, R. Whitaker, D. M. Gershenson, G. B. Mills, R. C. Bast, Jr. and K. H. Lu (2005). "Patterns of gene expression in different histotypes of epithelial ovarian cancer correlate with those in normal fallopian tube, endometrium, and colon." *Clin Cancer Res* **11**(17): 6116–26.
19. Zorn, K. K., T. Bonome, L. Gangi, G. V. Chandramouli, C. S. Awtrey, G. J. Gardner, J. C. Barrett, J. Boyd and M. J. Birrer (2005). "Gene expression profiles of serous, endometrioid, and clear cell subtypes of ovarian and endometrial cancer." *Clin Cancer Res* **11**(18): 6422–30.
20. Schwartz, D. R., S. L. Kardia, K. A. Shedden, R. Kuick, G. Michailidis, J. M. Taylor, D. E. Misek, R. Wu, Y. Zhai, D. M. Darrah, H. Reed, L. H. Ellenson, T. J. Giordano, E. R. Fearon, S. M. Hanash and K. R. Cho (2002). "Gene expression in ovarian cancer reflects both morphology and biological behavior, distinguishing clear cell from other poor-prognosis ovarian carcinomas." *Cancer Res* **62**(16): 4722–9.
21. Sugiyama, T., T. Kamura, J. Kigawa, N. Terakawa, Y. Kikuchi, T. Kita, M. Suzuki, I. Sato and K. Taguchi (2000). "Clinical characteristics of clear cell carcinoma of the ovary: a distinct histologic type with poor prognosis and resistance to platinum-based chemotherapy." *Cancer* **88**(11): 2584–9.
22. Itamochi, H., J. Kigawa, T. Sugiyama, Y. Kikuchi, M. Suzuki and N. Terakawa (2002). "Low proliferation activity may be associated with chemoresistance in clear cell carcinoma of the ovary." *Obstet Gynecol* **100**(2): 281–7.

23. Han, E. K., S. K. Tahir, S. P. Cherian, N. Collins and S. C. Ng (2000). "Modulation of paclitaxel resistance by annexin IV in human cancer cell lines." *Br J Cancer* **83**(1): 83–8.
24. Gagne, J. F., V. Montminy, P. Belanger, K. Journault, G. Gaucher and C. Guillemette (2002). "Common human UGT1A polymorphisms and the altered metabolism of irinotecan active metabolite 7-ethyl-10-hydroxycamptothecin (SN-38)." *Mol Pharmacol* **62**(3): 608–17.
25. Garrett, A. P., K. R. Lee, C. R. Colitti, M. G. Muto, R. S. Berkowitz and S. C. Mok (2001). "k-ras mutation may be an early event in mucinous ovarian tumorigenesis." *Int J Gynecol Pathol* **20**(3): 244–51.
26. Wamunyokoli, F. W., T. Bonome, J. Y. Lee, C. M. Feltmate, W. R. Welch, M. Radonovich, C. Pise-Masison, J. Brady, K. Hao, R. S. Berkowitz, S. Mok and M. J. Birrer (2006). "Expression profiling of mucinous tumors of the ovary identifies genes of clinicopathologic importance." *Clin Cancer Res* **12**(3 Pt 1): 690–700.
27. Chan, A. M., S. Takai, K. Yamada and T. Miki (1996). "Isolation of a novel oncogene, *NET1*, from neuroepithelioma cells by expression cDNA cloning." *Oncogene* **12**(6): 1259–66.
28. Sithanandam, G., L. W. Fornwald, J. Fields and L. M. Anderson (2005). "Inactivation of ErbB3 by siRNA promotes apoptosis and attenuates growth and invasiveness of human lung adenocarcinoma cell line A549." *Oncogene* **24**(11): 1847–59.
29. Ohishi, Y., Y. Oda, T. Uchiumi, H. Kobayashi, T. Hirakawa, S. Miyamoto, N. Kinukawa, H. Nakano, M. Kuwano and M. Tsuneyoshi (2002). "ATP-binding cassette superfamily transporter gene expression in human primary ovarian carcinoma." *Clin Cancer Res* **8**(12): 3767–75.
30. Hall, A. (1998). "Rho GTPases and the actin cytoskeleton." *Science* **279**(5350): 509–14.
31. Sahai, E. and C. J. Marshall (2002). "RHO-GTPases and cancer." *Nat Rev Cancer* **2**(2): 133–42.

Epigenetic Markers of Ovarian Cancer

Caroline A. Barton, Susan J. Clark, Neville F. Hacker,
and Philippa M. O'Brien

1 Introduction

Ovarian cancer is the fourth leading cause of cancer death in women, and has the highest mortality rate of the reproductive cancers (1). Ovarian cancer is often asymptomatic in its early stages and because of lack of early detection strategies, most patients are diagnosed with disseminated disease, for whom the 5-year overall survival rate is only 20% (2). In the absence of an early detection test, improved therapies for advanced disease are critical to improving the survival for women with ovarian cancer. Most patients receive cytotoxic chemotherapy following surgical resection of their tumor; however, although the majority of patients are initially responsive to chemotherapy, most of them eventually develop drug-resistant disease, that is essentially incurable. A better understanding of the molecular pathogenesis underlying ovarian cancer is the key to identifying markers for early detection and novel therapeutics.

Both genetic (changes in DNA sequence such as deletions/amplifications and mutations) and epigenetic changes, defined as heritable changes in gene expression that occur without changes to the DNA sequence (3), contribute to malignant transformation and progression. Commonly occurring epigenetic events include DNA methylation, the addition of a methyl group to the 5´-carbon of cytosine in CpG sequences, and chromatin remodeling via histone protein acetylation and methylation (4). The human genome is not methylated uniformly, containing regions of unmethylated segments interspersed with methylated regions (5). Although spontaneous deamination of methylated cytosine through evolution has decreased the proportion of CpG dinucleotides in the genome, there are regions ranging from 0.5 to 5 kb that contain clusters of CpG dinucleotides called CpG islands (6). These CpG-rich regions are often located in the 5´ region of genes and are associated with the promoters of genes. In contrast to the bulk of DNA, the CpG sites within CpG islands are almost always methylation free. This appears to be a prerequisite for active transcription of the genes under their control (7). The relationship between DNA methylation and posttranslational modification of histones appears to be complex but they collectively result in transcriptional silencing through effects on transcription factor binding and repression of gene expression in normal cellular processes (4, 8, 9).

The methylation patterns in cancer cells are significantly altered compared with those of normal cells. Cancer cells undergo changes in 5-methylcytosine distribution including global DNA hypomethylation (10) as well as hypermethylation of CpG islands (11–15). Genome hypomethylation, mainly due to hypomethylation of normally silenced repetitive sequences such as long interspersed nuclear elements, is present in most cancer cells compared with the normal tissue from which it originated (16, 17). Hypomethylation has been hypothesized to contribute to oncogenesis by transcriptional activation of oncogenes, activation of latent transposons, or by chromosome instability (18–22). At the same time, aberrant CpG island DNA methylation and histone modification, leading to transcriptional activation and gene silencing, is a common phenomenon in human cancer cells and an early event in carcinogenesis (4). In particular, hypermethylation of CpG islands in gene promoter regions is a frequent mechanism of inactivation of tumor suppressor genes (4, 10, 14, 15), and has been proposed as one of the two hits in Knudson's two hit hypothesis for oncogenic transformation (15).

Epigenetic alterations, including CpG island DNA methylation, occur in ovarian cancer (23, 24), and the identification of specific genes that are altered by these epigenetic events is an area of intense research. Although there are CpG islands that become methylated in multiple tumor types, differential patterns of methylation of specific genes can vary amongst neoplasms (25), with certain CpG islands only methylated in specific tumor types (26–29). Neoplasm-specific events may be useful as molecular biomarkers for early detection and prognostic significance. Large scale screening to identify ovarian cancer-specific epigenetic fingerprints is underway.

In this review, we discuss genes identified as being deregulated by epigenetic mechanisms in ovarian cancer, with a focus on genes with potential as diagnostic markers or markers of disease progression and therapeutic response.

2 Epigenetically Regulated Genes in Ovarian Cancer

Most studies to date have focused on candidate gene approaches to identify epigenetically regulated genes in ovarian cancer, in particular, methylated and silenced candidate tumor suppressor genes. Selected targets include genes with downregulated expression in ovarian cancer, genes in regions with known loss of heterozygosity (LOH) in ovarian cancer and thus where tumor suppressors likely reside, and genes that have been shown to be epigenetically regulated in other cancers. There is also a small but significant literature on hypomethylated genes in ovarian cancer. Although in their infancy, genome-wide array-based approaches to epigenetically regulated gene discovery are also beginning to emerge.

2.1 Hypermethylated and Silenced Genes

Epigenetic regulation of *BRCA1* (breast cancer susceptibility gene 1) has been studied extensively because of its known role in inherited forms of ovarian cancer.

BRCA1, a breast and ovarian cancer susceptibility gene, is involved in the maintenance of genome integrity. Carriers of *BRCA1* germline mutations develop predominantly breast and ovarian tumors. *BRCA1* promoter methylation only occurs in breast and ovarian cancers (30, 31) and mirrors the classical genetic mutation studies of familial cancers. Studies investigating *BRCA1* hypermethylation report methylation in 5–24% of epithelial ovarian cancers (24, 32–38) in association with loss of *BRCA1* expression (32, 39, 40). No correlation of methylation with histological subtypes or grade or stage has been found (40); however, *BRCA1* silencing is detectable in early (stage 1A) tumors (40). LOH at the *BRCA1* locus occurs in a significant proportion of sporadic ovarian cancers (30, 41); moreover, *BRCA1* hypermethylation is predominantly detected in cancers that exhibit LOH at the *BRCA1* locus (31, 39). Thus, silencing of *BRCA1* expression by methylation likely acts as the second hit required for tumor suppressor gene inactivation in Knudson's two hit hypothesis (15).

Several other genes located at regions of LOH are methylated and silenced in ovarian cancer. *ARHI* (Ras homologue member 1), a maternally imprinted tumor suppressor of the ras superfamily, expressed monoallelically from the paternal allele, maps to chromosome 1p31, which is associated with LOH in 40% of ovarian carcinomas. *ARHI* expression is lost in ovarian cancers (42, 43) compared with normal ovarian tissue including the ovarian surface epithelium (OSE). The remaining paternal allele is silenced by methylation in 10–15% of cases (44).

DLEC1 (deleted in lung and esophageal cancer 1) is located on 3p22.3 in a region of frequent LOH in cancer, and is a putative tumor suppressor in lung and other cancers. DLEC1 expression is downregulated in ovarian cancer cell lines and primary invasive epithelial ovarian cancers, where its expression is correlated with hypermethylation of the *DLEC1* promoter (45).

p16 *(CDKN2A)* encodes a cyclin-dependant kinase inhibitor involved in the regulation of the cell cycle. p16 expression is frequently disrupted in cancer. The p16 locus at chromosome 9p21 is located in a region of LOH documented in a wide variety of cancers, including ovarian carcinoma (46, 47). Hypermethylation of the *p16* promoter region is important in a subset of human carcinomas including lung, head and neck, and pancreatic cancer (30, 48, 49), but there is conflicting evidence whether *p16* methylation plays a role in ovarian carcinogenesis (24, 37, 38, 50–54).

OPCML, located at 11q25, is hypermethylated in 33–83% of epithelial ovarian cancers (38, 55, 56). *SFRP1*, a Wnt antagonist located at chromosome 8p11.2, is methylated in 5–12% of primary ovarian cancers (38, 57). *MYO18B*, located at 22q121.1, is methylated and silenced in primary ovarian cancer (58). The TRAIL receptor *DR4*, located at 8p21.1, is methylated in 28% of ovarian cancers and is associated with loss of expression (59). Each of these genes is located in chromosomal regions associated with LOH in ovarian cancer.

Certain epigenetically regulated genes postulated to act as tumor suppressors in other carcinomas are also methylated and transcriptionally silenced in ovarian cancer. For example, de novo methylation and inactivation of the ras homologue *RASSF1A* tumor suppressor gene is one of the most frequently detected epigenetic events in human cancer (60). *RASSF1A* is methylated and silenced in 10–50% of primary ovarian cancers (36–38, 61, 62). Hypermethylation and loss of expression of insulin-like growth factor binding protein 3 (IGFBP-3), a member of the IGFBP family, which regulates mitogenic and apoptotic effects of insulin-like growth factors, occurs in nonsmall cell lung cancer and hepatocellular carcinoma (63, 64). *IGFBP-3* promoter methylation is detected in 44% of samples from epithelial ovarian cancer patients (65). *ARLTS1*, a tumor suppressor previously described as a low penetrance cancer gene, shows downregulated expression in ovarian carcinomas because of DNA methylation in its promoter region (66).

Gene methylation patterns can also represent molecular characterizations of pathological and clinical features of ovarian carcinomas. For example, 14-3-3sigma (*SFN*), an inhibitor of cell cycle progression that is epigenetically deregulated in other cancers (67, 68), is methylated at a higher frequency in ovarian clear cell carcinomas than in other histological types of ovarian cancer (69). Similarly, aberrant methylation of *TMS1* (target of methylation-induced silencing) and the *WT1* (Wilms tumor suppressor 1 gene) sense and antisense promoters is more frequent in clear cell ovarian tumors than in other histological types (38, 70–72). Methylation profiles can also differentiate between ovarian low malignant potential (LMP) tumors and invasive ovarian carcinoma, for example, *RASSF1A, APC, GSTP1,* and *MGMT* show aberrant methylation exclusively in invasive ovarian carcinomas (53).

There are numerous other genes that exhibit promoter methylation and decreased expression in ovarian cancer. For example, loss of expression of TCEAL7 (Bex4) in primary cancers correlates with methylation of CpG sites in the promoter (73). Decreased levels of FANCF expression found in most ovarian cancers are in part due to promoter hypermethylation (28%) (74). Promoter methylation of $p33^{ING1b}$, the inhibitor of growth *1b* gene, is found in 24% of ovarian cancers and are significantly correlated with loss of mRNA expression (75). The CpG islands of the suppressor of cytokine signaling genes *SOCS1* and *SOCS2* are hypermethylated in 23% and 14% primary ovarian cancers, respectively (76). *TCF2*, which encodes the transcription factor HNF1β, is methylated in 26% of primary ovarian cancers (77). Finally, methylation of *MLH1* (6–13%) (24, 38, 78, 79), *HIC1* (16–35%) (24, 37, 38), *hTR* (24%) (24), *p73* (10%) (24), and *MINT25* (12–16%) (24, 38) have been reported in primary ovarian cancers.

2.2 Chromatin Modifications Influencing Gene Expression

Histone modification leads to changes in chromatin structure and results in alteration of transcriptional activity of a gene locus (80–82). Multiple acetylations at both histone H3 and H4 subunits are associated with transcriptionally active sequences and a lack of histone acetylation (hypoacetylation) correlates with transcriptional silencing (82). The removal of acetyl groups can lead to chromatin condensation and result in repression of transcription. Additionally, the unmodified histone lysine residues can be mono, di, or trimethylated. Histone H3 di or trimethylation at lysine

K4 is also associated with active transcription (82). In contrast, histone H3 di and trimethylation at lysine K9 are enriched in transcriptionally silenced, densely packed heterochromatin and are thus associated with gene silencing (82–84).

The emergence of chromatin immunoprecipitation (ChIP) analysis as a method to identify histone modifications has aided the identification of genes silenced by chromatin remodeling in ovarian cancer. For example, there is an enhanced association between acetylated histone H3 and H4 and the *DLEC1* promoter in cells that have lost DLEC1 expression (45), indicating that histone hypoacetylation is used to suppress DLEC1 expression in ovarian cancers. Acetylation and methylation of chromatin in the promoter region of *ARHI* is associated with loss of expression in ovarian cancer cells (85). *GATA4* and *GATA6* gene silencing, via an alteration of chromatin conformation, correlates with hypoacetylation of histones H3 and H4 and loss of histone H3 lysine K4 trimethylation at their promoters in ovarian cancer cell lines (86). In contrast to *DLEC1* where gene silencing is also associated with DNA hypermethylation (45), *GATA4* and *GATA6* silencing is independent of promoter methylation (86).

In addition to gene silencing, increased histone acetylation can lead to reexpression of genes in ovarian cancer cells. For example, high histone H3 acetylation and an open chromatin conformation, in addition to reduced DNA methylation, are important in claudin-4 (CLDN4) overexpression in ovarian cancer (87–92).

2.3 Hypomethylated Genes

Global DNA hypomethylation increases with malignancy in ovarian epithelial neoplasms (93). There are, however, limited examples of specific gene activation by hypomethylation in ovarian cancer. Demethylation of the maspin (*SERPINB5*) promoter in ovarian cancer cells is associated with a gain of maspin mRNA expression (94). Demethylation is also important in the abnormal expression of the metastasis-related gene synuclein-γ *(SNCG)*, a member of a family of small cytoplasmic proteins. Synuclein-γ is not normally expressed in OSE because of dense methylation in the promoter region but is hypomethylated and reexpressed in aggressive ovarian cancer cell lines (95) and in a substantial proportion of malignant ovarian carcinoma samples (96).

Similarly, little is known about activation of latent retrotransposons, though hypomethylation of the CpG dinucleotides associated with the L1 and HERV-W retrotransposons occurs in malignant relative to nonmalignant ovarian tissue consistent with an elevation in expression levels (97).

Hypomethylation and rearrangements in heterochromatin in the vicinity of the centromeres of chromosomes 1 and 16 are frequent in many types of cancer, including ovarian epithelial carcinomas (98). Satellite 2 (Sat2) DNA is the main sequence in the heterochromatin region adjacent to the centromere of these chromosomes. In all normal tissues, Sat2 DNA is highly methylated but in ovarian carcinomas there is significantly more hypomethylation in Sat2 DNA sequences in the juxtacentromeric heterochromatin of chromosome 1 (Chr1 Sat2) and chromosome 16 compared with borderline (LMP) ovarian tumors and cystadenomas (99). Thus degree of malignancy significantly correlates with the extent of Sat2 DNA hypermethylation (99). In addition, the study by Widschwendter et al. (100, 101) of 115 ovarian cancers and 26 nonneoplastic ovarian specimens demonstrated a highly significant difference in levels of satellite hypomethylation in the major DNA component of all the human centromeres, satellite α (Satα), in ovarian cancer. Advanced stage of disease and tumor grade were associated significantly with frequent hypomethylation of Chr1 Sat2 or Chr1 Satα, and serous and endometriod ovarian cancers had significantly higher hypomethylation levels than LMP or mucinous tumors (100). Finally, the methylation status of *NBL2*, a complex tandem DNA repeat in acrocentric chromosomes, is significantly related to degree of malignancy of ovarian epithelial carcinomas, with hypomethylation seen only in the carcinomas (102).

3 Clinical Epigenetic Markers

Alterations in epigenetic patterns, including changes in DNA methylation, have several advantages as a means to detect and classify cancer: (1) methylation analysis utilizes DNA, a more chemically stable molecule than RNA or protein; (2) aberrant DNA methylation is a "positive" signal that can be detected in a background of excess normal DNA molecules by sensitive assays that depend on signal amplification by PCR. Such assays include methylation-specific PCR (MSP) (103) and quantitative MSP (104) including the fluorescence-based real-time PCR-based MethyLight technique (105) and headloop suppression PCR, which allows the detection of a single methylated allele in 10,000 unmethylated alleles (106). These assays have sufficient signal-to-noise ratio and throughput capacity to sensitively analyze a broad spectrum of markers and thus may be useful in the clinical setting (101, 107, 108); (3) assay design can focus on a single amplifiable region (eg., CpG island) rather than scanning an entire gene for mutations. Moreover the detection of methylation of multiple genes can be combined in a high throughput manner to improve the specificity of cancer detection; (4) aberrant methylation is frequently observed in early cancer development, and hence has applicability to the detection of early stage disease; and (5) methylation biomarkers are present in patient serum/plasma and other bodily fluids (109), and hence may have application as the basis of noninvasive detection tests. To this end, a number of studies have shown the feasibility of detecting hypermethylation of multiple genes in circulating DNA from patients with a broad spectrum of tumors (110–112).

3.1 *Prognostic Markers*

Several epigenetically regulated genes have been assessed for their positive prognostic potential in ovarian cancer. For example, *IGFBP-3* methylation is associated with disease progression and death in ovarian cancer, particularly in patients with early-stage disease, where methylation was associated with a threefold higher risk of disease progression and a fourfold higher risk of death (65). Similarly, hypermethylation of 18S and 28S rDNA is associated with prolonged progression-free survival of ovarian cancer patients (113). Patients who demonstrated little or no hypomethylation of Chr1 Sat2 or Chr1 Satα had a significantly longer relapse-free survival compared with patients with strong hypomethylation of these regions (100). However, the small sample sizes used in these studies require these results to be confirmed by large independent studies.

It is quite likely that determining the methylation status of multiple genes simultaneously rather than individual genes will provide a more sensitive and specific assay for molecular classification and prognosis of ovarian cancer patients. To this end, genome-wide array-based approaches are being utilized to identify prognostic "methylation signatures" that can predict patient outcome. For example, using differential methylation hybridization, Wei et al. could stratify late-stage of ovarian tumors into two distinct groups with significantly different outcome on the basis of methylation profiling of 956 CpG island-containing loci (114, 115). This study was recently extended to identify 112 discriminatory methylated gene loci capable of predicting progression-free survival with 95% accuracy using rigorous classifying algorithms (116). Hence, although in its infancy, the identification of a prognostic panel of hypermethylated DNA markers for ovarian cancer remains a realistic possibility.

3.2 Markers of Therapeutic Responsiveness

Variations in patterns of methylation can occur within the same tumor types and in addition to providing prognostic information, methylation patterns are associated with response to chemotherapy. Epigenetic gene regulation plays a prominent role in both intrinsic and acquired drug resistance in cancer (117), and epigenetic markers may therefore prove useful in predicting chemotherapy response and outcome in patients with ovarian cancer. Methylated genes implicated in drug resistance are those involved in processes known to influence chemosensitivity, such as DNA repair and damage response pathways, cell cycle control, and apoptosis (118). For example, Teodoridis et al. (38) showed that methylation of at least one of the three genes involved in DNA repair/drug detoxification (*BRCA1, GSTP1,* and *MGMT*) is associated with improved response to chemotherapy of patients with late-stage epithelial ovarian tumors (38).

Chemotherapy itself can exert a positive selective pressure on subpopulations of cells in an initially chemoresponsive tumor. A number of recent studies suggest a direct role for epigenetic inactivation of genes underlying acquired chemoresistance at disease relapse. For example, matched cell line models of acquired resistance have shown that common patterns of CpG island methylation can be identified as being selected for chemotherapy in vitro (119). There is an increasing volume of evidence from clinical studies that supports this hypothesis. In the study by Wei et al., discussed in the above Prognostic Markers section, patients stratified as having a short progression-free survival (with a high degree of CpG island methylation) have a poorer response to second-line cytotoxic therapies when compared with patients with a longer progression-free survival (and low CpG island methylation), suggesting that patients with high CpG island methylation more readily acquire resistance to chemotherapy (115).

Silencing of *hMLH1*, a DNA mismatch repair gene, by hypermethylation of its promoter CpG island (24) has been linked with acquired resistance to platinum-based drugs in ovarian cell line models (24, 120, 121). Methylation of *MLH1* is increased at relapse in epithelial ovarian cancer patients; 25% (34/138) of plasma samples from relapsed patients showed methylation of *MLH1*, which is not evident in matched prechemotherapy plasma samples but consistent with acquisition of methylation after chemotherapy. Moreover, acquisition of *MLH1* methylation at relapse predicts poor overall patient survival and is associated with drug resistance (122).

FANCF is crucial for the activation of the DNA repair complex containing BRCA1 and BRCA2. Methylation-induced inactivation of *FANCF* is observed in ovarian cancer cells with a defective BRCA2 pathway, associated with increased sensitivity to cisplatin. Demethylation and reexpression of FANCF is associated with acquisition of cisplatin resistance in ovarian cancer cell lines (123). It has been proposed that inactivation of *FANCF* occurs early in tumor progression but chemotherapy selects for cells in which *FANCF* methylation has been reversed and therefore displays higher resistance to platinum-based chemotherapy (123). Methylation of *FANCF* has been observed in primary ovarian cancers (123), but its relevance to clinical outcome following chemotherapy is yet to be established.

Although no functional role has yet been assigned, methylation-controlled DNAJ (*MCJ*) was identified as a gene that rendered epithelial cells more sensitive to cisplatin and paclitaxel, the mainstay of chemotherapy for ovarian cancer patients (124). Unusual for a CpG island-associated gene, cell-type-specific DNA methylation and gene silencing of *MCJ* are observed in normal cells, including OSE (125). The majority of late-stage ovarian cancers also exhibit *MCJ* methylation; however, many of these have undergone a partial demethylation of the *MCJ* gene promoter, with only 17% of cancers maintaining very high (>90%) methylation, which is correlated with a poor response to chemotherapy and decreased survival (125, 126). Hence, *MCJ* methylation may be a useful marker of response to chemotherapy in ovarian cancer.

These data remain to be validated in large prospective studies; nonetheless, the identification of ovarian cancer-specific epigenetic changes clearly has promise in disease stratification and treatment individualization (118, 122).

3.3 Early Diagnostic Markers

Early diagnosis is critical for the successful treatment of many types of cancer, including ovarian cancer. The detection of cancer at early stages by noninvasive methods may be aided by the identification of cancer-specific biomarkers detectable in body fluids. It has been known for many years that tumors appear to "shed" DNA into the circulation (127). Moreover, specific methylated DNA markers can be detected in the serum/plasma and peritoneal fluid of ovarian cancer patients (36). The challenge remains to identify methylated markers that are commonly found in patients with ovarian cancer and would be suitable for diagnostic purposes. Unlike prostate cancer, in which GSTP1 is methylated in over 90% of cancers (128, 129), no single gene in ovarian cancer has been identified as being methylated in more than a relatively small proportion of cancers. Although new genome-wide approaches may discover such a gene(s), it is quite likely that a panel of methylated genes will be required to detect ovarian cancer at sufficient specificity and sensitivity. A combination of genes that are commonly methylated in cancer and genes that are methylated specifically in ovarian cancer is the most likely methylation signature capable of distinguishing ovarian cancers from neoplasms of other organs and from benign disease.

The detection of *RASSF1A* methylation in body fluids promises to be a useful marker for early cancer detection (60). In a recent feasibility study, tumor-specific hypermethylation of at least one of a panel of six tumor suppressor gene promoters, including *RASSF1A, BRCA1, APC, p14, p16*, and *DAPK*, could be detected in the serum or plasma of ovarian cancer patients with 100% specificity and 82% sensitivity (36), including 13/17 cases of stage I disease. Methylation was observed in only one peritoneal fluid sample from 15 stage IA or B patients, but 11/15 paired sera were positive for methylation (36). In addition to proof of principle, these data indicate that circulating ovarian tumor DNA is more readily accessible in the bloodstream than in the peritoneum, consistent with previous studies (127).

Although several limitations still exist, including the sensitivity of methylation assays relative to the amount of circulating tumor DNA, in principal, detection of specific epigenetic markers in the circulation of patients appears a promising candidate for the detection of early stage ovarian cancer.

4 Conclusions

There is an ever-increasing literature detailing epigenetically regulated genes in ovarian cancer, in particular, hypermethylated and silenced genes. However, many of these reported changes remain unverified in independent studies. Moreover, the frequency of methylation detection for individual genes can vary widely between studies. This variability in detection is likely a result of disparate tumor cohorts, DNA integrity, and assay platform and design. There are now several validated assays to assess DNA methylation, including high-throughput quantitative approaches; however, it should be emphasized that regardless of the approach chosen, careful assay design and correct interpretation of the results are critical in determining the true methylation frequency of a given chromosomal region.

To date, no gene(s) have been identified that are methylated and silenced in a high proportion of ovarian cancer cells. However, only a fraction of potential methylation targets have been examined. It is quite likely that a shift from candidate gene to genome-wide array-based approaches will aid in the discovery of methylated genes (23, 59, 130). We would anticipate from earlier studies that some of these methylation targets will be specific to the majority of cells in a particular stratified group of ovarian cancers, such as histological phenotype or cancers with acquired resistance to chemotherapy. Like other array-based data discovery platforms, methylated genes identified by high-throughput screening approaches will require careful analysis and validation. In particular, it will be important to concurrently analyze expression and DNA sequence changes in matched clinical samples to allow accurate analysis of methylation profiling data. Data analysis will also rely on prior knowledge of normal levels of DNA methylation in the ovary to establish a baseline from which to identify alterations in ovarian cancer which will, in part, be aided by the establishment of the Human Epigenomic Project (131).

In combination with genetic changes, it is clear that a distinct set of epigenetic changes underlie ovarian cancer initiation and development. Identifying the methylation signature of ovarian cancer cells will likely lead to a greater understanding of the molecular pathways causing ovarian cancer progression (115). Precise functional and genetic studies will be necessary to determine which epigenetic events are critical to tumorigenesis and thus have biological consequences, when compared with "bystander" genes that are methylated and selected during tumor development, perhaps due to epigenetic silencing of large chromosomal regions containing tumor suppressor genes (132), despite having no immediate effect on tumor phenotype (117).

Finally, the identification of epigenetic changes that correlate with clinicopathological parameters and patient outcome may provide new markers of clinical benefit. There is now accumulating evidence that epigenetic biomarkers offer great potential in the detection of cancer in its earliest stages and accurate assessment of individual risk. Moreover, epigenetically silenced cancer genes offer new targets for therapeutic approaches based on reexpression of tumor suppressor genes via demethylation and deacetylating drugs (133). The next decade will determine whether the promise of epigenetic markers holds true.

Acknowledgments Research in the authors' laboratories is supported by the Gynaecological Oncology (GO) Fund of the Royal Hospital for Women Foundation, Sydney, Australia; the Cancer Institute New South Wales; and the National Health and Medical Research Council (NH&MRC) of Australia. Philippa M. O'Brien is the recipient of a Cancer Institute NSW Career Development and Support Fellowships. Susan J. Clark is a NH&MRC Principal Research Fellow.

References

1. Jemal, A., Murray, T., Ward, E., Samuels, A., Tiwari, R.C., Ghafoor, A., Feuer, E.J., and Thun, M.J. (2005) Cancer statistics, 2005. CA Cancer J Clin 55, 10–30.

2. Barnholtz-Sloan, J.S., Schwartz, A.G., Qureshi, F., Jacques, S., Malone, J., and Munkarah, A.R. (2003) Ovarian cancer: changes in patterns at diagnosis and relative survival over the last three decades. Am J Obstet Gynecol 189, 1120–1127.
3. Wolffe, A.P., and Matzke, M.A. (1999) Epigenetics: regulation through repression. Science 286, 481–486.
4. Jones, P.A., and Baylin, S.B. (2002) The fundamental role of epigenetic events in cancer. Nat Rev Genet 3, 415–428.
5. Bird, A. (1986) CpG-rich islands and the function of DNA methylation. Nature 321, 209–213.
6. Gardiner-Garden, M., and Frommer, M. (1987) CpG islands in vertebrate genomes. J Mol Biol 196, 261–282.
7. Clark, S.J., and Melki, J. (2002) DNA methylation and gene silencing in cancer: which is the guilty party? Oncogene 21, 5380–5387.
8. Lund, A.H., and van Lohuizen, M. (2004) Epigenetics and cancer. Genes Dev 18, 2315–2335.
9. Stirzaker, C., Song, J.Z., Davidson, B., and Clark, S.J. (2004) Transcriptional gene silencing promotes DNA hypermethylation through a sequential change in chromatin modifications in cancer cells. Cancer Res 64, 3871–3877.
10. Ehrlich, M. (2002) DNA methylation in cancer: too much, but also too little. Oncogene 21, 5400–5413.
11. Bird, A. (1996) The relationship of DNA methylation to cancer. Cancer Surv 28, 87–101.
12. Egger, G., Liang, G., Aparicio, A., and Jones, P.A. (2004) Epigenetics in human disease and prospects for epigenetic therapy. Nature 429, 457–463.
13. Esteller, M. (2002) CpG island hypermethylation and tumor suppressor genes: a booming present, a brighter future. Oncogene 21, 5427–5440.
14. Herman, J.G., and Baylin, S.B. (2003) Gene silencing in cancer in association with promoter hypermethylation. N Engl J Med 349, 2042–2054.
15. Jones, P.A., and Laird, P.W. (1999) Cancer epigenetics comes of age. Nat Genet 21, 163–167.
16. Feinberg, A.P., and Tycko, B. (2004) The history of cancer epigenetics. Nat Rev Cancer 4, 143–153.
17. Walsh, C.P., Chaillet, J.R., and Bestor, T.H. (1998) Transcription of IAP endogenous retroviruses is constrained by cytosine methylation. Nat Genet 20, 116–117.
18. Alves, G., Tatro, A., and Fanning, T. (1996) Differential methylation of human LINE-1 retrotransposons in malignant cells. Gene 176, 39–44.
19. Costello, J.F., and Plass, C. (2001) Methylation matters. J Med Genet 38, 285–303.
20. Eden, A., Gaudet, F., Waghmare, A., and Jaenisch, R. (2003) Chromosomal instability and tumors promoted by DNA hypomethylation. Science 300, 455.
21. Gaudet, F., Hodgson, J.G., Eden, A., Jackson-Grusby, L., Dausman, J., Gray, J.W., Leonhardt, H., and Jaenisch, R. (2003) Induction of tumors in mice by genomic hypomethylation. Science 300, 489–492.
22. Tuck-Muller, C.M., Narayan, A., Tsien, F., Smeets, D.F., Sawyer, J., Fiala, E.S., Sohn, O.S., and Ehrlich, M. (2000) DNA hypomethylation and unusual chromosome instability in cell lines from ICF syndrome patients. Cytogenet Cell Genet 89, 121–128.
23. Ahluwalia, A., Yan, P., Hurteau, J.A., Bigsby, R.M., Jung, S.H., Huang, T.H., and Nephew, K.P. (2001) DNA methylation and ovarian cancer. I. Analysis of CpG island hypermethylation in human ovarian cancer using differential methylation hybridization. Gynecol Oncol 82, 261–268.
24. Strathdee, G., Appleton, K., Illand, M., Millan, D.W., Sargent, J., Paul, J., and Brown, R. (2001) Primary ovarian carcinomas display multiple methylator phenotypes involving known tumor suppressor genes. Am J Pathol 158, 1121–1127.
25. Grady, W.M. (2005) Epigenetic events in the colorectum and in colon cancer. Biochem Soc Trans 33, 684–688.
26. Costello, J.F., Fruhwald, M.C., Smiraglia, D.J., Rush, L.J., Robertson, G.P., Gao, X., Wright, F.A., Feramisco, J.D., Peltomaki, P., Lang, J.C., Schuller, D.E., Yu, L., Bloomfield, C.D.,

Caligiuri, M.A., Yates, A., Nishikawa, R., Su Huang, H., Petrelli, N.J., Zhang, X., O'Dorisio, M.S., Held, W.A., Cavenee, W.K., and Plass, C. (2000) Aberrant CpG-island methylation has non-random and tumour-type-specific patterns. Nat Genet 24, 132–138.
27. Esteller, M., and Herman, J.G. (2002) Cancer as an epigenetic disease: DNA methylation and chromatin alterations in human tumours. J Pathol 196, 1–7.
28. Melki, J.R., and Clark, S.J. (2002) DNA methylation changes in leukaemia. Semin Cancer Biol 12, 347–357.
29. Melki, J.R., Vincent, P.C., and Clark, S.J. (1999) Concurrent DNA hypermethylation of multiple genes in acute myeloid leukemia. Cancer Res 59, 3730–3740.
30. Esteller, M., Corn, P.G., Baylin, S.B., and Herman, J.G. (2001) A gene hypermethylation profile of human cancer. Cancer Res 61, 3225–3229.
31. Esteller, M., Silva, J.M., Dominguez, G., Bonilla, F., Matias-Guiu, X., Lerma, E., Bussaglia, E., Prat, J., Harkes, I.C., Repasky, E.A., Gabrielson, E., Schutte, M., Baylin, S.B., and Herman, J.G. (2000) Promoter hypermethylation and BRCA1 inactivation in sporadic breast and ovarian tumors. J Natl Cancer Inst 92, 564–569.
32. Baldwin, R.L., Nemeth, E., Tran, H., Shvartsman, H., Cass, I., Narod, S., and Karlan, B.Y. (2000) BRCA1 promoter region hypermethylation in ovarian carcinoma: a population-based study. Cancer Res 60, 5329–5333.
33. Buller, R.E., Shahin, M.S., Geisler, J.P., Zogg, M., De Young, B.R., and Davis, C.S. (2002) Failure of BRCA1 dysfunction to alter ovarian cancer survival. Clin Cancer Res 8, 1196–1202.
34. Catteau, A., Harris, W.H., Xu, C.F., and Solomon, E. (1999) Methylation of the BRCA1 promoter region in sporadic breast and ovarian cancer: correlation with disease characteristics. Oncogene 18, 1957–1965.
35. Geisler, J.P., Hatterman-Zogg, M.A., Rathe, J.A., and Buller, R.E. (2002) Frequency of BRCA1 dysfunction in ovarian cancer. J Natl Cancer Inst 94, 61–67.
36. Ibanez de Caceres, I., Battagli, C., Esteller, M., Herman, J.G., Dulaimi, E., Edelson, M.I., Bergman, C., Ehya, H., Eisenberg, B.L., and Cairns, P. (2004) Tumor cell-specific BRCA1 and RASSF1A hypermethylation in serum, plasma, and peritoneal fluid from ovarian cancer patients. Cancer Res 64, 6476–6481.
37. Rathi, A., Virmani, A.K., Schorge, J.O., Elias, K.J., Maruyama, R., Minna, J.D., Mok, S.C., Girard, L., Fishman, D.A., and Gazdar, A.F. (2002) Methylation profiles of sporadic ovarian tumors and nonmalignant ovaries from high-risk women. Clin Cancer Res 8, 3324–3331.
38. Teodoridis, J.M., Hall, J., Marsh, S., Kannall, H.D., Smyth, C., Curto, J., Siddiqui, N., Gabra, H., McLeod, H.L., Strathdee, G., and Brown, R. (2005) CpG island methylation of DNA damage response genes in advanced ovarian cancer. Cancer Res 65, 8961–8967.
39. Chan, K.Y., Ozcelik, H., Cheung, A.N., Ngan, H.Y., and Khoo, U.S. (2002) Epigenetic factors controlling the *BRCA1* and *BRCA2* genes in sporadic ovarian cancer. Cancer Res 62, 4151–4156.
40. Wilcox, C.B., Baysal, B.E., Gallion, H.H., Strange, M.A., and DeLoia, J.A. (2005) High-resolution methylation analysis of the BRCA1 promoter in ovarian tumors. Cancer Genet Cytogenet 159, 114–122.
41. Futreal, P.A., Liu, Q., Shattuck-Eidens, D., Cochran, C., Harshman, K., Tavtigian, S., Bennett, L.M., Haugen-Strano, A., Swensen, J., Miki, Y., and et al. (1994) BRCA1 mutations in primary breast and ovarian carcinomas. Science 266, 120–122.
42. Yu, Y., Xu, F., Peng, H., Fang, X., Zhao, S., Li, Y., Cuevas, B., Kuo, W.L., Gray, J.W., Siciliano, M., Mills, G.B., and Bast, R.C., Jr. (1999) *NOEY2 (ARHI)*, an imprinted putative tumor suppressor gene in ovarian and breast carcinomas. Proc Natl Acad Sci USA 96, 214–219.
43. Yuan, J., Luo, R.Z., Fujii, S., Wang, L., Hu, W., Andreeff, M., Pan, Y., Kadota, M., Oshimura, M., Sahin, A.A., Issa, J.P., Bast, R.C., Jr., and Yu, Y. (2003) Aberrant methylation and silencing of *ARHI*, an imprinted tumor suppressor gene in which the function is lost in breast cancers. Cancer Res 63, 4174–4180.
44. Yu, Y., Luo, R., Lu, Z., Wei Feng, W., Badgwell, D., Issa, J.P., Rosen, D.G., Liu, J., and Bast, R.C., Jr. (2005) Biochemistry and biology of *ARHI (DIRAS3)*, an imprinted tumor suppressor gene whose expression is lost in ovarian and breast cancers. Methods Enzymol 407, 455–468.

45. Kwong, J., Lee, J.Y., Wong, K.K., Zhou, X., Wong, D.T., Lo, K.W., Welch, W.R., Berkowitz, R.S., and Mok, S.C. (2006) Candidate tumor-suppressor gene DLEC1 is frequently downregulated by promoter hypermethylation and histone hypoacetylation in human epithelial ovarian cancer. Neoplasia 8, 268–278.
46. Campbell, I.G., Foulkes, W.D., Beynon, G., Davis, M., and Englefield, P. (1995) LOH and mutation analysis of CDKN2 in primary human ovarian cancers. Int J Cancer 63, 222–225.
47. Schultz, D.C., Vanderveer, L., Buetow, K.H., Boente, M.P., Ozols, R.F., Hamilton, T.C., and Godwin, A.K. (1995) Characterization of chromosome 9 in human ovarian neoplasia identifies frequent genetic imbalance on 9q and rare alterations involving 9p, including CDKN2. Cancer Res 55, 2150–2157.
48. Furonaka, O., Takeshima, Y., Awaya, H., Ishida, H., Kohno, N., and Inai, K. (2004) Aberrant methylation of *p14(ARF)*, *p15(INK4b)* and *p16(INK4a)* genes and location of the primary site in pulmonary squamous cell carcinoma. Pathol Int 54, 549–555.
49. Wong, T.S., Man, M.W., Lam, A.K., Wei, W.I., Kwong, Y.L., and Yuen, A.P. (2003) The study of *p16* and *p15* gene methylation in head and neck squamous cell carcinoma and their quantitative evaluation in plasma by real-time PCR. Eur J Cancer 39, 1881–1887.
50. Brown, I., Milner, B.J., Rooney, P.H., and Haites, N.E. (2001) Inactivation of the *p16INK4A* gene by methylation is not a frequent event in sporadic ovarian carcinoma. Oncol Rep 8, 1359–1362.
51. Hashiguchi, Y., Tsuda, H., Yamamoto, K., Inoue, T., Ishiko, O., and Ogita, S. (2001) Combined analysis of p53 and RB pathways in epithelial ovarian cancer. Hum Pathol 32, 988–996.
52. Katsaros, D., Cho, W., Singal, R., Fracchioli, S., Rigault De La Longrais, I.A., Arisio, R., Massobrio, M., Smith, M., Zheng, W., Glass, J., and Yu, H. (2004) Methylation of tumor suppressor gene *p16* and prognosis of epithelial ovarian cancer. Gynecol Oncol 94, 685–692.
53. Makarla, P.B., Saboorian, M.H., Ashfaq, R., Toyooka, K.O., Toyooka, S., Minna, J.D., Gazdar, A.F., and Schorge, J.O. (2005) Promoter hypermethylation profile of ovarian epithelial neoplasms. Clin Cancer Res 11, 5365–5369.
54. Niederacher, D., Yan, H.Y., An, H.X., Bender, H.G., and Beckmann, M.W. (1999) *CDKN2A* gene inactivation in epithelial sporadic ovarian cancer. Br J Cancer 80, 1920–1926.
55. Sellar, G.C., Watt, K.P., Rabiasz, G.J., Stronach, E.A., Li, L., Miller, E.P., Massie, C.E., Miller, J., Contreras-Moreira, B., Scott, D., Brown, I., Williams, A.R., Bates, P.A., Smyth, J.F., and Gabra, H. (2003) OPCML at 11q25 is epigenetically inactivated and has tumor-suppressor function in epithelial ovarian cancer. Nat Genet 34, 337–343.
56. Zhang, J., Ye, F., Chen, H.Z., Ye, D.F., Lu, W.G., and Xie, X. (2006) Deletion of *OPCML* gene and promoter methylation in ovarian epithelial carcinoma. Zhongguo Yi Xue Ke Xue Yuan Xue Bao 28, 173–177.
57. Takada, T., Yagi, Y., Maekita, T., Imura, M., Nakagawa, S., Tsao, S.W., Miyamoto, K., Yoshino, O., Yasugi, T., Taketani, Y., and Ushijima, T. (2004) Methylation-associated silencing of the Wnt antagonist *SFRP1* gene in human ovarian cancers. Cancer Sci 95, 741–744.
58. Yanaihara, N., Nishioka, M., Kohno, T., Otsuka, A., Okamoto, A., Ochiai, K., Tanaka, T., and Yokota, J. (2004) Reduced expression of *MYO18B*, a candidate tumor-suppressor gene on chromosome arm 22q, in ovarian cancer. Int J Cancer 112, 150–154.
59. Horak, P., Pils, D., Haller, G., Pribill, I., Roessler, M., Tomek, S., Horvat, R., Zeillinger, R., Zielinski, C., and Krainer, M. (2005) Contribution of epigenetic silencing of tumor necrosis factor-related apoptosis inducing ligand receptor 1 (DR4) to TRAIL resistance and ovarian cancer. Mol Cancer Res 3, 335–343.
60. Pfeifer, G.P., Yoon, J.H., Liu, L., Tommasi, S., Wilczynski, S.P., and Dammann, R. (2002) Methylation of the *RASSF1A* gene in human cancers. Biol Chem 383, 907–914.
61. Agathanggelou, A., Honorio, S., Macartney, D.P., Martinez, A., Dallol, A., Rader, J., Fullwood, P., Chauhan, A., Walker, R., Shaw, J.A., Hosoe, S., Lerman, M.I., Minna, J.D., Maher, E.R., and Latif, F. (2001) Methylation associated inactivation of RASSF1A from region 3p21.3 in lung, breast and ovarian tumours. Oncogene 20, 1509–1518.
62. Yoon, J.H., Dammann, R., and Pfeifer, G.P. (2001) Hypermethylation of the CpG island of the *RASSF1A* gene in ovarian and renal cell carcinomas. Int J Cancer 94, 212–217.

63. Chang, Y.S., Wang, L., Liu, D., Mao, L., Hong, W.K., Khuri, F.R., and Lee, H.Y. (2002) Correlation between insulin-like growth factor-binding protein-3 promoter methylation and prognosis of patients with stage I non-small cell lung cancer. Clin Cancer Res 8, 3669–3675.
64. Hanafusa, T., Yumoto, Y., Nouso, K., Nakatsukasa, H., Onishi, T., Fujikawa, T., Taniyama, M., Nakamura, S., Uemura, M., Takuma, Y., Yumoto, E., Higashi, T., and Tsuji, T. (2002) Reduced expression of insulin-like growth factor binding protein-3 and its promoter hypermethylation in human hepatocellular carcinoma. Cancer Lett 176, 149–158.
65. Wiley, A., Katsaros, D., Fracchioli, S., and Yu, H. (2006) Methylation of the insulin-like growth factor binding protein-3 gene and prognosis of epithelial ovarian cancer. Int J Gynecol Cancer 16, 210–218.
66. Petrocca, F., Iliopoulos, D., Qin, H.R., Nicoloso, M.S., Yendamuri, S., Wojcik, S.E., Shimizu, M., Di Leva, G., Vecchione, A., Trapasso, F., Godwin, A.K., Negrini, M., Calin, G.A., and Croce, C.M. (2006) Alterations of the tumor suppressor gene *ARLTS1* in ovarian cancer. Cancer Res 66, 10287–10291.
67. Akahira, J., Sugihashi, Y., Suzuki, T., Ito, K., Niikura, H., Moriya, T., Nitta, M., Okamura, H., Inoue, S., Sasano, H., Okamura, K., and Yaegashi, N. (2004b) Decreased expression of 14-3-3 sigma is associated with advanced disease in human epithelial ovarian cancer: its correlation with aberrant DNA methylation. Clin Cancer Res 10, 2687–2693.
68. Mhawech, P., Benz, A., Cerato, C., Greloz, V., Assaly, M., Desmond, J.C., Koeffler, H.P., Lodygin, D., Hermeking, H., Herrmann, F., and Schwaller, J. (2005) Downregulation of 14-3-3sigma in ovary, prostate and endometrial carcinomas is associated with CpG island methylation. Mod Pathol 18, 340–348.
69. Kaneuchi, M., Sasaki, M., Tanaka, Y., Shiina, H., Verma, M., Ebina, Y., Nomura, E., Yamamoto, R., Sakuragi, N., and Dahiya, R. (2004) Expression and methylation status of 14-3-3 sigma gene can characterize the different histological features of ovarian cancer. Biochem Biophys Res Commun 316, 1156–1162.
70. Akahira, J., Sugihashi, Y., Ito, K., Niikura, H., Okamura, K., and Yaegashi, N. (2004a) Promoter methylation status and expression of TMS1 gene in human epithelial ovarian cancer. Cancer Sci 95, 40–43.
71. Kaneuchi, M., Sasaki, M., Tanaka, Y., Shiina, H., Yamada, H., Yamamoto, R., Sakuragi, N., Enokida, H., Verma, M., and Dahiya, R. (2005) WT1 and WT1-AS genes are inactivated by promoter methylation in ovarian clear cell adenocarcinoma. Cancer 104, 1924–1930.
72. Terasawa, K., Sagae, S., Toyota, M., Tsukada, K., Ogi, K., Satoh, A., Mita, H., Imai, K., Tokino, T., and Kudo, R. (2004) Epigenetic inactivation of TMS1/ASC in ovarian cancer. Clin Cancer Res 10, 2000–2006.
73. Chien, J., Staub, J., Avula, R., Zhang, H., Liu, W., Hartmann, L.C., Kaufmann, S.H., Smith, D.I., and Shridhar, V. (2005) Epigenetic silencing of TCEAL7 (Bex4) in ovarian cancer. Oncogene 24, 5089–5100.
74. Wang, Z., Li, M., Lu, S., Zhang, Y., and Wang, H. (2006) Promoter hypermethylation of FANCF plays an important role in the occurrence of ovarian cancer through disrupting Fanconi anemia-BRCA pathway. Cancer Biol Ther 5, 256–260.
75. Shen, D.H., Chan, K.Y., Khoo, U.S., Ngan, H.Y., Xue, W.C., Chiu, P.M., Ip, P., and Cheung, A.N. (2005) Epigenetic and genetic alterations of p33ING1b in ovarian cancer. Carcinogenesis 26, 855–863.
76. Sutherland, K.D., Lindeman, G.J., Choong, D.Y., Wittlin, S., Brentzell, L., Phillips, W., Campbell, I.G., and Visvader, J.E. (2004) Differential hypermethylation of *SOCS* genes in ovarian and breast carcinomas. Oncogene 23, 7726–7733.
77. Terasawa, K., Toyota, M., Sagae, S., Ogi, K., Suzuki, H., Sonoda, T., Akino, K., Maruyama, R., Nishikawa, N., Imai, K., Shinomura, Y., Saito, T., and Tokino, T. (2006) Epigenetic inactivation of TCF2 in ovarian cancer and various cancer cell lines. Br J Cancer 94, 914–921.
78. Helleman, J., van Staveren, I.L., Dinjens, W.N., van Kuijk, P.F., Ritstier, K., Ewing, P.C., van der Burg, M.E., Stoter, G., and Berns, E.M. (2006) Mismatch repair and treatment resistance in ovarian cancer. BMC Cancer 6, 201.

79. Strathdee, G., MacKean, M.J., Illand, M., and Brown, R. (1999) A role for methylation of the hMLH1 promoter in loss of hMLH1 expression and drug resistance in ovarian cancer. Oncogene 18, 2335–2341.
80. Berger, S.L. (2002) Histone modifications in transcriptional regulation. Curr Opin Genet Dev 12, 142–148.
81. Gregory, P.D., Wagner, K., and Horz, W. (2001) Histone acetylation and chromatin remodeling. Exp Cell Res 265, 195–202.
82. Rice, J.C., and Allis, C.D. (2001) Histone methylation versus histone acetylation: new insights into epigenetic regulation. Curr Opin Cell Biol 13, 263–273.
83. Fischle, W., Wang, Y., and Allis, C.D. (2003) Histone and chromatin cross-talk. Curr Opin Cell Biol 15, 172–183.
84. Lachner, M., O'Carroll, D., Rea, S., Mechtler, K., and Jenuwein, T. (2001) Methylation of histone H3 lysine 9 creates a binding site for HP1 proteins. Nature 410, 116–120.
85. Yu, Y., Fujii, S., Yuan, J., Luo, R.Z., Wang, L., Bao, J., Kadota, M., Oshimura, M., Dent, S.R., Issa, J.P., and Bast, R.C., Jr. (2003) Epigenetic regulation of ARHI in breast and ovarian cancer cells. Ann N Y Acad Sci 983, 268–277.
86. Caslini, C., Capo-chichi, C.D., Roland, I.H., Nicolas, E., Yeung, A.T., and Xu, X.X. (2006) Histone modifications silence the GATA transcription factor genes in ovarian cancer. Oncogene 25, 5446–5461.
87. Hibbs, K., Skubitz, K.M., Pambuccian, S.E., Casey, R.C., Burleson, K.M., Oegema, T.R., Jr., Thiele, J.J., Grindle, S.M., Bliss, R.L., and Skubitz, A.P. (2004) Differential gene expression in ovarian carcinoma: identification of potential biomarkers. Am J Pathol 165, 397–414.
88. Honda, H., Pazin, M.J., Ji, H., Wernyj, R.P., and Morin, P.J. (2006) Crucial roles of Sp1 and epigenetic modifications in the regulation of the CLDN4 promoter in ovarian cancer cells. J Biol Chem 281, 21433–21444.
89. Hough, C.D., Sherman-Baust, C.A., Pizer, E.S., Montz, F.J., Im, D.D., Rosenshein, N.B., Cho, K.R., Riggins, G.J., and Morin, P.J. (2000) Large-scale serial analysis of gene expression reveals genes differentially expressed in ovarian cancer. Cancer Res 60, 6281–6287.
90. Lu, K.H., Patterson, A.P., Wang, L., Marquez, R.T., Atkinson, E.N., Baggerly, K.A., Ramoth, L.R., Rosen, D.G., Liu, J., Hellstrom, I., Smith, D., Hartmann, L., Fishman, D., Berchuck, A., Schmandt, R., Whitaker, R., Gershenson, D.M., Mills, G.B., and Bast, R.C., Jr. (2004) Selection of potential markers for epithelial ovarian cancer with gene expression arrays and recursive descent partition analysis. Clin Cancer Res 10, 3291–3300.
91. Rangel, L.B., Agarwal, R., D'Souza, T., Pizer, E.S., Alo, P.L., Lancaster, W.D., Gregoire, L., Schwartz, D.R., Cho, K.R., and Morin, P.J. (2003) Tight junction proteins claudin-3 and claudin-4 are frequently overexpressed in ovarian cancer but not in ovarian cystadenomas. Clin Cancer Res 9, 2567–2575.
92. Zhu, Y., Brannstrom, M., Janson, P.O., and Sundfeldt, K. (2006) Differences in expression patterns of the tight junction proteins, claudin 1, 3, 4 and 5, in human ovarian surface epithelium as compared to epithelia in inclusion cysts and epithelial ovarian tumours. Int J Cancer 118, 1884–1891.
93. Cheng, P., Schmutte, C., Cofer, K.F., Felix, J.C., Yu, M.C., and Dubeau, L. (1997) Alterations in DNA methylation are early, but not initial, events in ovarian tumorigenesis. Br J Cancer 75, 396–402.
94. Rose, S.L., Fitzgerald, M.P., White, N.O., Hitchler, M.J., Futscher, B.W., De Geest, K., and Domann, F.E. (2006) Epigenetic regulation of maspin expression in human ovarian carcinoma cells. Gynecol Oncol 102, 319–324.
95. Gupta, A., Godwin, A.K., Vanderveer, L., Lu, A., and Liu, J. (2003) Hypomethylation of the synuclein gamma gene CpG island promotes its aberrant expression in breast carcinoma and ovarian carcinoma. Cancer Res 63, 664–673.
96. Czekierdowski, A., Czekierdowska, S., Wielgos, M., Smolen, A., Kaminski, P., and Kotarski, J. (2006) The role of CpG islands hypomethylation and abnormal expression of neuronal protein synuclein-gamma (SNCG) in ovarian cancer. Neuro Endocrinol Lett 27.

97. Menendez, L., Benigno, B.B., and McDonald, J.F. (2004) L1 and HERV-W retrotransposons are hypomethylated in human ovarian carcinomas. Mol Cancer 3, 12.
98. Narayan, A., Ji, W., Zhang, X.Y., Marrogi, A., Graff, J.R., Baylin, S.B., and Ehrlich, M. (1998) Hypomethylation of pericentromeric DNA in breast adenocarcinomas. Int J Cancer 77, 833–838.
99. Qu, G., Dubeau, L., Narayan, A., Yu, M.C., and Ehrlich, M. (1999) Satellite DNA hypomethylation vs. overall genomic hypomethylation in ovarian epithelial tumors of different malignant potential. Mutat Res 423, 91–101.
100. Widschwendter, M., Jiang, G., Woods, C., Muller, H.M., Fiegl, H., Goebel, G., Marth, C., Muller-Holzner, E., Zeimet, A.G., Laird, P.W., and Ehrlich, M. (2004a) DNA hypomethylation and ovarian cancer biology. Cancer Res 64, 4472–4480.
101. Widschwendter, M., Siegmund, K.D., Muller, H.M., Fiegl, H., Marth, C., Muller-Holzner, E., Jones, P.A., and Laird, P.W. (2004b) Association of breast cancer DNA methylation profiles with hormone receptor status and response to tamoxifen. Cancer Res 64, 3807–3813.
102. Nishiyama, R., Qi, L., Lacey, M., and Ehrlich, M. (2005) Both hypomethylation and hypermethylation in a 0.2-kb region of a DNA repeat in cancer. Mol Cancer Res 3, 617–626.
103. Herman, J.G., Graff, J.R., Myohanen, S., Nelkin, B.D., and Baylin, S.B. (1996) Methylation-specific PCR: a novel PCR assay for methylation status of CpG islands. Proc Natl Acad Sci USA 93, 9821–9826.
104. Laird, P.W. (2003) The power and the promise of DNA methylation markers. Nat Rev Cancer 3, 253–266.
105. Eads, C.A., Danenberg, K.D., Kawakami, K., Saltz, L.B., Blake, C., Shibata, D., Danenberg, P.V., and Laird, P.W. (2000) MethyLight: a high-throughput assay to measure DNA methylation. Nucleic Acids Res 28, E32.
106. Rand, K.N., Ho, T., Qu, W., Mitchell, S.M., White, R., Clark, S.J., and Molloy, P.L. (2005) Headloop suppression PCR and its application to selective amplification of methylated DNA sequences. Nucleic Acids Res 33, e127.
107. Muller, H.M., Oberwalder, M., Fiegl, H., Morandell, M., Goebel, G., Zitt, M., Muhlthaler, M., Ofner, D., Margreiter, R., and Widschwendter, M. (2004) Methylation changes in faecal DNA: a marker for colorectal cancer screening? Lancet 363, 1283–1285.
108. Muller, H.M., Widschwendter, A., Fiegl, H., Ivarsson, L., Goebel, G., Perkmann, E., Marth, C., and Widschwendter, M. (2003) DNA methylation in serum of breast cancer patients: an independent prognostic marker. Cancer Res 63, 7641–7645.
109. Cottrell, S.E., and Laird, P.W. (2003) Sensitive detection of DNA methylation. Ann N Y Acad Sci 983, 120–130.
110. Hoque, M.O., Begum, S., Topaloglu, O., Jeronimo, C., Mambo, E., Westra, W.H., Califano, J.A., and Sidransky, D. (2004) Quantitative detection of promoter hypermethylation of multiple genes in the tumor, urine, and serum DNA of patients with renal cancer. Cancer Res 64, 5511–5517.
111. Hoque, M.O., Feng, Q., Toure, P., Dem, A., Critchlow, C.W., Hawes, S.E., Wood, T., Jeronimo, C., Rosenbaum, E., Stern, J., Yu, M., Trink, B., Kiviat, N.B., and Sidransky, D. (2006) Detection of aberrant methylation of four genes in plasma DNA for the detection of breast cancer. J Clin Oncol 24, 4262–4269.
112. Leung, W.K., To, K.F., Man, E.P., Chan, M.W., Bai, A.H., Hui, A.J., Chan, F.K., and Sung, J.J. (2005) Quantitative detection of promoter hypermethylation in multiple genes in the serum of patients with colorectal cancer. Am J Gastroenterol 100, 2274–2279.
113. Chan, M.W., Wei, S.H., Wen, P., Wang, Z., Matei, D.E., Liu, J.C., Liyanarachchi, S., Brown, R., Nephew, K.P., Yan, P.S., and Huang, T.H. (2005) Hypermethylation of 18S and 28S ribosomal DNAs predicts progression-free survival in patients with ovarian cancer. Clin Cancer Res 11, 7376–7383.
114. Huang, T.H., Perry, M.R., and Laux, D.E. (1999) Methylation profiling of CpG islands in human breast cancer cells. Hum Mol Genet 8, 459–470.
115. Wei, S.H., Chen, C.M., Strathdee, G., Harnsomburana, J., Shyu, C.R., Rahmatpanah, F., Shi, H., Ng, S.W., Yan, P.S., Nephew, K.P., Brown, R., and Huang, T.H. (2002) Methylation

microarray analysis of late-stage ovarian carcinomas distinguishes progression-free survival in patients and identifies candidate epigenetic markers. Clin Cancer Res 8, 2246–2252.
116. Wei, S.H., Balch, C., Paik, H.H., Kim, Y.S., Baldwin, R.L., Liyanarachchi, S., Li, L., Wang, Z., Wan, J.C., Davuluri, R.V., Karlan, B.Y., Gifford, G., Brown, R., Kim, S., Huang, T.H., and Nephew, K.P. (2006) Prognostic DNA methylation biomarkers in ovarian cancer. Clin Cancer Res 12, 2788–2794.
117. Balch, C., Huang, T.H., Brown, R., and Nephew, K.P. (2004) The epigenetics of ovarian cancer drug resistance and resensitization. Am J Obstet Gynecol 191, 1552–1572.
118. Teodoridis, J.M., Strathdee, G., and Brown, R. (2004) Epigenetic silencing mediated by CpG island methylation: potential as a therapeutic target and as a biomarker. Drug Resist Updat 7, 267–278.
119. Wei, S.H., Brown, R., and Huang, T.H. (2003) Aberrant DNA methylation in ovarian cancer: is there an epigenetic predisposition to drug response? Ann N Y Acad Sci 983, 243–250.
120. Brown, R., Hirst, G.L., Gallagher, W.M., McIlwrath, A.J., Margison, G.P., van der Zee, A.G., and Anthoney, D.A. (1997) hMLH1 expression and cellular responses of ovarian tumour cells to treatment with cytotoxic anticancer agents. Oncogene 15, 45–52.
121. Plumb, J.A., Strathdee, G., Sludden, J., Kaye, S.B., and Brown, R. (2000) Reversal of drug resistance in human tumor xenografts by 2´-deoxy-5-azacytidine-induced demethylation of the *hMLH1* gene promoter. Cancer Res 60, 6039–6044.
122. Gifford, G., Paul, J., Vasey, P.A., Kaye, S.B., and Brown, R. (2004) The acquisition of hMLH1 methylation in plasma DNA after chemotherapy predicts poor survival for ovarian cancer patients. Clin Cancer Res 10, 4420–4426.
123. Taniguchi, T., Tischkowitz, M., Ameziane, N., Hodgson, S.V., Mathew, C.G., Joenje, H., Mok, S.C., and D'Andrea, A.D. (2003) Disruption of the Fanconi anemia-BRCA pathway in cisplatin-sensitive ovarian tumors. Nat Med 9, 568–574.
124. Shridhar, V., Bible, K.C., Staub, J., Avula, R., Lee, Y.K., Kalli, K., Huang, H., Hartmann, L.C., Kaufmann, S.H., and Smith, D.I. (2001) Loss of expression of a new member of the DNAJ protein family confers resistance to chemotherapeutic agents used in the treatment of ovarian cancer. Cancer Res 61, 4258–4265.
125. Strathdee, G., Davies, B.R., Vass, J.K., Siddiqui, N., and Brown, R. (2004) Cell type-specific methylation of an intronic CpG island controls expression of the *MCJ* gene. Carcinogenesis 25, 693–701.
126. Strathdee, G., Vass, J.K., Oien, K.A., Siddiqui, N., Curto-Garcia, J., and Brown, R. (2005) Demethylation of the *MCJ* gene in stage III/IV epithelial ovarian cancer and response to chemotherapy. Gynecol Oncol 97, 898–903.
127. Hickey, K.P., Boyle, K.P., Jepps, H.M., Andrew, A.C., Buxton, E.J., and Burns, P.A. (1999) Molecular detection of tumour DNA in serum and peritoneal fluid from ovarian cancer patients. Br J Cancer 80, 1803–1808.
128. Lee, W.H., Morton, R.A., Epstein, J.I., Brooks, J.D., Campbell, P.A., Bova, G.S., Hsich, W.S., Isaacs, W.B., and Nelson, W.G. (1994) Cytidine methylation of regulatory sequences near the pi-class glutathione S-transferase gene accompanies human prostatic carcinogenesis. Proc Natl Acad Sci USA 91, 11733–11737.
129. Song, J.Z., Stirzaker, C., Harrison, J., Melki, J.R., and Clark, S.J. (2002) Hypermethylation trigger of the glutathione-S-transferase gene (*GSTP1*) in prostate cancer cells. Oncogene 21, 1048–1061.
130. Ushijima, T. (2005) Detection and interpretation of altered methylation patterns in cancer cells. Nat Rev Cancer 5, 223–231.
131. Jones, P.A., and Martienssen, R. (2005) A blueprint for a Human Epigenome Project: the AACR Human Epigenome Workshop. Cancer Res 65, 11241–11246.
132. Frigola, J., Song, J., Stirzaker, C., Hinshelwood, R.A., Peinado, M.A., and Clark, S.J. (2006) Epigenetic remodeling in colorectal cancer results in coordinate gene suppression across an entire chromosome band. Nat Genet 38, 540–549.
133. Yoo, C.B., and Jones, P.A. (2006) Epigenetic therapy of cancer: past, present and future. Nat Rev Drug Discov 5, 37–50.

Role of Genetic Polymorphisms in Ovarian Cancer Susceptibility: Development of an International Ovarian Cancer Association Consortium

Andrew Berchuck, Joellen M. Schildkraut, C. Leigh Pearce, Georgia Chenevix-Trench, and Paul D. Pharoah

The value of identifying women with an inherited predisposition to ovarian cancer has become readily apparent with the identification of the *BRCA1* and *BRCA2* genes. Women who inherit a deleterious mutation in one of these genes have a very high lifetime risk of ovarian cancer (10–60%) and lesser risks of fallopian tube and peritoneal cancer. These highly lethal cancers are almost completely prevented by prophylactic salpingoophorectomy. BRCA1/BRCA2 mutation testing has become the accepted standard of care in families with a strong history of breast and/or ovarian cancer. This approach has the potential to reduce ovarian cancer mortality by about 10%.

Although the ability to perform genetic testing for BRCA1 and BRCA2 represents a significant clinical advance, the frequency of mutations in these high penetrance ovarian cancer susceptibility genes in the general population is low (about 1 in 500 individuals). There is evidence to suggest that ovarian cancer susceptibility is affected by low penetrance genetic polymorphisms that are much more common. Although such polymorphisms would increase risk to a lesser degree, they could contribute to the development of many ovarian cancers by virtue of their high frequency in the population. It has been shown that the most powerful approach to studying low penetrance genes is an association study rather than a linkage study (1). Several groups have obtained funding to initiate such studies and these generally have focused on polymorphisms in candidate genes purportedly involved in ovarian biology or carcinogenesis.

Over the last decade, initial reports from ovarian cancer association studies have been disappointing. Although numerous positive associations have been reported, in most cases these have not been confirmed by other groups. The accumulated experience to date has served to highlight how difficult it is to conduct statistically and methologically rigorous ovarian cancer association studies. The main issues are summarized below.

1. Association studies of genetic polymorphisms require large numbers of subjects to have adequate power to identify low penetrance effects; but because of the relative rarity of ovarian cancer, most studies include hundreds of subjects rather than the thousands that are needed.
2. Because of the large number of polymorphisms in the human genome (about 10 million), false-positive associations are inevitably more frequent than true-positive

associations even when studies are conducted in a scientifically rigorous fashion. For example, using a significance level of 0.05, one false-positive result would be expected for every 20 polymorphisms examined.
3. Epithelial ovarian cancer is composed of several histological types that are somewhat heterogeneous with respect to predisposing risk factors and somatic mutations, and likewise it is possible that a given polymorphism may not affect the risk of all histologic types. The power of analyses stratified by histology is limited because of the smaller numbers of cases in each group.
4. Careful attention must be paid to issues of population stratification because both ovarian cancer rates and allele frequencies vary with race/ethnicity leaving open the possibility of residual confounding by race/ethnicity. This issue is one possible explanation for false-positive associations in the literature.
5. Epidemiological risk factor data should be considered in association studies to allow for examination of interactions between known etiologic factors (e.g., ovulation, endometriosis) and genetic risk factors. Because large samples sizes are needed to detect interactions, the power of these types of analyses in association studies has been extremely limited.

In view of the above-noted issues, over the last few years, collaborations have been initiated between groups in the US, UK, Europe, and Australia that are performing ovarian cancer association studies. To continue and expand this collaborative momentum, a meeting was held in Cambridge, England, in April 2005 to review the results of ongoing ovarian cancer association studies. The above-noted methodological issues that have slowed progress in the field were reviewed in detail. Presently, despite significant efforts by the various groups, little real progress has been achieved in understanding the contribution of genetic polymorphisms to ovarian cancer susceptibility. There was a consensus that many of the challenges inherent in this field can best be addressed by cooperative efforts. In view of this, the group unanimously decided to establish an ovarian cancer association consortium (OCAC). Shortly after the Cambridge meeting, an invitation to join the OCAC was extended to other groups known to be performing ovarian cancer association studies, and this was met with an enthusiastic response. Presently, 16 groups that are performing ovarian cancer case–control genetic association studies have joined the OCAC (Table 1). Together, over 10,000 cases and 15,000 controls have been accrued in these studies.

The work of the OCAC was funded in October 2005 by a generous donation from the family and friends of Kathryn Sladek Smith to the Ovarian Cancer Reseach Fund (www.ocrf.org). Biannual group meetings have been held for the past 2 years. The immediate goal of the group is to work together collaboratively to reach definitive results regarding polymorphisms that have been previously studied and to plan for future high quality studies. The development over time of a track record of collaboration and joint accomplishments will lay the groundwork for future studies, such as whole genome scans of thousands of polymorphisms.

Table 1 The Ovarian Cancer Association Consortium

United States

Duke University – North Carolina Ovarian Cancer Association Study
University of Southern California – Los Angeles Ovarian Cancer Association Study
University of Pittsburgh – HOPE (Hormones and Ovarian Cancer Prediction)
University of Washigton, Fred Hutchinson Cancer Institute – DOVE Study (Diseases of the Ovary and their Evaluation), OvCARE Study (Ovarian Cancer Contraceptive and Reproductive Experiences Study)
Mayo Clinic – Mayo Clinic Ovarian Cancer Association Study
Stanford University – San Francisco Bay Area Ovarian Cancer Genetic Epidemiology Study
Harvard University – New England Ovarian Cancer Case Control Study
Yale University – Connecticut Ovarian Cancer Study
University of California, Irvine – Orange County California Ovarian Cancer Study
University of South Florida, Moffitt Cancer Center – Tampa Bay Ovarian Cancer Study
University of Hawaii – Hawaii Ovarian Cancer Study

International

Cambridge University, UK – SEARCH East Anglian and West Midlands Study
University College London, UK – UK Ovarian Cancer Study
Queensland University, Australia – Australian Ovarian Cancer Study and Australian Cancer Study
Denmark – The Danish Malignant Ovarian Tumor study ("MALOVA")
NCI/Poland – Warsaw and Lodz Ovarian Cancer Study
Poland – West-Pomerania Region Hereditary Ovarian Cancer Study

1 Clinical Utility of Ovarian Cancer Susceptibility Polymorphisms

Although epidemiological risk factors for ovarian cancer have been identified, they are not sufficiently powerful to direct risk stratification in the clinic. Presently, ovarian cancer risk stratification is *not* used to guide clinical surveillance or interventions in the vast majority of women, other than in those rare individuals with mutations in the *BRCA* or *HNPCC* genes. The long-term goal of the OCAC is to identify a panel of ovarian cancer susceptibility polymorphisms that can be used in combination with known epidemiological risk factors such as family history, parity, and oral contraceptive use to better stratify ovarian cancer risk. We envision a future in which reduction of ovarian cancer incidence and mortality will be accomplished by implementation of screening and prevention interventions that focus on women defined as high risk, based on genetic and epidemiological risk factors. Such a focused approach likely will be more feasible and cost-effective than population-based approaches, given the relative rarity of ovarian cancer. Identifying genetic risk factors will also likely lead to improved understanding of the underlying biology and etiology of ovarian cancer and ultimately results in better ways of treating the disease.

Ovarian cancer is a highly lethal disease because most cases are detected at an advanced stage. Several obstacles to early detection of ovarian cancer exist, including its relative rarity, the occult location of the ovaries, and the lack of a well-defined preinvasive lesion. Despite these challenges, intensive efforts aimed at the development of a screening test are ongoing. In addition to screening strategies, the protective effect of oral contraceptives, pregnancy, and NSAIDs against ovarian cancer provides evidence that risk reduction through preventive approaches may be possible. In view of the relative rarity of ovarian cancer, both screening and prevention approaches likely would be most cost effective if focused on populations at increased risk, based on epidemiological and genetic risk factors.

The next sections summarize the present understanding of the contributions of epidemiological risk factors and genetic susceptibility to ovarian cancer risk.

2 Epidemiology of Ovarian Cancer

In addition to genetic susceptibility, reproductive behaviors are the other main risk factors for ovarian cancer. Both pregnancy and use of oral contraceptives (OCs) dramatically reduce ovarian cancer incidence (2). Women who have three children or use OCs for more than 5 years have more than a 50% risk reduction. It is thought that reductions in numbers of lifetime ovulations due to pregnancy, OC use, and breastfeeding may decrease risk by reducing gonadotropin levels, oxidative stress, DNA replication errors, and inclusion cyst formation in the ovarian epithelium. In addition, both pregnancy and use of OC are characterized by a protective progestagenic hormonal milieu (2, 3), and it has been suggested that this may reduce ovarian cancer risk by stimulating apoptosis of genetically damaged ovarian epithelial cells that otherwise might eventually evolve a fully transformed phenotype (4, 5). This may account for the observation that the protective effect of pregnancy and OCs is far greater than the extent to which lifetime ovulatory cycles are reduced (2). It has been suggested that combination OCs with high progestin potency were associated with a greater ovarian cancer risk reduction than those with low progestin potency (6, 7).

Additional risk factors apart from those that affect hormonal events and ovulation have been identified. Most notably, it has been shown that tubal ligation and hysterectomy reduce ovarian cancer risk by about 20–50% (2), perhaps by interrupting the access of perineal carcinogens such as talc to the ovary. In addition, endometriosis is associated with a two to threefold increased risk, particularly for clear cell and endometrioid cancers (8). Ovarian cancer incidence also has been noted to be higher in Northern regions with lower sunlight exposure (9). Finally, there is evidence that NSAIDs and other antiinflammatory drugs reduce ovarian cancer risk, as has also been noted for colon and breast cancer (10).

3 Genetic Susceptibility

Population-based case–control studies have described a two to threefold increased risk in first degree relatives of ovarian cancer patients. In principle, the familial aggregation of ovarian cancer may be the result of genetic or nongenetic factors that are shared within families. Twin studies that compare the concordance of ovarian cancer between monozygotic and dizygotic twins have shown that most of the excess familial risk of ovarian cancer is due to genetic factors (11). About 10% of invasive epithelial ovarian cancers are attributable to inherited mutations in high penetrance genes: *BRCA1* (3–6%), *BRCA2* (1–3%), *HNPCC* DNA mismatch repair genes (1–2%) (12, 13). Most deleterious BRCA mutations encode truncated protein products, although missense mutations that alter a single amino acid in BRCA1 or BRCA2 have been found to segregate with disease in a handful of familial ovarian cancer clusters (14, 15). Inheritance of a BRCA mutation increases lifetime risk of ovarian cancer from a baseline of 1.5% to about 15–25% in BRCA2 carriers and 20–40% in BRCA1 carriers (16–18). Highly penetrant germline BRCA mutations are rare, however, and are carried by less than 1 in 500 individuals in most populations, with the notable exception of Ashkenazi Jews (1 in 40 carrier rate) (19). The ability to identify BRCA mutation carriers is an exciting advance, as these women can consider oophorectomy and other approaches aimed at decreasing ovarian cancer mortality (12, 20). On the other hand, because BRCA mutations are rare, the overall impact on mortality will be inevitably small.

Rare, high penetrance susceptibility alleles for many cancer types have been cloned by focusing on families with multiple and/or early onset cases. More recently, it has been hypothesized that common, weakly penetrant alleles may exist, which contribute to the burden of cancers classified as sporadic. Several million common genetic variants (polymorphisms) have been identified in the human genome (21–25). The most common of these polymorphisms involves substitution of a single nucleotide (SNP). Many of these SNPs are located either outside genes, in introns, or in the coding sequence of genes, and are "silent" because they do not alter the amino acid encoded. However, some SNPs that change a single amino acid may significantly alter the activity of a protein or its interactions with other molecules. SNPs that arise in introns or promoter regions may also alter expression of the protein by affecting transcription. In addition, insertion/deletion polymorphisms may occur in repetitive DNA sequences. Some trinucleotide repeats encode a stretch of a single amino acid, and variant alleles may alter the number of amino acid residues.

All genes have numerous polymorphisms, and current estimates suggest that on average there is one common SNP for every 300 bp across the genome. Identification of common polymorphisms that predispose more weakly to cancer involves association studies using groups of individuals with a given type of cancer and unaffected controls (1, 25). Although the potential effects of these polymorphisms on risk are less striking than seen with BRCA mutations, they could account for a larger fraction of ovarian cancer cases by virtue of their high prevalence. There are two approaches that can be taken to association studies – direct and indirect. In the direct approach, putative functional variants are studied in the expectation that they are causally related to the disease of interest. Alternatively, the indirect approach takes advantage of the fact that polymorphisms in physical proximity are often inherited together as a haplotype block. The elucidation of the haplotype structure of genes is facilitating association studies by reducing the number of SNPs that must be examined in each gene (http://www.hapmap.org/) because of the correlated nature of the SNPs.

4 Link Between Epidemiological and Genetic Risk Factors

Ovarian cancer risk is quite likely determined by a complex interaction between various inherited and acquired factors. For example, it has been proposed that ovulation may increase ovarian cancer risk by increasing mutations in the epithelium that occur due to spontaneous errors in DNA synthesis or oxidative stress at the ovulatory site. If so, polymorphisms in genes involved in DNA repair or metabolism of free radicals could affect ovarian cancer risk. Similarly, any increased risk of ovarian cancer associated with talc use and other exogenous carcinogens could be modified by genes that affect xenobiotic metabolism. It has been proposed that high levels of gonadotropins associated with ovulation may stimulate sex steroid hormone production, which may enhance proliferation and transformation in the ovarian epithelium. Thus, polymorphisms in genes, which regulate and facilitate these processes, such as gonadotropin releasing hormone, the androgen receptor, and genes involved in sex steroid hormone biosynthesis and metabolism could affect ovarian cancer susceptibility. In addition, it is thought that the progestagenic milieu of pregnancy and OCs may have a protective effect by virtue of increasing apoptosis of ovarian epithelial cells that have undergone genetic damage. Thus, polymorphisms in the progesterone receptor or its downstream effectors could affect ovarian cancer risk. Likewise, the relationship between low sunlight exposure and increased ovarian cancer risk could be attributable to vitamin D activity, and polymorphisms in genes involved in its action could be a determinant of risk.

5 Review of Prior Ovarian Cancer Association Studies

Prior reports have examined the relationship between polymorphisms in several candidate genes and ovarian cancer risk. This includes the progesterone receptor (26–33), androgen receptor (34, 35), CYP17 (36, 37), p53 (38, 39), prohibitin (40), epoxide hydrolase (41, 42), BRCA1 and BRCA2 (43, 44), and others. Positive associations reported by some groups have not been confirmed by others, and this is attributable to chance; however, methodological weaknesses including using hospital- rather than population-based controls and employing controls that are poorly matched with respect to the presence of ovaries, age, and race (45). A few illustrative examples of some of these studies are described later.

5.1 Brca1/Brca2

Polymorphisms in *BRCA* genes are high priority ovarian cancer susceptibility candidates, since inactivation of these proteins strikingly increases ovarian cancer risk. Several members of the OCAC have examined common polymorphisms in BRCA1/

BRCA2. Initially, Ponder et al. reported that homozygosity for the H allele of the N372H polymorphism in *BRCA2* gene conferred a 1.3-fold increased risk of breast cancer (46). This is the only BRCA2 polymorphism with a rare allele frequency greater than 5% that results in an amino acid change. Dr. Chenevix-Trench examined N372H in UK and Australian ovarian cancer cases and controls and found a 1.7-fold increased risk (43). This polymorphism was also examined in the North Carolina Ovarian Cancer study, but no association was found between the H allele and risk of ovarian cancer (44). The overall odds ratio for HH homozygotes was 0.8 (95% CI = 0.4–1.5) and was similar in all subsets including invasive serous cases.

With regard to BRCA1, five amino acid changing polymorphisms have minor allele frequencies greater than 5% (Q356R, L871P, E1038g, K1183R, S1613G) (47). With the exception of Q356R, the others are highly correlated and only three haplotypes occur with a frequency of greater than 1.3% (48). In a population-based study of BRCA1 sequence variants in Southern California by Anton-Culver et al., the Q356R polymorphism was significantly associated with a family history among cases, suggesting that this polymorphism may influence risk (49). However, in the North Carolina Ovarian Cancer Study, neither the BRCA1 Q356R (OR = 0.9, 95% CI 0.5–1.4) nor P871L (OR = 0.9, 95% CI 0.6–1.9) polymorphisms were associated with ovarian cancer risk (44). A significant racial difference in allele frequencies was noted for the P871L polymorphism ($P = 0.64$ in Caucasians, $L = 0.76$ in African Americans, $p < 0.0001$).

5.2 Progesterone Receptor

In view of the protective effect of a progestin-dominant hormonal milieu (OC use, pregnancy), progesterone receptor variants with altered biological activity might affect ovarian cancer susceptibility. Polymorphisms in this gene have been studied in greater depth than those of any other gene, yet it remains unclear whether specific variants affect risk of ovarian cancer.

A German group reported that an insertion polymorphism in intron G of the progesterone receptor was associated with a 2.1-fold increased ovarian cancer risk (26, 27). It subsequently was shown that this intronic *Alu* insertion is in linkage disequilibrium with polymorphisms across the locus, including an amino acid changing SNP in exon 4 and a silent SNP in exon 5. However, several subsequent studies have failed to confirm an association between these polymorphisms and ovarian cancer risk (28–31). In addition, the evidence that this complex of polymorphisms, termed PROGINS, alters progesterone receptor function remains uncertain (50).

More recently, sequencing of the progesterone receptor gene by Pearce et al. at USC revealed the presence of four major haploytpe blocks within the gene (32). In this study, the association of PROGINS with ovarian cancer was explained by its cosegregation with the minor allele of the SNP rs608995. Homozygosity for the minor allele was seen in 4% of 387 controls compared with 11.2% of 267 cases (OR = 3.0; 95% CI = 1.63–5.89).

In addition to polymorphisms in the exons and introns of the progesterone receptor gene, additional polymorphisms have been identified in the promoter region (51). The A allele of the +331SNP creates an unique transcriptional start site that favors production of the progesterone receptor B (PR-B) isoform over progesterone receptor A (PR-A) (51). The PR-A and PR-B isoforms are ligand-dependent members of the nuclear receptor family that are structurally identical except for an additional 164 amino acids at the N-terminus of PR-B, but their actions are distinct. The full length PR-B functions as a transcriptional activator and in the tissues where it is expressed, it is a mediator of various responses, including the proliferative response to estrogen or the combination of estrogen and progesterone (52). PR-A is a transcriptionally inactive dominant-negative repressor of steroid hormone transcription activity that is thought to oppose estrogen-induced proliferation. An association has been reported between the +331A allele of the progesterone receptor promoter polymorphism and increased susceptibility to endometrial (51) and breast cancers (53), although the breast cancer association has not been confirmed in two subsequent studies (32, 54). It was postulated that upregulation of PR-B in carriers of the +331A allele might enhance formation of these cancers because of an increased proliferative response.

Through collaborative efforts between two members of the OCAC (Duke and Australia), convincing evidence of association has been found between the +331A allele and ovarian cancer risk (33). Analyses involving the combined data set between these two studies showed a significant association between the +331A allele and decreased risk of endometrioid/clear cell cases (OR = 0.46, 95% CI = 0.23–0.92) (P = 0.027). The example underscores the importance of working together because the major finding is present among the less common endometrioid and clear cell subtypes of ovarian cancer that represent 21% of invasive cases. Endometriosis is known to increase risk of endometrioid and clear cell ovarian cancers, many of which may arise in ovarian deposits of endometriosis (8).

The literature is fraught with false-positive association studies of genetic susceptibility polymorphisms (25, 45), but several features mitigate the likelihood of this in the present study. First, the known protective benefit of progestins against ovarian cancer provides a preexisting biologic plausibility for the observed association. In addition, this collaborative effort showed a consistent effect across both the Duke and Australian study populations. Lastly, we have been able to combine the results from two additional OCAC members to provide further evidence of a true-positive association (Fig. 1). This is one of the first three variants that will be studied by the OCAC. Although the results appear convincingly positive, we cannot rule out publication bias as an issue for this association and therefore the combined efforts of the OCAC are necessary.

Finally, there is evidence to suggest that steroid hormones other than progesterone play a major role in ovarian carcinogenesis, both via affects on ovulation and direct effects on the ovarian epithelium. In view of this, polymorphisms in genes that comprise the estrogen, progesterone, androgen, and vitamin D receptor pathways also are high priority candidates

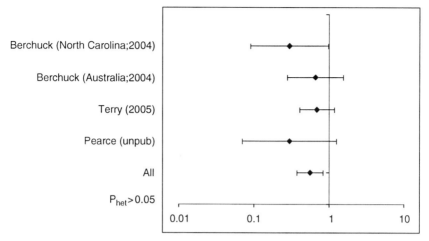

Fig. 1 Metaanalysis of the +331G/A progesterone receptor promoter polymorphism in endometrioid and clear cell ovarian cancers. The X axis represents the relative risk on a log scale. The *diamond* for each study represents the relative risk and the *error bar* the confidence intervals. A value of 1 indicates no association with values to the *right* representing increased risk and values to the *left* decreased risk. With all of the data combined there are 479 cases and 2,158 controls. The odds ratio for all of the data combined is 0.56 (95% CI 0.37–0.83)

5.3 DNA Repair Genes

The very strong association between mutated forms of *BRCA1*, *BRCA2*, and *HNPCC* mismatch repair genes and cancer underscores that DNA damage response pathways may be critical in the development of ovarian cancer. It is possible that variants in the genes that encode other proteins in the BRCA1 or BRCA2-associated complexes may adversely affect the efficiency of DNA repair and increase the risk of cancer, even if there are no high penetrance mutations in *BRCA1*, *BRCA2*, or *HNPCC* genes (e.g., *FANCD2, PMS2, BACH1, BARD1, GADD45, XPD, XRCC1*). In addition, polymorphisms associated with DNA damage response and the p53 DNA damage checkpoint may be important in the pathogenesis of ovarian cancer and affect the frequency of p53 overexpression and/or spectrum of p53 mutations (e.g., *p21, MDM2, ARF,* and *PIG3*). Genes involved in apoptosis also are appealing candidates, as failure to undergo cell death when DNA repair is not adequate may play a role in the development of some cancers.

5.4 Inflammation Pathways

Many of the established risk factors for ovarian cancer including ovulation and endometriosis have a link with inflammatory processes. Furthermore, it has been shown that analgesic use is associated with decreased ovarian cancer risk. In view of this, polymorphisms in genes involved in inflammation pathways could affect ovarian cancer risk. This includes genes that encode cytokines or other molecules related to cytokine activity (e.g., TNF-α, IL-1β and IL-6, IL-1RA and IL-10). In addition, polymorphisms in genes involved in analgesic drug metabolism, drug effects (e.g., cyclooxygenases), and those that mediate the actions of arachidonic acid metabolites affect ovarian cancer risk (e.g., CYP2C9, CYP3A4, PTGS1, PTGS1) and could modify the protective effect of analgesics against ovarian cancer.

5.5 Other Pathways

Additional pathways to consider include those involved in methylation and acetylation of genes and chromatin remodeling as well as DNA replication and cell cycle regulation. Genes that regulate angiogenesis, invasion and metastasis, stromal–epithelial interactions, and those shown to be overexpressed in ovarian cancers using genomic approaches also represent appealing candidates.

6 Ovarian Cancer Association Consortium

6.1 Candidate Gene Approaches

The OCAC will work together to validate initial associations between single SNPs in candidate genes and ovarian cancer that are reported by individual groups. Data will be pooled for joint analyses. It is becoming increasingly desirable to study genes using a comprehensive approach to "rule out" the involvement of a given gene with a given phenotype (e.g., ovarian cancer). Taking this approach a step further, studies of complete biological pathways (e.g., DNA repair) involving multiple genes are being conducted at increasing frequency (55–57). In future, the OCAC will increasingly attempt to examine a set of SNPs that capture as completely as possible the underlying population variability within the chosen genetic loci. Although a locus or gene will contain many SNPs, a few "tag" SNPs can provide most of the information on its pattern of genetic variation such that all of the variation in the locus is marked by the tag SNPs. This is the principle of the indirect approach through which it is unnecessary to identify the "key" SNP in a gene as long as it is coinherited or in linkage disequilibrium (LD) with a representative genotyped SNP. The International HapMap Project (HapMap) was organized to provide extensive genotype data to the scientific community for the purposes of identifying disease associations (http://www.hapmap.org/) and this resource can be used to select tag SNPs. We will select tag SNPs using standard methods (http://www.broad.mit.edu/mpg/tagger/) (58, 59).

There remain many important methodological issues to address in analyzing SNP data. For example, the optimal design strategy with regard to determining the number of samples that need to be genotyped is unclear in the context of preserving both financial and biological material resources. In addition, because both allele frequency and ovarian cancer rates vary by race/ethnicity, it will be important to consider the issue of population substructure among studies. In addition, the best way to explore the effect of multiple genes in a pathway is an area of active research (60).

The goal of genetic association studies is to determine whether a specific variant is associated with the risk of developing a given disease. However, the phenotypic expression of genetic variants is affected by environmental and behavioral factors. Information regarding known epidemiological risk factors should be incorporated into genetic association studies. For example, it is possible that the effect of certain polymorphisms on risk may only be manifest in women who are nulliparous or in those with a history of endometriosis. Because of the moderate size of most ovarian cancer association studies, it has not been possible for individual groups to perform meaningful analyses of gene–environment interactions. One of the aims of the consortium will be to establish a common data sheet that includes basic information relating to the major epidemiological risk factors. This will focus mainly on family history and reproductive risk factors. Analyses will be performed to examine interactions between specific risk factors, genetic polymorphisms, and ovarian cancer risk.

6.2 Whole Genome Studies

The discussion above focuses on association studies of polymorphisms in candidate genes that are selected based on a biological or epidemiological link to ovarian cancer. A potential pitfall of studies aimed at identification of ovarian cancer susceptibility polymorphisms using a candidate gene approach is the large number of polymorphisms in the genome. In addition, because our understanding of ovarian carcinogenesis is incomplete, many of the relevant genes may still be unidentified. An alternative strategy involves nonhypothesis-based high throughput approaches that examine thousands of polymorphisms across the genome to look for linkage. These whole genome approaches are arduous and generate many regions across the genome that must be studied further. Also, the optimal design for whole genome association scans is still an area of extensive debate. The infrastructure and working relationships established as the OCAC matures will lay the ground work for whole genome association studies in ovarian cancer.

Acknowledgments The Ovarian Cancer Association is funded by a generous donation from the family and friends of Kathryn Sladek Smith to the Ovarian Cancer Research Fund.

References

1. Risch, N. and Merikangas, K. The future of genetic studies of complex human diseases, Science, *273:* 1516–1517, 1996.
2. Whittemore, A.S., Harris, R. and Itnyre, J. Characteristics relating to ovarian cancer risk. Collaborative analysis of twelve US case-control studies. II. Invasive epithelial ovarian cancers in white women, Am J Epidemiol, *136:* 1184–1203, 1992.
3. Risch, H.A. Hormonal etiology of epithelial ovarian cancer, with a hypothesis concerning the role of androgens and progesterone, J Natl Cancer Inst, *90:* 1774–1786, 1998.
4. Rodriguez, G.C., Walmer, D.K., Cline, M., Krigman, H.R., Lessey, B.A., Whitaker, R.S., Dodge, R.K. and Hughes, C.L. Effect of progestin on the ovarian epithelium of macaques: Cancer prevention through apoptosis? J Soc Gynecol Invest, *5:* 271–276, 1998.
5. Rodriguez, G.C., Nagarsheth, N.P., Lee, K.L., Bentley, R.C., Walmer, D.K., Cline, M., Whitaker, R.S., Isner, P., Berchuck, A., Dodge, R.K. and Hughes, C.L. Progestin-induced apoptosis in the macaque ovarian epithelium: Differential regulation of transforming growth factor-beta, J Natl Cancer Inst, *94:* 50–60, 2002.
6. Schildkraut, J.M., Calingaert, B., Marchbanks, P.A., Moorman, P.G. and Rodriguez, G.C. Impact of progestin and estrogen potency in oral contraceptives on ovarian cancer risk, J Natl Cancer Inst, *94:* 32–38, 2002.
7. Pike, M.C., Pearce, C.L., Peters, R., Cozen, W., Wan, P. and Wu, A.H. Hormonal factors and the risk of invasive ovarian cancer: A population-based case-control study, Fertil Steril, *82:* 186–95, 2004.
8. Ness, R.B. Endometriosis and ovarian cancer: Thoughts on shared pathophysiology, Am J Obstet Gynecol, *189:* 280–294, 2003.
9. Lefkowitz, E.S. and Garland, C.F. Sunlight, vitamin D, and ovarian cancer mortality rates in US women, Int J Epidemiol, *23:* 1133–1136, 1994.
10. Fairfield, K.M., Hunter, D.J., Fuchs, C.S., Colditz, G.A. and Hankinson, S.E. Aspirin, other NSAIDs, and ovarian cancer risk, Cancer Causes Control, *13:* 535–542, 2002.
11. Lichtenstein, P., Holm, N.V., Verkasalo, P.K., Iliadou, A., Kaprio, J., Koskenvuo, M., Pukkala, E., Skytthe, A. and Hemminki, K. Environmental and heritable factors in the causation of cancer – analyses of cohorts of twins from Sweden, Denmark, and Finland, N Engl J Med, *343:* 78–85, 2000.
12. Berchuck, A., Schildkraut, J.M., Marks, J.R. and Futreal, P.A. Managing hereditary ovarian cancer risk, Cancer, *86:* 2517–2524, 1999.
13. Frank, T.S., Manley, S.A., Olopade, O.I., Cummings, S., Garber, J.E., Bernhardt, B., Antman, K., Russo, D., Wood, M.E., Mullineau, L., Isaacs, C., Peshkin, B., Buys, S., Venne, V., Rowley, P.T., Loader, S., Offit, K., Robson, M., Hampel, H., Brener, D., Winer, E.P., Clark, S., Weber, B., Strong, L.C. and Thomas, A. Sequence analysis of BRCA1 and BRCA2: Correlation of mutations with family history and ovarian cancer risk, J Clin Oncol, *16:* 2417–2425, 1998.
14. Couch, F.J. and Weber, B.L. Mutations and polymorphisms in the familial early-onset breast cancer *(BRCA1)* gene, Hum Mutat, *8:* 8–18, 1996.
15. Shattuck-Eidens, D., Oliphant, A., McClure, M., McBride, C., Gupte, J., Rubano, T., Pruss, D., Tavtigian, S.V., Teng, D.H., Adey, N., Staebell, M., Gumpper, K., Lundstrom, R., Hulick, M., Kelly, M., Holmen, J., Lingenfelter, B., Manley, S., Fujimura, F., Luce, M., Ward, B., Cannon-Albright, L., Steele, L., Offit, K., Thomas, A., et al. BRCA1 sequence analysis in women at high risk for susceptibility mutations. Risk factor analysis and implications for genetic testing, JAMA, *278:* 1242–1250, 1997.
16. Whittemore, A.S., Gong, G. and Itnyre, J. Prevalence and contribution of BRCA1 mutations in breast cancer and ovarian cancer: Results from three U.S. population-based case-control studies of ovarian cancer, Am J Hum Genet, *60:* 496–504, 1997.
17. Struewing, J.P., Hartge, P., Wacholder, S., Baker, S.M., Berlin, M., McAdams, M., Timmerman, M.M., Brody, L.C. and Tucker, M.A. The risk of cancer associated with specific mutations of BRCA1 and BRCA2 among Ashkenazi Jews, N Engl J Med, *336:* 1401–1408, 1997.

18. Risch, H.A., McLaughlin, J.R., Cole, D.E., Rosen, B., Bradley, L., Kwan, E., Jack, E., Vesprini, D.J., Kuperstein, G., Abrahamson, J.L., Fan, I., Wong, B. and Narod, S.A. Prevalence and penetrance of germline BRCA1 and BRCA2 mutations in a population series of 649 women with ovarian cancer, Am J Hum Genet, *68:* 700–710, 2001.
19. Szabo, C.I. and King, M.C. Invited editorial: Population genetics of BRCA1 and BRCA2, Am J Hum Genet, *60:* 1013–1020, 1997.
20. Guillem, J., Wood, W., Berchuck, Karlan, B., Mutch, D., Gagel, R., Weitzel, J., Morrow, M., Weber, B., Giardiello, F., Rodriguez-Bigas, M., Church, J., Gruber, S. and Offit, K. ASCO/SSO review of current role of risk-reducing surgery in common hereditary cancer syndromes, J Clin Oncol, *24:* 4642–60, 2006.
21. Cargill, M., Altshuler, D., Ireland, J., Sklar, P., Ardlie, K., Patil, N., Shaw, N., Lane, C.R., Lim, E.P., Kalyanaraman, N., Nemesh, J., Ziaugra, L., Friedland, L., Rolfe, A., Warrington, J., Lipshutz, R., Daley, G.Q. and Lander, E.S. Characterization of single-nucleotide polymorphisms in coding regions of human genes, Nat Genet, *22:* 231–8, 1999.
22. Halushka, M.K., Fan, J.B., Bentley, K., Hsie, L., Shen, N., Weder, A., Cooper, R., Lipshutz, R. and Chakravarti, A. Patterns of single-nucleotide polymorphisms in candidate genes for blood-pressure homeostasis, Nat Genet, *22:* 239–47, 1999.
23. Sachidanandam, R., Weissman, D., Schmidt, S.C., Kakol, J.M., Stein, L.D., Marth, G., Sherry, S., Mullikin, J.C., Mortimore, B.J., Willey, D.L., Hunt, S.E., Cole, C.G., Coggill, P.C., Rice, C.M., Ning, Z., Rogers, J., Bentley, D.R., Kwok, P.Y., Mardis, E.R., Yeh, R.T., Schultz, B., Cook, L., Davenport, R., Dante, M., Fulton, L., Hillier, L., Waterston, R.H., McPherson, J.D., Gilman, B., Schaffner, S., Van Etten, W.J., Reich, D., Higgins, J., Daly, M.J., Blumenstiel, B., Baldwin, J., Stange-Thomann, N., Zody, M.C., Linton, L., Lander, E.S. and Altshuler, D. A map of human genome sequence variation containing 1.42 million single nucleotide polymorphisms, Nature, *409:* 928–33, 2001.
24. The International HapMap Project, Nature, *426:* 789–96, 2003.
25. Carlson, C.S., Eberle, M.A., Kruglyak, L. and Nickerson, D.A. Mapping complex disease loci in whole-genome association studies, Nature, *429:* 446–452, 2004.
26. McKenna, N.J., Kieback, D.G., Carney, D.N., Fanning, M., McLinden, J. and Headon, D.R. A germline TaqI restriction fragment length polymorphism in the progesterone receptor gene in ovarian carcinoma, Br J Cancer, *71:* 451–455, 1995.
27. Rowe, S.M., Coughlan, S.J., McKenna, N.J., Garrett, E., Kieback, D.G., Carney, D.N. and Headon, D.R. Ovarian carcinoma-associated TaqI restriction fragment length polymorphism in intron G of the progesterone receptor gene is due to an Alu sequence insertion, Cancer Res, *55:* 2743–2745, 1995.
28. Spurdle, A.B., Webb, P.M., Purdie, D.M., Chen, X., Green, A. and Chenevix-Trench, G. No significant association between progesterone receptor exon 4 Val660Leu G/T polymorphism and risk of ovarian cancer, Carcinogenesis, *22:* 717–721, 2001.
29. Lancaster, J.M., Berchuck, A., Carney, M., Wiseman, R.W. and Taylor, J.A. Progesterone receptor gene and risk of breast and ovarian cancer, Br J Cancer, *78*: 277, 1997.
30. Manolitsas, T.P., Englefield, P., Eccles, D.M. and Campbell, I.G. No association of a 306-bp insertion polymorphism in the progesterone receptor gene with ovarian and breast cancer, Br J Cancer, *75:* 1398–1399, 1900.
31. Lancaster, J.M., Wenham, R.M., Halabi, S., Calingaert, B., Marks, J.R., Moorman, P.G., Bentley, R.C., Berchuck, A. and Schildkraut, J.M. No relationship between ovarian cancer risk and progesterone receptor gene polymorphism in a population-based, case-control study in North Carolina, Cancer Epidemiol Biomarkers Prev, *12:* 226–227, 2003.
32. Pearce, C.L., Hirschhorn, J.N., Wu, A.H., Burtt, N.P., Stram, D.O., Young, S., Kolonel, L.N., Henderson, B.E., Altshuler, D. and Pike, M.C. Clarifying the PROGINS allele association in ovarian and breast cancer risk: A haplotype-based analysis, J Natl Cancer Inst, *97:* 51–59, 2005.
33. Berchuck, A., Schildkraut, J.M., Wenham, R.M., Calingaert, B., Ali, S., Henriott, A., Halabi, S., Rodriguez, G.C., Gertig, D., Purdie, D.M., Kelemen, L., Spurdle, A.B., Marks, J. and Chenevix-Trench, G. Progesterone receptor promoter +331A polymorphism is associated with

increased risk of endometrioid and clear cell ovarian cancers, Cancer Epidemiol Biomarkers Prev, *13:* 2141–2147, 2004.
34. Spurdle, A.B., Webb, P.M., Chen, X., Martin, N.G., Giles, G.G., Hopper, J.L. and Chenevix-Trench, G. Androgen receptor exon 1 CAG repeat length and risk of ovarian cancer, Int J Cancer, *87:* 637–643, 2000.
35. Levine, D.A. and Boyd, J. The androgen receptor and genetic susceptibility to ovarian cancer: results from a case series, Cancer Res, *61:* 908–911, 2001.
36. Garner, E.I., Stokes, E.E., Berkowitz, R.S., Mok, S.C. and Cramer, D.W. Polymorphisms of the estrogen-metabolizing genes *CYP17* and catechol-*O*-methyltransferase and risk of epithelial ovarian cancer, Cancer Res, *62:* 3058–3062, 2002.
37. Spurdle, A.B., Chen, X., Abbazadegan, M., Martin, N., Khoo, S.K., Hurst, T., Ward, B., Webb, P.M. and Chenevix-Trench, G. CYP17 promotor polymorphism and ovarian cancer risk, Int J Cancer, *86:* 436–439, 2000.
38. Runnebaum, I.B., Tong, X.W., Konig, R., Zhao, H., Korner, K., Atkinson, E.N., Kreienberg, R., Kieback, D.G. and Hong, Z.c.t.Z.H. p53-based blood test for p53PIN3 and risk for sporadic ovarian cancer, Lancet, *345:* 994, 1995.
39. Lancaster, J.M., Brownlee, H.A., Wiseman, R.W. and Taylor, J. p53 polymorphism in ovarian and bladder cancer, Lancet, *346:* 182, 1995.
40. Spurdle, A.B., Purdie, D., Chen, X. and Chenevix-Trench, G. The prohibitin 3′ untranslated region polymorphism is not associated with risk of ovarian cancer, Cancer Res, *90:* 145–149, 2003.
41. Lancaster, J.M., Taylor, J.A., Brownlee, H.A., Bell, D.A., Berchuck, A. and Wiseman, R.W. Microsomal epoxide hydrolase polymorphism as a risk factor for ovarian cancer, Mol Carcinog, *17:* 160–162, 1996.
42. Spurdle, A.B., Purdie, D.M., Webb, P.M., Chen, X., Green, A. and Chenevix-Trench, G. The microsomal epoxide hydrolase Tyr113His polymorphism: Association with risk of ovarian cancer, Mol Carcinog, *30:* 71–78, 2001.
43. Auranen, A., Spurdle, A.B., Chen, X., Lipscombe, J., Purdie, D.M., Hopper, J.L., Green, A., Healey, C.S., Redman, K., Dunning, A.M., Pharoah, P.D., Easton, D.F., Ponder, B.A., Chenevix-Trench, G. and Novik, K.L. BRCA2 Arg372His polymorphism and epithelial ovarian cancer risk, Int J Cancer, *103:* 427–430, 2003.
44. Wenham, R.M., Schildkraut, J.M., McLean, K., Calingaert, B., Bentley, R.C., Marks, J.R. and Berchuck, A. Polymorphisms in BRCA1 and BRCA2 and risk of epithelial ovarian cancer, Clin Cancer Res, *9:* 4396–4403, 2003.
45. Thomas, D.C. and Witte, J.S. Point: Population stratification: A problem for case-control studies of candidate-gene associations? Cancer Epidemiol Biomarkers Prev, *11:* 505–512, 2002.
46. Healey, C.S., Dunning, A.M., Teare, M.D., Chase, D., Parker, L., Burn, J., Chang-Claude, J., Mannermaa, A., Kataja, V., Huntsman, D.G., Pharoah, P.D., Luben, R.N., Easton, D.F. and Ponder, B.A. A common variant in BRCA2 is associated with both breast cancer risk and prenatal viability, Nat Genet, *26:* 362–364, 2000.
47. Durocher, F., Shattuck-Eidens, D., McClure, M., Labrie, F., Skolnick, M.H., Goldgar, D.E. and Simard, J. Comparison of BRCA1 polymorphisms, rare sequence variants and/or missense mutations in unaffected and breast/ovarian cancer populations, Hum Mol Genet, *5:* 835–842, 1996.
48. Dunning, A.M., Chiano, M., Smith, N.R., Dearden, J., Gore, M., Oakes, S., Wilson, C., Stratton, M., Peto, J., Easton, D., Clayton, D. and Ponder, B.A. Common BRCA1 variants and susceptibility to breast and ovarian cancer in the general population, Hum Mol Genet, *6:* 285–289, 1997.
49. Janezic, S.A., Ziogas, A., Krumroy, L.M., Krasner, M., Plummer, S.J., Cohen, P., Gildea, M., Barker, D., Haile, R., Casey, G. and Anton-Culver, H. Germline BRCA1 alterations in a population-based series of ovarian cancer cases, Hum Mol Genet, *8:* 889–897, 1999.
50. Agoulnik, I.U., Tong, X.W., Fischer, D.C., Korner, K., Atkinson, N.E., Edwards, D.P., Headon, D.R., Weigel, N.L. and Kieback, D.G. A germline variation in the progesterone

receptor gene increases transcriptional activity and may modify ovarian cancer risk, J Clin Endocrinol Metab, *89:* 6340–6347, 2004.
51. DeVivo, I., Huggins, G.S., Hankinson, S.E., Lescault, P.J., Boezen, M., Colditz, G.A. and Hunter, D.J. A functional polymorphism in the promoter of the progesterone receptor gene associated with endometrial cancer risk, Proc Natl Acad Sci USA, *99:* 12263–12268, 2002.
52. Giangrande, P.H., Kimbrel, E.A., Edwards, D.P. and McDonnell, D.P. The opposing transcriptional activities of the two isoforms of the human progesterone receptor are due to differential cofactor binding, Mol Cell Biol, *20:* 3102–3115, 2000.
53. DeVivo, I., Hankinson, S.E., Colditz, G.A. and Hunter, D.J. A functional polymorphism in the progesterone receptor gene is associated with an increase in breast cancer risk, Cancer Res, *63:* 5236–5238, 2003.
54. Feigelson, H.S., Rodriguez, C., Jacobs, E.J., Diver, W.R., Thun, M.J. and Calle, E.E. No association between the progesterone receptor gene +331G/A polymorphism and breast cancer, Cancer Epidemiol Biomarkers Prev, *13:* 1084–5, 2004.
55. Lohmueller, K.E., Pearce, C.L., Pike, M., Lander, E.S. and Hirschhorn, J.N. Meta-analysis of genetic association studies supports a contribution of common variants to susceptibility to common disease, Nat Genet, *33:* 177–82, 2003.
56. Conti, D.V., Cortessis,V., Molitor, J. and Thomas, D.C. Bayesian modeling of complex metabolic pathways, Hum Hered, *56:* 83–93, 2003.
57. Conti, D.V. and Gauderman, W.J. SNPs, haplotypes, and model selection in a candidate gene region: The SIMPle analysis for multilocus data, Genet Epidemiol, *27:* 429–41, 2004.
58. Stram, D.O., Haiman, C.A., Hirschhorn, J.N., Altshuler, D., Kolonel, L.N., Henderson, B.E. and Pike, M.C. Choosing haplotype-tagging SNPS based on unphased genotype data using a preliminary sample of unrelated subjects with an example from the Multiethnic Cohort Study, Hum Hered, *55:* 27–36, 2003.
59. Johnson, G.C., Esposito, L., Barratt, B.J., Smith, A.N., Heward, J., Di Genova, G., Ueda, H., Cordell, H.J., Eaves, I.A., Dudbridge, F., Twells, R.C., Payne, F., Hughes, W., Nutland, S., Stevens, H., Carr, P., Tuomilehto-Wolf, E., Tuomilehto, J., Gough, S.C., Clayton, D.G. and Todd, J.A. Haplotype tagging for the identification of common disease genes, Nat Genet, *29:* 233–237, 2001.
60. Thomas, D.C. The need for a systematic approach to complex pathways in molecular epidemiology, Cancer Epidemiol Biomarkers Prev, *14:* 557–559, 2005.

MicroRNA in Human Cancer: One Step Forward in Diagnosis and Treatment

Lin Zhang, Nuo Yang, and George Coukos

Cancer is a disease involving multistep dynamic changes in the genome. However, studies to date on the cancer genome have focused most heavily on protein-coding genes, and our knowledge on alterations of the functional noncoding sequences in cancer is largely absent. MicroRNAs (miRNAs) are ~22 nucleotide (nt) noncoding RNAs, which regulate gene expression in a sequence-specific manner via translational inhibition or mRNA degradation. Mounting evidence shows that miRNAs may play an important role in tumor development, and a better understanding of their alteration in cancer genome and oncogenic property should contribute to the diagnosis and treatment of cancer.

1 miRNAs in Human Genome

miRNAs are endogenous ~22 nt noncoding small RNAs, which negatively regulate gene expression in a sequence-specific manner via mRNA degradation, transcriptional regulation, or translational repression (1–8). The human genome may contain ~1,000 miRNAs, and more than 300 of them have been identified by molecular cloning (8). Vertebrate miRNA targets are thought to be plentiful in number (9–13). Up to one-third of human mRNAs is predicted to be miRNA targets (12). Each miRNA can target about 200 transcripts directly or indirectly (14, 15), while more than one miRNA can converge on a single protein-encoding gene target (9–13). Therefore, the potential regulatory circuitry afforded by miRNA is enormous. Increasing evidence indicates that miRNAs may in fact be key regulators of processes such as development (16, 17), cell proliferation and death (18), apoptosis and fat metabolism (19), hematopoiesis (20), and stem cell division (21).

The expression of miRNAs is highly specific for tissues and developmental stage (5–7) and has recently allowed for molecular classification of tumors (22), but little is known about how these expression patterns are regulated. Most miRNA genes are located in regions of the genome distinct from previously known protein-encoding genes (2, 4) and primary transcripts (pri-miRNA) are generated by polymerase II (23). These transcripts are capped, polyadenylated, and are usually several thousand bases in length (24). Smaller portions are of miRNAs located within introns of pre-mRNAs and are likely transcribed together with the cognate protein-encoding genes (2, 10). Some miRNAs are clustered and transcribed as multicistronic primary transcripts, but the majority of human miRNAs are not clustered and are transcribed independently (5–7). The biogenesis and function of miRNA require a common set of proteins. Drosha, an RNase III endonuclease, is responsible for processing of pri-miRNAs in the nucleus and releasing 60- to 70-nt precursor miRNAs (pre-miRNAs) (25). Drosha associates with the double-stranded RNA-binding protein DGCR8 in human (26) or Pasha in flies (27) to form the microprocessor complex, which is required for directing the specific cleavage of pri-miRNA by Drosha (26, 27). Pre-miRNAs with hallmarks of Drosha-mediated cleavage and specific hairpin secondary structure are then transported to the cytoplasm by Exportin-5 (28, 29). The RNase III endonuclease Dicer further cleaves pre-miRNA, releasing a 22-nt mature double-stranded miRNA (30). One strand of the miRNA duplex is subsequently incorporated into an effector complex termed RNA-induced silencing complex (RISC), which mediates target gene expression. Argonaute 2 is the key component of RISC, and it may function as an endonulease that cleaves target mRNAs (31, 32).

2 Expression of miRNAs is Deregulated in Cancer

Increasing evidence shows that expression of miRNAs is deregulated in human cancer including leukemia (33–35), lymphoma (36–43), glioblastoma (44, 45), colon (22, 46–49), lung (22, 47, 50–53), breast (22, 47, 54, 55), prostate (55), thyroid (56, 57), liver (58), and ovarian cancer (59). Most recently, high-throughput miRNA quantification technologies, such as miRNA microarray (60–63), bead-based flow cytometric method (22), RNA-primed array-based Klenow enzyme (RAKE) assay (64), miRNA serial analysis of gene expression (miRAGE) (48), and real-time RT-PCR-based TaqMan miRNA assay (65, 66), have provided powerful tools to study the global miRNA profile in whole cancer genome. Some important questions on miRNA deregulation in cancer are being addressed with the advent of these new technologies.

First, is there a differential global expression profile between tumors and their corresponding normal tissues? According to high-throughput studies to date, global expression of miRNAs is ostensibly deregulated in most, if not all, cancer types (22, 47, 52, 54, 56, 58). Most interestingly, miRNA expression seems globally elevated more in normal tissues than in tumors, as revealed by a large-scale bead-based flow cytometric study with human samples of multiple tumor types (22). As the authors pointed out, global downregulation of miRNAs might reflect the state of cellular differentiation in cancer, suggesting that abrogation of miRNA at large may in fact be a hallmark of all human cancers (22). However, such exclusive downregulation of miRNAs in cancer was not observed in other studies based on microarrays (47, 52, 54, 56). In fact, a mixed pattern of downregulation and upregulation of select miRNA genes has been reported, which appears to be tumor specific

(47, 52, 54, 56). Second, is miRNA expression signature informative enough to identify and/or classify human cancers? Indeed, several recent studies have proven a surprisingly promising answer to this question. It is progressively becoming obvious that although the number of miRNAs (~300) is much smaller than the protein-coding genes (~22,000), miRNA expression signatures quite accurately reflect the developmental lineage and tissue origin of human cancers (22, 47). Third, can miRNA expression signature reflect the distinguished subtypes or predict biological and clinical behavior within the same cancer type? The dawning example is that a specific miRNA signature that was identified in B cell chronic lymphocytic leukemia (CLL) could predict the presence or absence of the 70-kDa zeta-associated protein (*ZAP-70*), one of the few known factors that can be used to predict early disease progression (34). Subsequent large-scale studies with human tumor specimens further demonstrated that miRNA expression signatures are associated with specific subtypes as well as clinical behaviors of the same cancer type (22, 35, 52, 54). Last but not least, is miRNA expression associated with the prognosis and progression of human cancer? The earliest insight comes from the expression of *let-7* miRNA family in lung cancer (50). Using a real-time RT-PCR-based method, Takamizawa et al. reported that certain *let-7* family numbers are downregulated in lung cancer, and the reduced expression is significantly associated with the poor outcome of those patients (50). Follow-up studies of large-scale global miRNA profiling further proved that expression signatures of miRNA genes have diagnostic and prognostic significance in leukemia (35) and lung cancer (52). For example, a unique miRNA expression signature composed of 13 genes can differentiate individual cases of CLL, and can predict disease progression (35).

Taken together, deregulated miRNA expression in human cancer may prove to be a powerful tool for diagnosis, classification, and prediction of clinical behavior. In addition, further studies of deregulated miRNA genes should contribute remarkably to our understanding of their involvement in tumorigenic mechanisms and in the development of new drug targets.

3 miRNA Expression Might be Regulated by Epigenetic Alterations in Cancer

Epigenetic changes such as DNA methylation and histone modification play important roles in chromatin remodeling and general regulation of protein-coding gene expression in human cancer (67). Likewise, such mechanisms may also function to affect miRNA expression in cancer. To test this hypothesis, several groups treated cancer cell lines with DNA-demethylating reagents and/or histone deactylase inhibitors in vitro and monitored miRNA expression by microarray analysis (52, 68–70). In SKBr3, a human breast cancer cell line, rapid alteration of miRNA levels was observed in response to histone deacetylase (HDAC) inhibitor LAQ824 (68). In T24, a human bladder cancer cell line, expression of 17 of 313 miRNAs was significantly upregulated upon simultaneous treatment with chromatin-modifying drugs, 5-aza-2'-deoxycytidine and 4-phenylbutyric acid (69). For example, *mir-127*, which is generally expressed in normal cells but absent in cancer cells, was markedly induced after treatment. Intriguingly, a predicted target of *mir-127*, *BCL6*, was translationally downregulated after treatment (69). These data suggest that epigenetic alteration might play a critical role in tuning miRNA expression in human cancers, and epigenetic treatment may provide a novel strategy for cancer therapy. By contrast, reports from other groups have shown that demethylation and HDAC inhibitors do not alter the expression of miRNAs in lung cancer cell lines A549 and NCI-H157 (52, 70). This discrepancy might be due to the unique epigenetic regulation on miRNAs in a tumor/tissue-type specific manner. Nevertheless, further investigation of the epigenetic regulatory mechanisms on miRNA expression is warranted in both malignant and in normal cells.

4 miRNAs Exhibit High Frequency DNA Copy Number Alterations in Cancer

Alterations in DNA copy number is one mechanism to modify gene expression and function, and DNA dosage alterations occurring in somatic cells are frequent contributors to cancer (71). In 2002, the first example of an miRNA gene with DNA copy number alteration in cancer was reported in CLL patients (Fig. 1). It was found that *mir-16-1* and *mir-15a* at 13q14 are deleted in more than 50% patients, with concurrent reduced expression in ~65% patients (33). Further studies demonstrated that these two miRNAs suppress *BCL2* expression and may serve as tumor suppressor genes in this disease (72). Deletion of *mir-16-1* and *mir-15a* was also identified in epithelial tumors, such as pituitary adenomas (73), ovarian, and breast cancers (59). In 2004, amplification of *C13orf25* at 13q31-32 was first reported in lymphoma patients (37). Most interestingly, this amplified region contains seven miRNAs as a polycistronic cluster, and the expression of primary and mature miRNAs derived from this locus is increased in this type of lymphoma (39, 40). We now know that this miRNA cluster actually serves as an oncogene in human cancer (39,

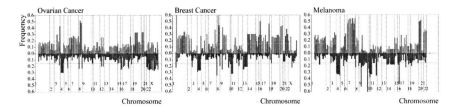

Fig. 1 High frequency miRNA gene copy number alterations in human cancer. Genome-wide gains (*upper*) and losses (*lower*) of miRNA genes in human cancer. Copy number alterations observed in >15% tumors are considered significant. Data for miRNA alterations on sex chromosomes in melanoma is not available

53, 74) through disturbed balance between cell death and proliferation via the protooncogene *c-Myc*-mediated pathway (74, 75).

Using public database-retrieval and bioinformatics-based approaches, Calin et al. compared 186 miRNA loci to the sequences of previously reported nonrandom genetic alterations and discovered for the first time that miRNA genes are frequently residing in fragile sites, as well as in minimal regions of loss of heterozygosity, minimal regions of amplification, or common breakpoint regions. On the basis of their study, 98 of 186 (52.5%) miRNA genes are in the cancer-associated genomic regions (76). Most recently, this result was further confirmed by an array-based comparative genomic hybridization (aCGH) study in 227 human tumors. Zhang et al. analyzed 283 known miRNA genes using a high-resolution aCGH (~1 Mbp). It was found that a large proportion of miRNA gene-containing genomic loci exhibit DNA copy number alterations in ovarian cancer (37.1%), breast cancer (72.8%), and melanoma (85.9%) (59). These findings support the notion that copy number alterations of miRNAs are highly prevalent in cancer and may account partly for the frequent miRNA gene deregulation. At this point, the mechanisms underlying the high-frequency DNA copy number alteration of miRNA genes observed in cancer genome remain unclear. One potential explanation is that genomic aberrations preferentially involve the fragile regions that contain miRNA genes at a high density. Alternatively, clones with miRNA amplifications or deletions are selected because of the biological advantage that is afforded by these miRNA expression changes.

5 Mutations of miRNAs are Identified in Cancer

Germ-line and/or somatic mutations of protein-coding genes efficiently render gain- or loss-of-function of the important proteins involved in tumorigenesis. At present, the information on the mutation and polymorphism of miRNAs in cancer is just emerging. Recently, germ-line or somatic mutations of miRNA genes were identified in CLL samples. Sequencing of 42 miRNA genes in 75 patients revealed mutations in 5 of the 42 analyzed genes (34). C to T germ-line mutation (or rare polymorphism) in primary *mir-16-1* and *mir-15a* sequences was found in 2 of the 75 patients. Most interestingly, this mutation is associated with a lower expression of the mature *mir-16-1* and *mir-15a* (34). Sequence variations of miRNAs have also been reported in solid human tumors in which 15 cancer-associated miRNAs were analyzed in 91 cancer cell lines of epithelial-origin (70). One sequence variation in an miRNA precursor and 15 variations in primary miRNAs were identified. However, no functional consequence (e.g., obscured processing of miRNA) was observed as a result of those aberrations (70).

Because the ultimate function of miRNAs is mediated by the miRNA/mRNA duplex, in addition to the mutation of miRNA gene per se, mutations in the 3-terminal untranslated region (3' UTR) of protein-coding genes might also interfere with the recognition of their target sequence by miRNAs. One interesting example was recently reported in human papillary thyroid carcinoma (PTC) (56). *Mir-221*, *mir-22*, and *mir-146* are significantly upregulated in this disease, whereas the expression of their predicted target gene *KIT* is lost concurrently with the miRNA upregulation. Germ-line single-nucleotide changes were identified in the 3′ UTR of *KIT* and these SNPs seem to lead to the alterations in the miRNA/target mRNA duplex conformation (56). In summary, mutation or polymorphism in the miRNA-targeted genes might also disturb the miRNA/mRNA interaction and thus contribute to cancer development.

6 miRNA: One Step Forward for Cancer Diagnosis and Treatment

The pivotal role of miRNAs in development has been widely investigated and recently their involvement in pathological processes, such as cancer, are beginning to be understood (1–8). Some miRNAs function as tumor suppressor genes by regulating critical oncogenic events. For example, *let-7*, which regulates *Ras* (51), is downregulated in lung cancer (50, 51). *Mir-15* and *mir-16*, which target *BCL2* (72), are deleted or downregulated in leukemia (33). On the other hand, other miRNAs seem to play a more direct role in tumorigenesis. Strong clinical and experimental evidence indicates that the miRNA polycistron, *mir-17-92* (75), may serve as an oncogene in lymphoma (39) and in lung cancer (53), while *mir-372* and *mir-373* may be novel oncogenes in testicular germ cell tumors (77). Because of the imperfect complementarity between miRNA and its target sequence, the mechanism of those so-called "oncomirs" (78) in cancer is being painstakingly studied. A better understanding of their role in tumor development may provide invaluable cues for cancer diagnosis and treatment.

Acknowledgments This work was supported by the OCRF, the Abramson Family Cancer Research Institute, the Pennsylvania Department of Health and Ovarian SPORE P01-CA83638.

References

1. Lee RC, Feinbaum RL, Ambros V. The *C. elegans* heterochronic gene lin-4 encodes small RNAs with antisense complementarity to lin-14. Cell 1993; 75: 843–54.
2. Lagos-Quintana M, Rauhut R, Lendeckel W, Tuschl T. Identification of novel genes coding for small expressed RNAs. Science 2001; 294: 853–8.
3. Lau NC, Lim LP, Weinstein EG, Bartel DP. An abundant class of tiny RNAs with probable regulatory roles in *Caenorhabditis elegans*. Science 2001; 294: 858–62.
4. Lee RC, Ambros V. An extensive class of small RNAs in *Caenorhabditis elegans*. Science 2001; 294: 862–4.
5. Bartel DP. MicroRNAs: genomics, biogenesis, mechanism, and function. Cell 2004; 116: 281–97.
6. Ambros V. The functions of animal microRNAs. Nature 2004; 431: 350–5.

7. He L, Hannon GJ. MicroRNAs: small RNAs with a big role in gene regulation. Nat Rev Genet 2004; 5: 522–31.
8. Zamore PD, Haley B. Ribo-gnome: the big world of small RNAs. Science 2005; 309: 1519–24.
9. Lewis BP, Shih IH, Jones-Rhoades MW, Bartel DP, Burge CB. Prediction of mammalian microRNA targets. Cell 2003; 115: 787–98.
10. John B, Enright AJ, Aravin A, Tuschl T, Sander C, Marks DS. Human microRNA targets. PLoS Biol 2004; 2: e363.
11. Kiriakidou M, Nelson PT, Kouranov A, Fitziev P, Bouyioukos C, Mourelatos Z, et al. A combined computational-experimental approach predicts human microRNA targets. Genes Dev 2004; 18: 1165–78.
12. Lewis BP, Burge CB, Bartel DP. Conserved seed pairing, often flanked by adenosines, indicates that thousands of human genes are microRNA targets. Cell 2005; 120: 15–20.
13. Krek A, Grun D, Poy MN, Wolf R, Rosenberg L, Epstein EJ, et al. Combinatorial microRNA target predictions. Nat Genet 2005; 37: 495–500.
14. Bartel DP, Chen CZ. Micromanagers of gene expression: the potentially widespread influence of metazoan microRNAs. Nat Rev Genet 2004; 5: 396–400.
15. Lim LP, Lau NC, Garrett-Engele P, Grimson A, Schelter JM, Castle J, et al. Microarray analysis shows that some microRNAs downregulate large numbers of target mRNAs. Nature 2005; 433: 769–73.
16. Reinhart BJ, Slack FJ, Basson M, Pasquinelli AE, Bettinger JC, Rougvie AE, et al. The 21-nucleotide let-7 RNA regulates developmental timing in *Caenorhabditis elegans*. Nature 2000; 403: 901–6.
17. Giraldez AJ, Cinalli RM, Glasner ME, Enright AJ, Thomson MJ, Baskerville S, et al. MicroRNAs regulate brain morphogenesis in Zebrafish. Science 2005.
18. Brennecke J, Hipfner DR, Stark A, Russell RB, Cohen SM. Bantam encodes a developmentally regulated microRNA that controls cell proliferation and regulates the proapoptotic gene hid in *Drosophila*. Cell 2003; 113: 25–36.
19. Xu P, Vernooy SY, Guo M, Hay BA. The *Drosophila* microRNA Mir-14 suppresses cell death and is required for normal fat metabolism. Curr Biol 2003; 13: 790–5.
20. Chen CZ, Li L, Lodish HF, Bartel DP. MicroRNAs modulate hematopoietic lineage differentiation. Science 2004; 303: 83–6.
21. Hatfield SD, Shcherbata HR, Fischer KA, Nakahara K, Carthew RW, Ruohola-Baker H. Stem cell division is regulated by the microRNA pathway. Nature 2005; 435: 974–8.
22. Lu J, Getz G, Miska EA, Alvarez-Saavedra E, Lamb J, Peck D, et al. MicroRNA expression profiles classify human cancers. Nature 2005; 435: 834–8.
23. Lee Y, Kim M, Han J, Yeom KH, Lee S, Baek SH, et al. MicroRNA genes are transcribed by RNA polymerase II. Embo J 2004; 23: 4051–60.
24. Kim VN, Nam JW. Genomics of microRNA. Trends Genet 2006; 22: 165–73.
25. Lee Y, Ahn C, Han J, Choi H, Kim J, Yim J, et al. The nuclear RNase III Drosha initiates microRNA processing. Nature 2003; 425: 415–9.
26. Gregory RI, Yan KP, Amuthan G, Chendrimada T, Doratotaj B, Cooch N, et al. The Microprocessor complex mediates the genesis of microRNAs. Nature 2004; 432: 235–40.
27. Denli AM, Tops BB, Plasterk RH, Ketting RF, Hannon GJ. Processing of primary microRNAs by the Microprocessor complex. Nature 2004; 432: 231–5.
28. Yi R, Qin Y, Macara IG, Cullen BR. Exportin-5 mediates the nuclear export of pre-micro RNAs and short hairpin RNAs. Genes Dev 2003; 17: 3011–6.
29. Lund E, Guttinger S, Calado A, Dahlberg JE, Kutay U. Nuclear export of microRNA precursors. Science 2004; 303: 95–8.
30. Bernstein E, Caudy AA, Hammond SM, Hannon GJ. Role for a bidentate ribonuclease in the initiation step of RNA interference. Nature 2001; 409: 363–6.
31. Tabara H, Sarkissian M, Kelly WG, Fleenor J, Grishok A, Timmons L, et al. The *rde-1* gene, RNA interference, and transposon silencing in *C. elegans*. Cell 1999; 99: 123–32.
32. Hammond SM, Boettcher S, Caudy AA, Kobayashi R, Hannon GJ. Argonaute2, a link between genetic and biochemical analyses of RNAi. Science 2001; 293: 1146–50.

33. Calin GA, Dumitru CD, Shimizu M, Bichi R, Zupo S, Noch E, et al. Frequent deletions and down-regulation of micro-RNA genes *miR15* and *miR16* at 13q14 in chronic lymphocytic leukemia. Proc Natl Acad Sci USA 2002; 99: 15524–9.
34. Calin GA, Liu CG, Sevignani C, Ferracin M, Felli N, Dumitru CD, et al. MicroRNA profiling reveals distinct signatures in B cell chronic lymphocytic leukemias. Proc Natl Acad Sci USA 2004; 101: 11755–60.
35. Calin GA, Ferracin M, Cimmino A, Di Leva G, Shimizu M, Wojcik SE, et al. A MicroRNA signature associated with prognosis and progression in chronic lymphocytic leukemia. N Engl J Med 2005; 353: 1793–801.
36. Metzler M, Wilda M, Busch K, Viehmann S, Borkhardt A. High expression of precursor microRNA-155/BIC RNA in children with Burkitt lymphoma. Genes Chromosomes Cancer 2004; 39: 167–9.
37. Ota A, Tagawa H, Karnan S, Tsuzuki S, Karpas A, Kira S, et al. Identification and characterization of a novel gene, *C13orf25*, as a target for 13q31-q32 amplification in malignant lymphoma. Cancer Res 2004; 64: 3087–95.
38. Eis PS, Tam W, Sun L, Chadburn A, Li Z, Gomez MF, et al. Accumulation of miR-155 and BIC RNA in human B cell lymphomas. Proc Natl Acad Sci USA 2005; 102: 3627–32.
39. He L, Thomson JM, Hemann MT, Hernando-Monge E, Mu D, Goodson S, et al. A microRNA polycistron as a potential human oncogene. Nature 2005; 435: 828–33.
40. Tagawa H, Seto M. A microRNA cluster as a target of genomic amplification in malignant lymphoma. Leukemia 2005; 19: 2013–6.
41. Kluiver J, Poppema S, de Jong D, Blokzijl T, Harms G, Jacobs S, et al. BIC and miR-155 are highly expressed in Hodgkin, primary mediastinal and diffuse large B cell lymphomas. J Pathol 2005; 207: 243–9.
42. Jiang J, Lee EJ, Schmittgen TD. Increased expression of microRNA-155 in Epstein-Barr virus transformed lymphoblastoid cell lines. Genes Chromosomes Cancer 2006; 45: 103–6.
43. Costinean S, Zanesi N, Pekarsky Y, Tili E, Volinia S, Heerema N, et al. Pre-B cell proliferation and lymphoblastic leukemia/high-grade lymphoma in E(mu)-miR155 transgenic mice. Proc Natl Acad Sci USA 2006; 103: 7024–9.
44. Chan JA, Krichevsky AM, Kosik KS. MicroRNA-21 is an antiapoptotic factor in human glioblastoma cells. Cancer Res 2005; 65: 6029–33.
45. Ciafre SA, Galardi S, Mangiola A, Ferracin M, Liu CG, Sabatino G, et al. Extensive modulation of a set of microRNAs in primary glioblastoma. Biochem Biophys Res Commun 2005; 334: 1351–8.
46. Michael MZ, SM OC, van Holst Pellekaan NG, Young GP, James RJ. Reduced accumulation of specific microRNAs in colorectal neoplasia. Mol Cancer Res 2003; 1: 882–91.
47. Volinia S, Calin GA, Liu CG, Ambs S, Cimmino A, Petrocca F, et al. A microRNA expression signature of human solid tumors defines cancer gene targets. Proc Natl Acad Sci USA 2006; 103: 2257–61.
48. Cummins JM, He Y, Leary RJ, Pagliarini R, Diaz LA, Jr., Sjoblom T, et al. The colorectal microRNAome. Proc Natl Acad Sci USA 2006.
49. Bandres E, Cubedo E, Agirre X, Malumbres R, Zarate R, Ramirez N, et al. Identification by Real-time PCR of 13 mature microRNAs differentially expressed in colorectal cancer and non-tumoral tissues. Mol Cancer 2006; 5: 29.
50. Takamizawa J, Konishi H, Yanagisawa K, Tomida S, Osada H, Endoh H, et al. Reduced expression of the let-7 microRNAs in human lung cancers in association with shortened postoperative survival. Cancer Res 2004; 64: 3753–6.
51. Johnson SM, Grosshans H, Shingara J, Byrom M, Jarvis R, Cheng A, et al. RAS is regulated by the let-7 microRNA family. Cell 2005; 120: 635–47.
52. Yanaihara N, Caplen N, Bowman E, Seike M, Kumamoto K, Yi M, et al. Unique microRNA molecular profiles in lung cancer diagnosis and prognosis. Cancer Cell 2006; 9: 189–98.
53. Hayashita Y, Osada H, Tatematsu Y, Yamada H, Yanagisawa K, Tomida S, et al. A polycistronic microRNA cluster, miR-17-92, is overexpressed in human lung cancers and enhances cell proliferation. Cancer Res 2005; 65: 9628–32.

54. Iorio MV, Ferracin M, Liu CG, Veronese A, Spizzo R, Sabbioni S, et al. MicroRNA gene expression deregulation in human breast cancer. Cancer Res 2005; 65: 7065–70.
55. Mattie MD, Benz CC, Bowers J, Sensinger K, Wong L, Scott GK, et al. Optimized high-throughput microRNA expression profiling provides novel biomarker assessment of clinical prostate and breast cancer biopsies. Mol Cancer 2006; 5: 24.
56. He H, Jazdzewski K, Li W, Liyanarachchi S, Nagy R, Volinia S, et al. The role of microRNA genes in papillary thyroid carcinoma. Proc Natl Acad Sci USA 2005; 102: 19075–80.
57. Pallante P, Visone R, Ferracin M, Ferraro A, Berlingieri MT, Troncone G, et al. MicroRNA deregulation in human thyroid papillary carcinomas. Endocr Relat Cancer 2006; 13: 497–508.
58. Murakami Y, Yasuda T, Saigo K, Urashima T, Toyoda H, Okanoue T, et al. Comprehensive analysis of microRNA expression patterns in hepatocellular carcinoma and non-tumorous tissues. Oncogene 2006; 25: 2537–45.
59. Zhang L, Huang J, Yang N, Greshock J, Megraw MS, Giannakakis A, et al. microRNAs exhibit high frequency genomic alterations in human cancer. Proc Natl Acad Sci USA 2006; 103: 9136–41.
60. Liu CG, Calin GA, Meloon B, Gamliel N, Sevignani C, Ferracin M, et al. An oligonucleotide microchip for genome-wide microRNA profiling in human and mouse tissues. Proc Natl Acad Sci USA 2004; 101: 9740–4.
61. Goff LA, Yang M, Bowers J, Getts RC, Padgett RW, Hart RP. Rational probe opetimization and enhanced detection strategy for microRNAs using microarrays. RNA Biol 2005; 2: e9–e16.
62. Thomson JM, Parker J, Perou CM, Hammond SM. A custom microarray platform for analysis of microRNA gene expression. Nat Methods 2004; 1: 47–53.
63. Castoldi M, Schmidt S, Benes V, Noerholm M, Kulozik AE, Hentze MW, et al. A sensitive array for microRNA expression profiling (miChip) based on locked nucleic acids (LNA). RNA 2006; 12: 913–20.
64. Nelson PT, Baldwin DA, Scearce LM, Oberholtzer JC, Tobias JW, Mourelatos Z. Microarray-based, high-throughput gene expression profiling of microRNAs. Nat Methods 2004; 1: 155–61.
65. Jiang J, Lee EJ, Gusev Y, Schmittgen TD. Real-time expression profiling of microRNA precursors in human cancer cell lines. Nucleic Acids Res 2005; 33: 5394–403.
66. Chen C, Ridzon DA, Broomer AJ, Zhou Z, Lee DH, Nguyen JT, et al. Real-time quantification of microRNAs by stem-loop RT-PCR. Nucleic Acids Res 2005; 33: e179.
67. Egger G, Liang G, Aparicio A, Jones PA. Epigenetics in human disease and prospects for epigenetic therapy. Nature 2004; 429: 457–63.
68. Scott GK, Mattie MD, Berger CE, Benz SC, Benz CC. Rapid alteration of microRNA levels by histone deacetylase inhibition. Cancer Res 2006; 66: 1277–81.
69. Saito Y, Liang G, Egger G, Friedman JM, Chuang JC, Coetzee GA, et al. Specific activation of microRNA-127 with downregulation of the proto-oncogene BCL6 by chromatin-modifying drugs in human cancer cells. Cancer Cell 2006; 9: 435–43.
70. Diederichs S, Haber DA. Sequence variations of microRNAs in human cancer: alterations in predicted secondary structure do not affect processing. Cancer Res 2006; 66: 6097–104.
71. Pinkel D, Albertson DG. Array comparative genomic hybridization and its applications in cancer. Nat Genet 2005; 37 Suppl: S11–S17.
72. Cimmino A, Calin GA, Fabbri M, Iorio MV, Ferracin M, Shimizu M, et al. miR-15 and miR-16 induce apoptosis by targeting BCL2. Proc Natl Acad Sci USA 2005; 102: 13944–9.
73. Bottoni A, Piccin D, Tagliati F, Luchin A, Zatelli MC, degli Uberti EC. miR-15a and miR-16-1 down-regulation in pituitary adenomas. J Cell Physiol 2005; 204: 280–5.
74. Dews M, Homayouni A, Yu D, Murphy D, Sevignani C, Wentzel E, et al. Augmentation of tumor angiogenesis by a Myc-activated microRNA cluster. Nat Genet 2006.
75. O'Donnell KA, Wentzel EA, Zeller KI, Dang CV, Mendell JT. c-Myc-regulated microRNAs modulate E2F1 expression. Nature 2005; 435: 839–43.
76. Calin GA, Sevignani C, Dumitru CD, Hyslop T, Noch E, Yendamuri S, et al. Human microRNA genes are frequently located at fragile sites and genomic regions involved in cancers. Proc Natl Acad Sci USA 2004; 101: 2999–3004.

77. Voorhoeve PM, le Sage C, Schrier M, Gillis AJ, Stoop H, Nagel R, et al. A genetic screen implicates miRNA-372 and miRNA-373 as oncogenes in testicular germ cell tumors. Cell 2006; 124: 1169–81.
78. Esquela-Kerscher A, Slack FJ. Oncomirs – microRNAs with a role in cancer. Nat Rev Cancer 2006; 6: 259–69.

Ovarian Carcinogenesis: An Alternative Hypothesis

Jurgen M.J. Piek, Paul J. van Diest, and René H.M. Verheijen

1 Introduction

Cancer of the ovary is among the most common female genital tract cancers and has the worst prognosis. This is largely caused by the fact that these cancers are detected at late stage of disease, because of absence of early clinical symptoms. Consequently, early events in ovarian carcinogenesis remain remarkably unknown. Therefore, the precursor cell of these tumours remains a matter of debate (1, 2).

Three explanations as to the origin of serous ovarian adenocarcinomas have been put forward. The first one points towards the ovarian surface epithelium (OSE) as tissue of origin (3). Second, remnants of the embryologic Müllerian duct, the secondary Müllerian system, have been suggested as possible tissue of origin (4). The third is the Fallopian tube inner surface epithelium (TSE) (=oviduct epithelium) (1, 5). In this chapter, the third possible origin is highlighted.

2 Women at Risk to Develop Ovarian Carcinoma

Women harbouring a mutation in one of the breast cancer (BRCA) 1 or BRCA2 genes are at high risk to develop breast and/or female adnexal (ovarian and Fallopian tube) carcinoma (6). Several studies highlighted the occurrence of Fallopian tube carcinoma in women harbouring these mutations (Table 1).

Lifetime risk of female adnexal carcinomas in mutation carriers is in the order of 60% (17, 18). To reduce the risk of adnexal cancer, prophylactic bilateral salpingo-oophorectomy is advised to women who have completed their families (19). These adnexes are a potential source for studies into early steps of carcinogenesis, since pre-malignant and early malignant lesions can be expected to be present in these tissues. Reports on findings of pre-malignant changes in prophylactically removed ovaries are inconclusive whether such lesions do exist (Table 2). Some of these studies indicate that cortical inclusion cysts, papillomatosis, cortical invaginations, nuclear enlargement, and stromal activity are more common in ovaries from women with hereditary predisposition for female adnexal

Table 1 Studies indicating Fallopian tube carcinoma to be part of the BRCA-related cancer spectrum

Author	BRCA germline mutation
Schubert et al. (16)	-BRCA2 3034delAAAc
	-BRCA2 3034delAAAc
Tong et al. (16)	-BRCA1 Cys61Gly
Sobol et al. (16)	-BRCA1
	-N/A
Rose et al. (16)	-BRCA2 6563delGA
Zweemer et al. (16)	-BRCA1 1410insT
	-BRCA1 2804delAA
Hartley et al. (16)	-BRCA1 N/A
Colgan et al. (16)	-BRCA?
Aziz et al. (16)	-BRCA1 185delAG
	-BRCA2 2024del5
	-BRCA1 5083del19
	-BRCA1 C61G
	-BRCA1 5382insC
	-BRCA1 R1495M
	-BRCA2 6174delT
Hébert-Blouin et al. (16)	-BRCA1 K679X
Agoff et al. (16)	-BRCA1 2800delAA
	-BRCA1 2800delAA
	-BRCA2 2558insA
	-N/A
	-N/A
Scheuer et al. (16)	-BRCA1 Q563X
Leeper et al. (16)	-BRCA1 2800delAA
	-BRCA1 2800delAA
	-BRCA2 2558insA
Peyton-Jones et al. (16)	-BRCA2
Dijkhuizen et al. (16)	*BRCA1*
Levine et al. (16)	$-4 \times$ *BRCA1* 185delAG
	-BRCA1 5382insC
	-BRCA2 6174delT
Baudi et al. (16)	-BRCA2 q3034R
Olivier et al. (7)	-BRCA1 3875del4
	-BRCA1
	-BRCA12312del5
Casarsa et al. (8)	$-3 \times$ BRCA2
Powell et al. (9)	-BRCA2 6174delT
	-BRCA2 6174delT
	-BRCA1 Y1563X
	-BRCA1 5382insC
Meeuwissen (10)	-BRCA1
Cass et al. (11)	$-11 \times$ BRCA1
	$-1 \times$ BRCA2
Carcangiu et al. (12)	-BRCA1 1207delA
Finch et al. (13)	-BRCA2
	-BRCA2
	-BRCA1
Damayanti et al. (14)	-BRCA1 2845insA
Medeiros et al. (15)	$-3 \times$ BRCA2
	$-2 \times$ BRCA1

Ovarian Carcinogenesis: An Alternative Hypothesis

Table 2 Studies comparing ovaries prophylactically removed from patients at high hereditary risk of adnexal cancer with ovaries from women without such hereditary risk

	Kerner et al. (20)	Piek et al. (16)	Casey et al. (16)	Barakat et al. (16)	Stratton et al. (16)	Werness et al. (16)	Deligdisch et al. (16)	Salazar et al. (16)	Gusberg et al. (16)
Inclusion cysts				NS	NS	0.016		0.006	
Papillomatosis			0.039	NS	NS	NS		0.005	
Cortical invaginations	0.042			NS	NS	NS		0.0004	
Nuclear enlargement			Ns			0.006	A distinction could be made between normal-, dysplastic- and neoplastic nuclei		Nuclear enlargement in ovaries of twins. None in control ovaries
Stromal activity						NS		0.00017	
Epithelial hyperplasia		NS	Ns					NS	
Metaplasia		NS			NS				
Pseudo stratification			NS						
Psammoma bodies			NS						
Atypical changes	0.014								

adenocarcinomas. Other studies contradict these findings. Three recent studies on the expression of cell cycle and differentiation related markers in OSE in vivo indicate no differences between OSE from women with and without this hereditary predisposition (21–23). However, reports on prophylactically removed Fallopian tubes do invariably show a high incidence of preneoplastic lesions (Table 3 and Fig. 1). All these studies signify that tubal epithelium is prone to undergo (pre)malignant changes, this in contrast with OSE.

3 Sporadic Ovarian Cancer

Studies in which women were screened for ovarian carcinomas by CA125 levels, to diagnose disease in early stage, resulted in detection of tubal carcinomas 25 times more often than the expected (31, 32). Neoplastic lesions in the Fallopian tube may be expected to be present in up to 8% of patients with serous epithelial

Table 3 Studies highlighting preneoplastic lesions within Fallopian tubes prophylactically removed from patients at high hereditary risk of adnexal cancer

Author	Percentage dysplasia/atypical hyperplasia
Leunen et al. (24)	4% (2/52)
Lamb et al. (25)	3.5% (4/113)
Hermsen et al. (26)	32% (27/85)
McEwen et al. (27)	Case report
Olivier et al. (7)	3.3% (3/90)
Carcangiu (28)	15% (4/26)
Agoff et al. (29)	Case report
Piek et al. (30)	50% (6/12)

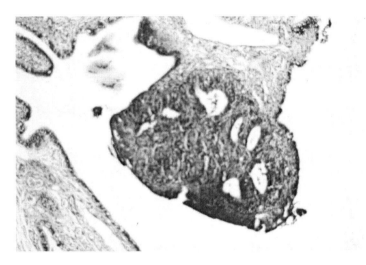

Fig. 1 Preneoplastic lesion within a Fallopian tube, prophylactically removed from a BRCA1 mutation carrier

malignancy of the ovary (33). Therefore, the true incidence of tubal carcinoma is probably highly underestimated. These studies indicate that at present most Fallopian tube carcinomas are quite likely to be misdiagnosed as ovarian carcinomas, since they do not fulfil all the criteria of Hu (34) (These criteria require that (a) the main tumour is in the Fallopian tube and arises from the endosalpinx, (b) histological features reflect a tubal pattern, (c) if the tubal wall is involved, the transition between malignant and benign tubal epithelium should be detectable, and (d) the Fallopian tube contains more tumour than the ovary or endometrium), although they may yet originate from the Fallopian tube.

4 Animal Models

A relative high incidence of serous tubal carcinomas is observed in hens (35–37) especially when kept under conditions in which they ovulate daily (38). In 1975, Ilchmann et al. proposed oviduct epithelium to be the tissue of origin for serous peritoneal and serous ovarian carcinomas in hens, as all tumours encountered in their study were of serous histotype and in all cases preneoplastic lesions were detected in the oviduct (39).

5 The Cell of Origin

It has been hypothesised that most ovarian carcinomas originate de novo, in the sense that the malignant tumour does not arise from a pre-existing benign epithelial lesion (40). Apart from OSE, the epithelium of cortical inclusion cysts, inclusion cyst epithelium (ICE), has been proposed as precursor (40). Inclusion cysts are found from birth till old age; however, serous changes are only detected in women after menarche (41, 42). Postmenarchal ICE occasionally expresses E-cadherin, normally expressed in serous tubal epithelium, but rarely or not in OSE (43, 44). Moreover, in contrast to OSE, bcl-2 positive cells often line ovarian inclusion cysts. Recently, it has been demonstrated that bcl-2 is a differentiation marker, also of serous tubal epithelial cells (45). The origin of these cortical inclusion cysts is a matter of debate. However, most investigators believe ICE to be included as metaplastic OSE cells. It has been hypothesized that OSE-lined cysts develop either during ingrowth of OSE into a stigma, formed after ovulation (46) (see Fig. 2), or that they are caused by interplay between OSE and the underlying ovarian stroma (47). However, some of these cysts are lined by cells that are indistinguishable from epithelial cells lining the Fallopian tube. Moreover, their morphological arrangement resembles Fallopian tube epithelium architecture (41, 46, 48–54) (see Fig. 3).

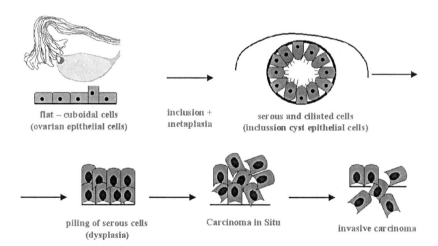

Fig. 2 Possible route to serous ovarian cancer (metaplasia theory): ovarian surface epithelial cells get entrapped within the ovarian stroma during the ovulatory process, undergo metaplastic changes, and eventually form a (pre)malignant lesion due to genetic changes occurring during mitosis

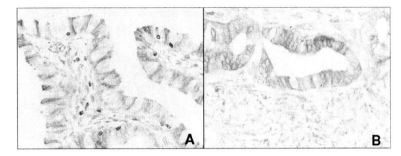

Fig. 3 (a) Fallopian tube stained for bcl-2, which is a differentiation marker for serous tubal cells. (b) Ovary stained for bcl-2; inclusion cyst lined by serous and cilliated cells. Morphology similar to that of the Fallopian tube. Reproduced from Histopatholgy May 2001, with permission from Blackwell Publishing

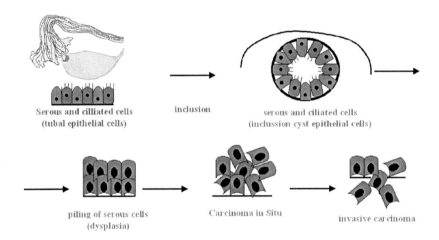

Fig. 4 Possible route to serous ovarian cancer (exfoliation theory): exfoliated tubal epithelial cells get entrapped within an ovarian stigma during the ovulatory process and eventually form a (pre)malignant lesion due to genetic changes occurring during mitosis

Since the fimbrial part of the tube is involved in ovum pick-up and is in close contact with the ovary (55–57), also exfoliated tubal epithelial cells included within the ovarian stroma are a source of ICE and consequently can form a (pre)malignant lesion (see Fig. 4). Additionally, during menstruation, retrograde flow through patent Fallopian tubes occurs carrying endometrial cells (58), but also TSE (1, 59).

6 Summary

Observations indicate three different tissues of origin for ovarian carcinoma. The ovarian surface epithelium, oviduct epithelium (TSE), and derivates of the Müllerian duct. This chapter discusses the TSE-related ovarian carcinogenesis (exfoliation theory). Recent evidence from prophylactic removed ovaries and Fallopian tubes shows (pre)neoplastic lesions primarily within the tubes and not in the ovaries, putting another light on the malignant potential of the tubal epithelium and possibly ovarian carcinogenesis.

References

1. Piek, J. M.; van Diest, P. J.; Zweemer, R. P.; Kenemans, P.; Verheijen, R. H. *Lancet* **2001**, *358*, 844.
2. Foulkes, W. D. *Cancer Cell* **2002**, *1*, 11–12.
3. Auersperg, N.; Maines-Bandiera, S. L.; Dyck, H. G. *J. Cell Physiol* **1997**, *173*, 261–265.
4. Dubeau, L. *Gynecol. Oncol.* **1999**, *72*, 437–442.
5. Goldgar, D.; Eeles, R. A.; Easton, D.; Piver, M. S.; Piek, J. M.; Diest P.J.; Verheijen, R. H.; Szabo, C.; Monteiro, A. N.; Devilee, P.; Narod, S.; Meijers-Heijboer, E. H. Inherited tumour syndromes; In *Pathology and Genetics of Tumours of the Breast and Female Genital Organs*; Tavassoli, F. A., Stratton, J. F., eds. IARC: Lyon, 2003.
6. Piek, J. M.; Dorsman, J. C.; Zweemer, R. P.; Verheijen, R. H.; Diest P.J.; Colgan, T. J. *Int. J. Gynecol. Pathol.* **2003**, *22*, 315–316.
7. Olivier, R. I.; Van Beurden, M.; Lubsen, M. A.; Rookus, M. A.; Mooij, T. M.; van de Vijver, M. J.; Van't Veer, L. J. *Br. J. Cancer* **2004**, *90*, 1492–1497.
8. Casarsa, S.; Puglisi, F.; Baudi, F.; De Paola, L.; Venuta, S.; Piga, A.; Di Loreto, C.; Marchesoni, D.; D'Elia, A. V.; Damante, G. *Oncol. Rep.* **2004**, *12*, 313–316.
9. Powell, C. B.; Kenley, E.; Chen, L. M.; Crawford, B.; McLennan, J.; Zaloudek, C.; Komaromy, M.; Beattie, M.; Ziegler, J. *J. Clin. Oncol.* **2005**, *23*, 127–132.
10. Meeuwissen, P. A.; Seynaeve, C.; Brekelmans, C. T.; Meijers-Heijboer, H. J.; Klijn, J. G.; Burger, C. W. *Gynecol. Oncol.* **2005**, *97*, 476–482.
11. Cass, I.; Holschneider, C.; Datta, N.; Barbuto, D.; Walts, A. E.; Karlan, B. Y. *Obstet. Gynecol.* **2005**, *106*, 1327–1334.
12. Carcangiu, M. L.; Peissel, B.; Pasini, B.; Spatti, G.; Radice, P.; Manoukian, S. *Am. J. Surg. Pathol.* **2006**, *30*, 1222–1230.
13. Finch, A.; Beiner, M.; Lubinski, J.; Lynch, H. T.; Moller, P.; Rosen, B.; Murphy, J.; Ghadirian, P.; Friedman, E.; Foulkes, W. D.; Kim-Sing, C.; Wagner, T.; Tung, N.; Couch, F.; Stoppa-Lyonnet, D.; Ainsworth, P.; Daly, M.; Pasini, B.; Gershoni-Baruch, R.; Eng, C.; Olopade, O. I.; McLennan, J.; Karlan, B.; Weitzel, J.; Sun, P.; Narod, S. A. *JAMA* **2006**, *296*, 185–192.
14. Damayanti, Z.; Ali, A. B.; Iau, P. T.; Ilancheran, A.; Sng, J. H. *Int. J. Gynecol. Cancer* **2006**, *16 Suppl 1*, 362–365.
15. Medeiros, F.; Muto, M. G.; Lee, Y.; Elvin, J. A.; Callahan, M. J.; Feltmate, C.; Garber, J. E.; Cramer, D. W.; Crum, C. P. *Am. J. Surg. Pathol.* **2006**, *30*, 230–236.
16. Piek, J. M.; Kenemans, P.; Verheijen, R. H. *Am. J. Obstet. Gynecol.* **2004**, *191*, 718–732.
17. Easton, D. F.; Ford, D.; Bishop, D. T. *Am. J. Hum. Genet.* **1995**, *56*, 265–271.

18. Sutcliffe, S.; Pharoah, P. D.; Easton, D. F.; Ponder, B. A. *Int. J. Cancer* **2000**, *87*, 110–117.
19. Verheijen, R. H.; Boonstra, H.; Menko, F. H.; de Graaff, J.; Vasen, H. F.; Kenter, G. G. *Ned. Tijdschr. Geneeskd.* **2002**, *146*, 2414–2418.
20. Kerner, R.; Sabo, E.; Gershoni-Baruch, R.; Beck, D.; Ben Izhak, O. *Gynecol. Oncol.* **2005**, *99*, 367–375.
21. Werness, B. A.; Afify, A. M.; Eltabbakh, G. H.; Huelsman, K.; Piver, M. S.; Paterson, J. M. *Int. J. Gynecol. Pathol.* **1999**, *18*, 338–343.
22. Sherman, M. E.; Lee, J. S.; Burks, R. T.; Struewing, J. P.; Kurman, R. J.; Hartge, P. *Int. J. Gynecol. Pathol.* **1999**, *18*, 151–157.
23. Piek, J. M.; Verheijen, R. H.; Menko, F. H.; Jongsma, A. P.; Weegenaar, J.; Gille, J. J.; Pals, G.; Kenemans, P.; van Diest, P. J. *Histopathology* **2003**, *43*, 26–32.
24. Leunen, K.; Legius, E.; Moerman, P.; Amant, F.; Neven, P.; Vergote, I. *Int. J. Gynecol. Cancer* **2006**, *16*, 183–188.
25. Lamb, J. D.; Garcia, R. L.; Goff, B. A.; Paley, P. J.; Swisher, E. M. *Am. J. Obstet. Gynecol.* **2006**, *194*, 1702–1709.
26. Hermsen, B. B.; van Diest, P. J.; Berkhof, J.; Menko, F. H.; Gille, J. J.; Piek, J. M.; Meijer, S.; Winters, H. A.; Kenemans, P.; Mensdorff-Pouilly, S.; Verheijen, R. H. *Int. J. Cancer* **2006**, *119*, 1412–1418.
27. McEwen, A. R.; McConnell, D. T.; Kenwright, D. N.; Gaskell, D. J.; Cherry, A.; Kidd, A. M. *Gynecol. Oncol.* **2004**, *92*, 992–994.
28. Carcangiu, M. L.; Radice, P.; Manoukian, S.; Spatti, G.; Gobbo, M.; Pensotti, V.; Crucianelli, R.; Pasini, B. *Int. J. Gynecol. Pathol.* **2004**, *23*, 35–40.
29. Agoff, S. N.; Mendelin, J. E.; Grieco, V. S.; Garcia, R. L. *Am. J. Surg. Pathol.* **2002**, *26*, 171–178.
30. Piek, J. M.; van Diest, P. J.; Zweemer, R. P.; Jansen, J. W.; Poort-Keesom, R. J.; Menko, F. H.; Gille, J. J.; Jongsma, A. P.; Pals, G.; Kenemans, P.; Verheijen, R. H. *J. Pathol.* **2001**, *195*, 451–456.
31. Woolas, R.; Jacobs, I.; Davies, A. P.; Leake, J.; Brown, C.; Grudzinskas, J. G.; Oram, D. *Int. J. Gynecol. Cancer* **1994**, *4*, 384–388.
32. Woolas, R.; Smith, J.; Paterson, J. M.; Sharp, F. Fallopian tube carcinoma: an under-recognized primary neoplasm. *Int. J. Gynecol. Cancer* **1997**, *7*, 284–288.
33. Bannatyne, P.; Russell, P. *Diagn. Gynecol. Obstet.* **1981**, *3*, 49–60.
34. Hu, C. Y.; Taymor, M. L.; Hertig, A. T. Primary carcinoma of the Fallopian tube. *Am. J. Obstet. Gynecol.* **1950**, *59*, 58.
35. Campbell, J. G. Some unusual gonadal tumours of the fowl. *Br. J. Cancer* **1951**, *5* (1), 69–84.
36. Fredrickson, T. N. Ovarian tumors of the Hen. *Environ. Health Perspect.* **1987**, *73*, 33–51.
37. Rodriguez-Burford, C.; Barnes, M. N.; Berry, W.; Partridge, E. E.; Grizzle, W. E. *Gynecol. Oncol.* **2001**, *81*, 373–379.
38. Wilson, J. E. Adeno-carcinomata in hens kept in a constant environment. *Poult. Sci.* **1958**, *37*, 1253.
39. Ilchmann, G.; Bergmann, V. *Arch. Exp. Veterinarmed.* **1975**, *29*, 897–907.
40. Scully, R. E. *Int. J. Gynaecol. Obstet.* **1995**, *49 Suppl*, S9–S15.
41. Blaustein, A.; Kantius, M.; Kaganowicz, A.; Pervez, N.; Wells, J. *Int. J. Gynecol. Pathol.* **1982**, *1*, 145–153.
42. Mulligan, R. M. *J. Surg.Oncol.* **1976**, *8*, 61–66.
43. Sundfeldt, K.; Piontkewitz, Y.; Ivarsson, K.; Nilsson, O.; Hellberg, P.; Brannstrom, M.; Janson, P. O.; Enerback, S.; Hedin, L. *Int. J. Cancer* **1997**, *74*, 275–280.
44. Davies, B. R.; Worsley, S. D.; Ponder, B. A. *Histopathology* **1998**, *32*, 69–80.
45. McCluggage, W. G.; Maxwell, P. *Histopathology* **2002**, *40*, 107–108.
46. Aoki, Y.; Kawada, N.; Tanaka, K. *J. Reprod. Med.* **2000**, *45*, 159–161.
47. Scully, R. E. *J. Cell Biochem. Suppl* **1995**, *23*, 208–218.
48. Fenoglio, C. M.; Castadot, M. J.; Ferenczy, A.; Cottral, G. A.; Richart, R. M. *Gynecol. Oncol.* **1977**, *5*, 203–218.

49. Stenback, F. *Pathol. Res. Pract.* **1981**, *172*, 58–72.
50. Resta, L.; De Benedictis, G.; Scordari, M. D.; Orlando, E.; Borraccino, V.; Milillo, F. *Tumori* **1987**, *73*, 249–256.
51. Rothacker, D. *Pathologe* **1991**, *12*, 266–269.
52. Resta, L.; Russo, S.; Colucci, G. A.; Prat, J. *Obstet. Gynecol.* **1993**, *82*, 181–186.
53. Maines-Bandiera, S. L.; Auersperg, N. *Int. J. Gynecol. Pathol.* **1997**, *16*, 250–255.
54. Piek, J. M.; Verheijen, R. H.; Menko, F. H.; Jongsma, A. P.; Weegenaar, J.; Gille, J. J.; Pals, G.; Kenemans, P.; van Diest, P. J. *Histopathology* **2003**, *43*, 26–32.
55. Brosens, I. A.; Vasquez, G. *J. Reprod. Med.* **1976**, *16*, 171–178.
56. Gordts, S.; Campo, R.; Rombauts, L.; Brosens, I. *Hum. Reprod.* **1998**, *13*, 1425–1428.
57. Ahmad-Thabet, S. M. *J. Obstet. Gynaecol. Res.* **2000**, *26*, 65–70.
58. Kruitwagen, R. F.; Poels, L. G.; Willemsen, W. N.; Jap, P. H.; de Ronde, I. J.; Hanselaar, T. G.; Rolland, R. *Eur. J. Obstet. Gynecol. Reprod. Biol.* **1991**, *41*, 215–223.
59. Poropatich, C.; Ehya, H. *Acta Cytol.* **1986**, *30*, 442–444.

BRCA1-Induced Ovarian Oncogenesis

Louis Dubeau

Women carrying a germline mutation in *BRCA1* have a 40% risk of developing ovarian cancer by the age of 70, and are also predisposed to cancers of the Fallopian tubes and breast (1). The molecular mechanisms responsible for cancer predisposition in these individuals remain unclear in spite of the huge effort focused on understanding the normal function of the BRCA1 protein since the encoding gene was first isolated. Particularly intriguing is the site specificity of the cancers that develop in such individuals. Indeed, although BRCA1 is expressed ubiquitously in most cell types, individuals carrying germline *BRCA1* mutations are predisposed primarily to cancers of the breast and female reproductive tract. This chapter focuses on observations with an experimental model that not only provide a potential explanation why germline BRCA1 mutations are associated almost exclusively with predisposition to breast and ovarian cancers, but also sheds light into an underlying mechanism contributing to such predisposition.

1 Evidence for and Against the Idea that BRCA1 Functions as a Classical Tumor Suppressor

The concept that certain genes act as suppressors of cancer development originated largely from observations made in the context of familial cancer predisposition. Over three decades ago, Knudsen proposed that two genetic hits are needed for retinoblastoma development, and further suggested that one of these two hits is inherited through the germline in individuals with familial predisposition to this disease (2). It has since been established that the two hits referred to in this hypothesis correspond to inactivation of the two alleles of RB, the first tumor suppressor gene ever identified. A similar scenario where two alleles of a tumor suppressor are inactivated independently, one from a germline mutation and the other from a somatic event, has been applied to other familial cancer predisposition syndromes and has become a central dogma in cancer genetics. This scenario implies that the loss of both alleles of a given tumor suppressor provides an inherent growth survival advantage to cells harboring such loss.

BRCA1 is involved in a variety of important cellular processes such as cell cycle regulation, control of apoptosis, DNA repair, chromatin remodeling, transcriptional regulation, X chromosome inactivation, and posttranslational protein modification (3–7). Functions associated with cell growth or DNA repair are especially supportive of the idea that this protein functions as a classical tumor suppressor on the basis of Knudsen's hypothesis. Earlier reports (8–10) that tumors developing in individuals with germline *BRCA1* mutations, if showing loss of heterozygosity, almost always show loss of the wild type allele provide further support for this notion, as these observations suggest that cancer cells that lack a normal BRCA1 allele have a survival advantage over those cells in which such an allele is present. A number of observations are not readily reconciled with the idea that BRCA1 functions as a classical tumor suppressor in spite of these arguments. Tumor cell lines lacking a functional *BRCA1* gene have been extremely difficult to establish from cancers arising in individuals with germline mutations in this gene. Only a handful of such cell lines is available, and these cell lines surprisingly have very long doubling times and are difficult to work with (11, 12). The idea that loss of BRCA1 function provides a survival advantage is hard to defend in light of these observations. Furthermore, primary cultures derived from mouse embryos with homozygous knockouts of the *Brca1* gene do not proliferate. Such cultures are only successful when derived from embryos carrying double *Brca1* and *p53* knockouts, and given that cells from such embryos grow only clonally, additional events must be needed to ensure viability of the cells (13, 14). In addition, cells from *Brca1* knock out embryos show evidence of cell cycle arrest at the G2/M phase (15), which is consistent with findings that the *BRCA1* gene product is important for regulation of progression through this cell cycle checkpoint (16). All these findings are at odds with the notion that this gene is a tumor suppressor gene. Finally, the idea that *BRCA1* functions as a classical tumor suppressor gene is difficult to reconcile with the observation that mutations in this ubiquitously expressed gene lead mainly to predisposition to breast and gynecological cancers.

2   Support for the Existence of a Link Between BRCA1 Expression and Menstrual Cycle Regulation

Several epidemiological studies have demonstrated that the normal menstrual cycle is an important risk factor for ovarian cancer (17). In fact, this cycle is probably the most important determinant of ovarian cancer risk in individuals who do not carry a genetic predisposition to this disease. Interruption of ovulatory activity protects against the development of this disease independently of whether such interruption is achieved through pregnancy or oral contraceptives. For example, use of oral contraceptives for 5 years results in an approximately 60% decrease in ovarian cancer risk, which is similar to the protective effect of five pregnancies after the first (18). More recent studies suggest that late pregnancies are more protective than those occurring at early ages (19).

We hypothesized that the molecular mechanisms underlying familial ovarian cancer predisposition in individuals carrying germline BRCA1 mutations could be directly linked to those mediating cancer predisposition associated with the ovulatory cycle. The fact that pregnancy or oral contraceptive use, both of which confer strong protection against ovarian cancer in the general population, also provides a similar protection in BRCA1 mutation carriers (20) is supportive of this hypothesis. We, therefore, reasoned that BRCA1 might, at least in part, influence ovarian tumorigenesis indirectly, by controlling an effector secreted by cells important for the control of menstrual cycle progression. In other words, loss of BRCA1 function could influence ovarian tumorigenesis cell nonautonomously, by disrupting interactions between cells that control the menstrual cycle and cells from which ovarian epithelial tumors originate. Thus, it is the role of BRCA1 as a regulator of transcription and in cell-to-cell signaling rather than its role in DNA repair that is the basis for our hypothesis, which is illustrated schematically in Fig. 1.

Given the central role of granulosa cells in regulating progression through the normal menstrual cycle and the role of this cycle in predisposition to ovarian cancer, we further hypothesized that these cells might interact with the cell of origin of ovarian tumors and influence their neoplastic transformation as suggested in the model shown in Fig. 1. Indeed, granulosa cells secrete a variety of hormones thought to influence growth and signal transduction in ovarian tumors. Such hormones include estrogens, progesterone, and the peptide hormone mullerian inhibiting

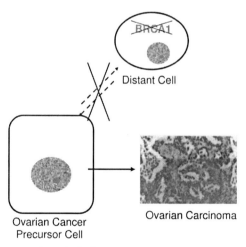

Fig. 1 Cell nonautonomous hypothesis for cancer predisposition in BRCA1 mutation carriers. This model stipulates that in the presence of normal BRCA1 function, the cell type from which ovarian epithelial tumors originate (ovarian cancer precursor cell) interacts with another cell type (distant cell) from a distance, either via endocrine or paracrine mechanisms. Loss of normal BRCA1 function in the distant cells leads to disruption of these normal intercellular interactions, resulting in predisposition to neoplastic transformation in the precursor cell

substance (MIS). MIS belongs to the TGF-beta family (21). It is secreted by Sertoli cells of the testes in male embryos, functioning to prevent the development of mullerian ducts, from which female reproductive organs other than the ovaries are derived (21). It is also secreted by granulosa cells in adult ovaries, resulting in detectable levels of MIS in the serum of premenopausal women (22–24). The function of this hormone in women of reproductive age is unknown, although a role in controlling follicular growth has recently been suggested (25). A possible role for MIS in controlling ovarian cancer development is suggested by the facts that (1) MIS prevents the development of mullerian ducts in the embryo and (2) that ovarian epithelial tumors bear a close resemblance to tissues derived from mullerian ducts, which include the Fallopian tubes, uterus, and endocervix (2 6). In support of this idea, MIS can inhibit ovarian epithelial tumor cell growth in vitro and in vivo (27, 28). The MIS receptor has a very limited normal tissue distribution, as it is present exclusively in the uterus, Fallopian tubes (or uterine horns in mice), granulosa cells of the ovaries, and Sertoli cells (29). Conolly et al. (30) recently took advantage of this tissue specificity to develop a transgenic mouse model for ovarian carcinoma.

3 Granulosa Cell-Specific Inactivation of BRCA1 in a Mouse Model

We used the *cre-lox* system to inactivate the *Brca1* gene in mouse granulosa cells specifically (31). This tissue-specific gene knock out was attempted by crossing mice carrying a floxed *Brca1* allele with mice carrying a *cre* recombinase transgene driven by a truncated form of the FSH receptor promoter, which was reported to drive expression exclusively in granulosa cells (32). Although the mutant mice indeed showed inactivation of Brca1 in secondary and tertiary ovarian follicles (31), further characterization also showed low frequency of Brca1 rearrangement in 10–20% of cells within the anterior pituitary gland, implying that the entire pituitary-gonadal axis might have been affected (unpublished observations from the author's laboratory). The exact significance of Brca1 inactivation in the pituitary gland remains unclear because of the small proportion of cells that are affected in that organ.

The mutant mice were fertile and their litters were of normal size, at least in the first 4 months of life. Two thirds of the mice developed epithelial cysts in their reproductive organs by the time they reached the age of 12–18 months (31). Some of those cysts involved the ovary and were very similar to human ovarian cystadenomas. These tumors were not confined to the ovary, but were seen along the entire mullerian tract in a manner reminiscent of para-ovarian and para-tubal epithelial cysts in humans (31). The finding of abnormalities in the uterine horns in addition to the ovaries is compatible with reports that women undergoing prophylactic oophorectomy for familial predisposition to ovarian cancer have a high incidence of preneoplastic lesions in the Fallopian tube epithelium (33–36). Although the tumors were benign, preliminary results suggest that crossing the mutant mice with mice carrying a homozygous knock out of p53 increases the rate of malignant transformation. The fact that the cystic tumors showed no evidence of rearrangement of Brca1, implying that they expressed a functional Brca1 protein, strongly supports the hypothesis that cells that control the ovulatory cycle, including ovarian granulosa cells and possibly a subset of cells from the anterior pituitary gland, use signaling pathways dependent on the presence of a normal *Brca1* gene product to influence the development of ovarian epithelial tumors.

The mutant mice would be expected to develop lesions in their mammary glands in addition to their reproductive tract, if this experimental model was relevant to familial cancer predisposition in human BRCA1 mutation carriers. Although the mammary glands of these animals has not yet been systematically examined, preliminary findings show that mutant animals have prominent large ectatic ducts, suggesting that the phenotypic consequences of Brca1 inactivation in the pituitary-gonadal axis include abnormalities in the breasts in addition to the ovaries and uterine horns (unpublished observations from the author's laboratory). Thus, the distribution of the lesions seen in this mouse model closely mimics that of cancers developing in human BRCA1 mutation carriers as indicated in Table 1.

4 Consequences of BRCA1 Inactivation on the Estrus Cycle

Much of the rationale for creating this mouse model was based on the idea that cancer predisposition in BRCA1 mutation carriers is mediated through mechanisms similar to those responsible for such predisposition in incessantly ovulating women. We, therefore, tested the hypothesis that the mutant mice showed differences in their estrus cycle and that such differences could be in part responsible for increased predisposition to epithelial cysts in their reproductive tract. Daily vaginal cytology specimens were obtained from mutant and littermate control mice over 3–5 weeks when the animals were 3–4 and 7–8 months old. Given that characteristic cytological changes are associated with each phase of the estrus cycle, microscopic examination of each sample allowed determination of the phase of the cycle present at each time point in each mouse. We used these data to calculate and compare the average length of each phase of the cycle in mutant vs. normal mice. There was a

Table 1 Comparison of lesions in mice with granulosa cell specific inactivation of Brca1 to cancers in human with germline BRCA1 mutations

Organ	Human	$Brca1^{flox/flox}$; *Fshr-Cre* mice
Ovary	Serous carcinoma	Serous cystadenoma
Fallopian tube/uterine horn	Serous carcinoma	Multiple serous cysts
Breast	Ductal carcinoma	Ductal ectasia

statistically significant elongation of the proestrus phase in mutant mice compared with wild type mice in both age groups that was most marked in the 7–8 month old group ($P = 0.003$). Given that proestrus is characterized by unopposed estrogens, these results support the idea that tumor predisposition in mutant mice is mediated, at least in part, by increased estrogen stimulation because of an increase in the average length of the proestrus phase. A recent report that downregulation of BRCA1 expression in human granulosa cells leads to increased expression of aromatase, the rate-limiting enzyme in estradiol biosynthesis, is well in line with this idea (37). These results also raise the possibility that women harboring germline BRCA1 mutations could similarly have differences in their menstrual cycle such as elongation of the follicular phase, which is the equivalent of the proestrus phase in the estrus cycle. Whether mice showing an increase in the length of their proestrus phase are more likely to develop epithelial cysts is still unclear because although the current data suggests that such an association indeed exists, the results do not reach statistical significance and have low statistical power due to the small number of animals so far examined.

5 Cancer Predisposition in BRCA1 Mutation Carriers

Given that the epithelial cysts that develop in mutant mice in our animal model do not harbor mutant *Brca1* alleles, a strong argument can be made that at least in this experimental model, a Brca1 mutation acts cell nonautonomously to cause proliferative lesions in the epithelium of the entire mullerian tract. Although the relevance of our animal model to cancer predisposition in humans is still unclear, the fact that mutant animals develop abnormalities in the same organs that are at risk in women harboring BRCA1 mutations (Table 1) argues in favor of such relevance. We therefore propose, based on this evidence, that predisposition to breast and gynecological cancers in women with germline BRCA1 mutations is mediated, at least in part, by an overall decrease in *BRCA1* gene dosage that is the direct result of this germline mutation. Such decrease in BRCA1 expression affects granulosa cells as well as perhaps other components of the pituitary-gonadal axis and interferes with endocrine or paracrine interactions take that normally occur between those cells and the epithelial cells lining the mullerian tract, resulting in predisposition to neoplastic transformation. This hypothesis does not rule out a cell autonomous mechanism based on the possibility that BRCA1 also functions as a classical tumor suppressor, as those two scenarios are not mutually exclusive and both could cooperate with each other to promote cancer development. However, the idea of a cell nonautonomous mechanism not only provides a straightforward explanation for the site specificity of the cancers that develop in individuals carrying germline BRCA1 mutations, but also accounts for the protective effect of surgical ablation of the ovaries, the site of granulosa cells, on breast cancer predisposition in these patients (38).

6 Implications for the Identification of the Cell of Origin of Ovarian Epithelial Tumors

Ovarian epithelial tumors are thought to arise from the mesothelial layer that covers the ovarian surface, according to the favored hypothesis. An argument has been made that these tumors could instead originate from derivatives of the mullerian tract on the basis of their morphological and functional characteristics (26). Indeed, serous ovarian carcinoma, which is the ovarian tumor subtype that typically develops in BRCA1 mutation carriers, is morphologically indistinguishable from neoplasms of the Fallopian tubes, which are part of the mullerian tract. This resemblance is so striking that pathologists, by convention, have for decades diagnosed all serous tumors from the tubo-ovarian area as serous ovarian neoplasms unless they were dealing with lesions small enough to be confined to the tubes or distributed in such a way that an origin from the tubes could clearly be demonstrated. It seems unlikely that cells that are as different in their function and embryological origin as the ovarian surface mesothelium and the Fallopian tube epithelium could give rise to identical tumors. In addition, if serous ovarian tumors indeed developed in the cell layer lining the ovarian surface, these tumors would be the only example of a tumor of somatic cells that shows a greater degree of differentiation than the cell type from which it originates. I have argued that all tumors currently classified as ovarian epithelial tumors originate in components of the mullerian tract, either the Fallopian tube or the numerous mullerian derivatives found within and around the ovary such as endosalpingiosis, endometriosis, and endocervicosis, which have also been referred to as secondary mullerian system (26). The fact that the epithelial cysts that develop in mice lacking a functional Brca1 in their pituitary–gonadal axis are not confined to the ovary, but distributed along the entire mullerian tract is not only supportive of this hypothesis, but provides an attractive experimental model to test it further.

7 Concluding Remarks

Our results strongly suggest that a circulating factor secreted by granulosa cells and under the control of Brca1 can influence predisposition to tumor development in the ovary as well as in components of the mullerian tract in rodents. Our hypotyhesis, in line with our views regarding the site of origin of ovarian epithelial tumors in humans is that the lesions involving the ovary in this animal model originate in cells derived from the mullerian tract. At this point, it is not clear whether the mechanism of tumor predisposition in this model is similar to that in humans with germline *BRCA1* mutations. Even if the mechanisms are not identical, it is quite likely that there are significant overlaps because the tumors in both species involve similar organs and tissues and are driven by inactivation of a similar gene. This has potentially important translational implications because knowledge of a circulating factor secreted by granulosa cells or other components of the pituitary–gonadal axis and associated with ovarian cancer predisposition could lead to the development of a novel approach, possibly based on a simple blood test, for screening for ovarian cancer predisposition. This knowledge may also form the basis for novel strategies on the basis of manipulations of the levels of the factor in question, for ovarian cancer prevention in individuals with familial predisposition to this disease.

Acknowledgments The writing of this article was aided by grant #W81XWH-04-1-0125 from the US Department of Defense and by grant #RO1CA119078 from the US National Institutes of Health.

References

1. Brose MS, Rebbeck TR, Calzone KA, Stopfer JE, Nathanson KL, Weber BL. Cancer risk estimates for BRCA1 mutation carriers identified in a risk evaluation program. J Natl Cancer Inst 2002;94:1365–72.
2. Knudsen AG. Mutation and cancer: Statistical study of retinoblastoma. Proc Natl Acad Sci USA 1971;68:820–3.
3. Deng C-X. BRCA1: Cell cycle checkpoint, genetic instability, DNA damage response and cancer evolution. Nucl Acids Res 2006;34:1416–26.
4. Deng CX, Scott F. Role of the tumor suppressor gene *Brca1* in genetic stability and mammary gland tumor formation. Oncogene 2000;19:1059–64.
5. Scully R, Livingston DM. In search of the tumour-suppressor functions of BRCA1 and BRCA2. Nature 2000;408:429–32.
6. Venkitaraman AR. Cancer susceptibility and the functions of BRCA1 and BRCA2. Cell 2002;108:171–82.
7. Zheng L, Li S, Boyer TG, Lee WH. Lessons learned from BRCA1 and BRCA2. Oncogene 2000;19:6159–75.
8. Cornelis RS, Neuhausen SL, Johansson O, et al. High allele loss rates at 17q12-q21 in breast and ovarian tumors from BRCA1-linked families. The Breast Cancer Linkage Consortium. Genes Chromosomes Cancer 1995;13:203–10.
9. Neuhausen SL, Marshall CJ. Loss of heterozygosity in familial tumors from three BRCA1-linked kindreds. Cancer Res 1994;54:6069–72.
10. Smith SA, Easton DF, Evans DG, Ponder BA. Allele losses in the region 17q12-21 in familial breast and ovarian cancer involve the wild-type chromosome. Nature Genet 1992;2:128–31.
11. Gowen LC, Johnson BL, Latour AM, Sulik KK, Koller BH. BRCA1 deficiency results in early embryonic lethality characterized by neuroepithelial abnormalities. Nat Genet 1996;12:191–4.
12. Shen SX, Waeaver Z, Xu XL, et al. A targeted disruption of the murine *BRCA1* gene causes y-radiation hypersensitivity and genetic instability. Oncogene 1998;17:3117–24.
13. Rice J, Massey-Brown K, Futscher B. Aberrant methylation of the BRCA1 CpG island promoter is associated with decreased BRCA1 mRNA in sporadic breast cancer cells. Oncogene 1998;1998(17):1807–12.
14. Tomlinson GE, Chen TT-L, Stastny VA, et al. Characterization of a breast cancer cell line derived from a germ-line BRCA1 mutation carrier. Cancer Res 1998;58:3237–42.
15. Xu X, Weaver Z, Linke SP, et al. Centrosome amplification and a defective G2/M cell cycle checkpoint induce genetic instability in BRCA1 exon 11 isoform deficient cells. Mol Cell 1999;3:389–95.
16. Ouchi M, Fujiuchi N, Sasai K, et al. BRCA1 phosphorylation by aurora in the regulation of G2 to M transition. J Biol Chem 2004;279:19643–8.

17. Whittemore AS, Harris R, Itnyre J. Characteristics relating to ovarian cancer risk: Collaborative analysis of 12 US case-control studies. II. Invasive epithelial ovarian cancers in white women. Collaborative Ovarian Cancer Group [see comments]. Am J Epidemiol 1992;136(10):1184–203.
18. Pike MC. Age-related factors in cancers of the breast, ovary, and endometrium. J Chronic Dis 1987;40(Suppl 2):59S–69S.
19. Pike MC, Pearce CL, Peters R, Cozen W, Wan P, Wu AH. Hormonal factors and the risk of invasive ovarian cancer: A population-based case-control study. Fertil Steril 2004;82:186–95.
20. Narod SA, Risch H, Moslehi R, et al. Oral contraceptives and the risk of hereditary ovarian cancer. Hereditary Ovarian Cancer Clinical Study Group. New Engl J Med 1998;339:424–8.
21. Josso N, di Clemente N, Gouedart L. Anti-mullerian hormone and its receptors. Molec Cell Endocrinol 2001;179:25–32.
22. de Vet A, Laven JSE, de Jong FH, Themmen APN, Fauser BCJM. Antimullerian hormone serum levels: A putative marker for ovarian aging. Fertil Steril 2002;77:357–62.
23. Gustafson ML, Lee MM, Scully RE, et al. Mullerian inhibiting substance as a marker for ovarian sex-cord tumor. New Engl J Med 1992;326:466–71.
24. Long W-Q, Ranchin V, Pautier P, et al. Detection of minimal levels of serum anti-mullerian hormone during follow-up of patients with ovarian granulosa cell tumor by means of a highly sensitive enzyme-linked immunosorbent assay. J Clin Endocrinol Metab 2000;85:540–4.
25. Durlinger AMJG, Gruijters MJG, Kramer P, Karels B, Grootegoed JA, Themmen PN. Anti-mullerian hormone inhibits initiation of primordial follicle growth in the mouse ovary. Endocrinology 2002;143:1076–84.
26. Dubeau L. The cell of origin of ovarian epithelial tumors and the ovarian surface epithelium dogma: Does the emperor have no clothes? Gynecol Oncol 1999;72:437–42.
27. Masiakos PT, MacLaughlin DT, Maheswaran S, et al. Human ovarian cancer, cell lines, and primary ascites cells express the human Mullerian inhibiting substance (MIS) type II receptor, bind, and are responsive to MIS. Clin Cancer Res 1999;5:3488–99.
28. Stephen AE, Pearsall LA, Christian BP, Donahoe PK, Vacanti JP, MacLaughlin DT. Highly purified mullerian inhibiting substance inhibits human ovarian cancer in vivo. Clin Cancer Res 2002;8:2640–6.
29. Dutertre M, Gouedart L, Xavier F, et al. Ovarian granulosa cell tumors express a functional membrane receptor for anti-mullerian hormone in transgenic mice. Endocrinology 2001;142:4040–6.
30. Conolly DC, Bao R, Nikitin AY, et al. Female mice chimeric for expression of the simian virus 40 TAg under control of the MISIIR promoter develop epithelial ovarian cancer. Cancer Res 2003;63:1389–97.
31. Chodankar R, Kwang S, Sangiorgi F, et al. Cell-nonautonomous induction of ovarian and uterine serous cystadenomas in mice lacking a functional Brca1 in ovarian granulosa cells. Curr Biol 2005;15:561–5.
32. Griswold MD, Heckert L, Linder C. The molecular biology of the FSH receptor. J Steroid Biochem Mol Biol 1995;53:215–8.
33. Colgan TJ, Murphy J, Cole DE, Narod S, Rosen B. Occult carcinoma in prophylactic oophorectomy specimens: Prevalence and association with BRCA germline mutation status. Am J Surg Pathol 2001;25:1283–9.
34. Leeper K, Garcia R, Swisher E, Goff B, Greer B, Paley P. Pathologic findings in prophylactic oophorectomy specimens in high-risk women. Gynecol Oncol 2002;87:52–6.
35. Piek JM, van Diest PJ, Zweemer RP, et al. Dysplastic changes in prophylactically removed Fallopian tubes of women predisposed to developing ovarian cancer. J Pathol 2001;195:451–6.
36. Tonin P, Weber B, Offit K, et al. Frequency of recurrent BRCA1 and BRCA2 mutations in 222 Ashkenazi Jewish breast cancer families. Nature Med 1996;2:1179–83.
37. Hu Y, Ghosh S, Amleh A, et al. Modulation of aromatase expression by BRCA1: A possible link to tissue-specific tumor suppression. Oncogene 2005;24:8343–8.
38. Kramer JL, Velazquez IA, Chen BE, Rosenberg PS, Struewing JP, Greene MH. Prophylactic oophorectomy reduces breast cancer penetrance during prospective, long-term follow-up of BRCA1 mutation carriers. J Clin Oncol 2005;23:8629–35.

Role of p53 and Rb in Ovarian Cancer

David C. Corney, Andrea Flesken-Nikitin, Jinhyang Choi, and Alexander Yu. Nikitin

1 Introduction

Ovarian cancer is the second most common gynecological neoplasm with over 20,000 new cases and 15,000 deaths predicted in 2006 (1). Although significant decreases in mortality have been observed in cancers of the breast and cervix, mortality rates for cancer of the ovary has remained essentially constant over the past 30 years. The majority of cases present at advanced stages, at which point the disease is rarely curable using the existing treatment schemes. Accordingly, the 5-year survival rate for advanced ovarian cancer is 29%. In addition to asymptomatic development, a scarcity of accurate animal models has resulted in a marked lack of knowledge of how the disease progresses, which in turn has precluded the development of desperately needed treatment regimens and screening programs.

Ovarian cancer is a wide-ranging term that groups together a diverse set of neoplasms originating from the ovary, with carcinomas comprising 90% of ovarian cancers. On the basis of the morphological criteria, epithelial ovarian cancers (EOCs) are classified as serous, mucinous, endometrioid, clear cell, transitional cell, squamous cell, and mixed epithelial neoplasms (2). The ovarian surface epithelium (OSE) is a single layer of flat-to-cuboidal cells covering the ovary and is the presumed cell of origin for EOCs (3–6). Recent studies indicate that this layer may possess stem cell properties, and both tumors and cell lines of transformed mouse OSE cells contain a side population (7), which is considered by many investigators as an indicator of cancer stem cells in other tissues (8–11).

2 Disease Etiology

The etiology of EOC is poorly understood, and although several risk factors have been identified, their direct involvement remains largely unaddressed. Of all proposed risk factors, ovulation has received the widest attention. The theory that persistent ovulation increases ovarian cancer incidence was first proposed by Fathalla in 1971 (12, 13), and has been supported by numerous studies demonstrating that a reduction in ovulatory events by pregnancy and/or oral contraceptive decreases EOC risk (14–18). Advocates of this so-called incessant ovulation hypothesis argue that repeated rupture of the ovarian surface during ovulation and subsequent repair by OSE proliferation may increase the frequency at which mutations arise. However, some have deemed this model as too simplistic because neither the effects of reproductive hormones nor acute inflammation is taken into account, both of which may be mutagenic (19–26).

In a recent study using a serial transvaginal ultrasonography approach, approximately 50% of ovarian carcinomas were shown to develop from preexisting benign-appearing cysts or endometriotic cysts, while no preexisting lesions had been evident in the remaining cases 12 months prior to diagnosis (27). Strikingly, upon histopathological analysis, the majority of tumors that arose from preexisting lesions were mucinous, endometrioid, or clear cell carcinomas with adjacent benign- or borderline-like lesions in the vicinity of the carcinoma. In stark contrast, tumors with no evidence of preexisting lesions were mostly of a serous pathological nature. Although a minority of serous carcinomas were of low grade and located adjacent to borderline-like lesions, a majority of them were of high grade with no evidence of precursor lesions in the vicinity of the carcinoma. These observations give significant weight to the hypothesis that low grade serous carcinomas arise in a stepwise manner from benign lesions, while high grade serous carcinomas are distinct and arise de novo from the OSE (28).

3 Genetics of Ovarian Cancer

Although germline mutations in *BRCA1* and *BRCA2* are the most common genetic aberrations in hereditary ovarian carcinomas, by far the most frequent alterations in sporadic EOC are in the p53 and RB pathways. Defects in these two tumor suppressor pathways are present in over 80% of human cancers (29, 30), and have been associated with poor prognosis in ovarian carcinomas (31–37).

3.1 Mutations in the p53 Pathway

Mutation of the *p53* gene at the locus 17p13.1 is the most common single genetic alteration in sporadic human EOC. The p53 protein contains four functional domains – a transcriptional activation domain, a tetramerization domain, and two DNA-binding domains. In addition to possessing transcriptional activating properties, transcriptional repression has been described, although binding sites are less well-characterized (38–42).

Either loss of wild type p53 function, gain of oncogenic function, or the ability to activate p53 inappropriately severely compromises the capacity for controlled cellular proliferation and growth. Numerous stimuli have been demonstrated to activate p53, including UV irradiation-induced DNA damage, inappropriate protooncogene activation, mitogenic signaling, and hypoxia. Depending on the cellular context, one of several responses is implemented, such as cell cycle arrest, senescence, differentiation, or induction of the apoptotic cascade. Through its activity as a transcription factor, p53 executes each response by directly binding p53-binding sites in regulatory regions of target genes. Using bioinformatic approaches, over 4,000 putative target genes were identified (43). Validated target genes include the CDK inhibitor *p21*, members of the proapoptotic family *BCL-2*, the death receptor *FAS*, and p53 repressor *HDM2* (mdm2 in mice) (44–47).

The majority of *p53* mutations are missense mutations that cause single residue changes, largely occurring in the DNA-binding domain (48). Mutant p53 protein has the ability to form a tetramer with wild type p53, acting as a dominant negative to repress normal physiological processes of p53, possibly by inducing an inactive conformation of the DNA-binding domain and reducing the ability to transactivate/repress target genes (49–52). Normally, p53 exists in a negative feedback loop with HDM2, which tightly controls both p53 and HDM2 levels in the cell. Loss of transcriptional activity, however, may result in decreased HDM2, with the consequence of mutant p53 stabilization and therefore increased amount of nonfunctional/gain-of-function mutant p53 protein (53).

Although *p53* mutations have been detected in all histological types of EOC, a number of studies have demonstrated higher frequencies of such mutations in serous carcinomas (Table 1).

Table 1 Frequency of *p53* mutations in histological subtypes of epithelial ovarian carcinoma (EOC)

Type of EOC (average %)		Defective/total cases (%)	Reference
Serous	Low grade (16%)	4/22 (18)	(54)
		1/12 (8)	(55)
		33/190 (17)	(56)
		5/27 (19)	(57)
	High grade (66%)	30/47 (64)	(54)
		30/59 (51)	(55)
		167/180 (93)	(56)
		25/46 (54)	(57)
		33/46 (72)	(58)
		47/71 (66)	(59)
		16/26 (62)	(60)
	Grade not determined (64%)	14/31 (45)	(61)
		11/20 (55)	(62)
		18/23 (78)	(63)
		31/42 (74)	(64)
		73/126 (58)	(65)
Clear cell (8%)		6/38 (17)	(66)
		0/4 (0)	(67)
		1/12 (8)	(60)
Endometrioid (45%)		5/15 (33)	(64)
		7/13 (54)	(62)
		13/27 (48)	(60)
Mucinous (19%)		1/12 (8)	(64)
		3/12 (25)	(63)
		3/11 (27)	(62)
		3/21 (14)	(60)

Furthermore, a number of studies that have paid particular attention to histological criteria of malignancy of serous tumors have found that *p53* mutations are strongly associated with high grade serous carcinomas, but are rare in low grade or borderline serous carcinomas (68–71). In contrast, borderline/low grade tumors frequently harbor mutations in *K-ras*, which are very rare events in high grade serous adenocarcinomas (71–76). These observations have given strong support to the hypothesis that high grade and low grade serous carcinomas arise via discrete pathways (28). Lending further support to this hypothesis is the observation that *p53* is mutated in early stage high grade carcinomas as well as in adjacent dysplastic epithelium in prophylactically removed ovaries from *BRCA1* heterozygotes (77, 78). This supports a model in which *p53* mutation is not only required for carcinogenesis, but also is an early event in the pathogenesis of high grade serous carcinoma.

Of interest are the interactions between p53 and BRCA1 in ovarian carcinogenesis. $Brca1^{-/-}$ mouse embryos are embryonic lethal at embryonic (e) day 6.5; however, if embryos are compound null mutants for both *Brca1* and *p53*, lethality is delayed, leading to a "death by checkpoint" hypothesis (79). This stipulates that in order for accelerated tumor development, p53 function must be lost so that genome instability is tolerated. In one epidemiological study (80), no instance of *p53* loss was observed without simultaneous loss of *BRCA1*. To test this model in a more defined setting, Xing and Orsulic (81) generated a mouse model to study p53 and Brca1 interaction further. They observed that inactivation of *Brca1* and *p53* in mouse OSE cells of ovary explants did not lead to transformation unless the *Myc* oncogene is over expressed virally, while Clark-Knowles et al. reported increased proliferation in mouse OSE cells deficient for *Brca1* and *p53* but no increase if *Brca1* or *p53* was inactivated independently (82). Both of these studies are in good agreement with the observation that transformation of *p53* deficient mouse OSE cells requires multiple hits for transformation to occur (83).

3.2 Mutations in the RB Pathway

The *Retinoblastoma 1 (RB)* gene was originally identified as a tumor suppressor gene in hereditary and sporadic retinoblastoma in children (84–88). Mutations in either *RB* or its pathways are also common in neoplasms of adults (30).

RB is the founding member of a three-member family of tumor suppressors, which also contains p107 and p130. All three interact with a large number of proteins, yet their direct binding to the E2F family of transcription factors is fundamental to their roles as tumor suppressors (30). RB is only able to interact with E2F when hypophosphorylated. When RB is phosphorylated by cyclin D-dependant kinases, E2Fs are no longer bound and are free to bind regulatory regions of E2F-responsive genes leading to progression into S phase of the cell cycle. In addition to cell cycle effects through E2F, RB also has wide-ranging and frequently poorly understood functions in several cellular processes, including control of cell death and differentiation, and histone modification. For example, RB plays a role in the transition of proliferating myoblasts to differentiating myocytes (89) and differentiation of fetal liver macrophages by opposing inhibitory functions of Id2 on transcription factor PU.1 (90). Furthermore, inactivation of *Rb* results in p53-independent apoptotic death in the developing nervous system of the mouse (91). Rb is involved in epigenetic modifications (92–94) and most recently *Caenorhabditis elegans* homologs of the RB pathway have been implicated in repressing the RNA interference pathway (95).

Although loss of heterozygosity (LOH) of *RB* is well demonstrated in many somatic cancers, a specific role of RB in ovarian cancer has been difficult to determine given conflicting data. Liu et al. observed inactivation of *RB* in 60% of ovarian cancer samples (96), while a study by Gras et al. reported LOH of the *RB* locus in 17% of EOC samples and 30% of tumors with serous differentiation (97). However, because of a limited number of samples, statistical significance was not attained in the later study. In contrast, independent studies by Dodson et al. and Kim et al. show RB immunohistochemistry staining in over 90% of clinical EOC samples that showed LOH at the *RB* locus, suggesting the presence of a second tumor suppressor at this locus (98, 99). Unfortunately, no corroborating experiments such as Western blots or RT-PCR assays were performed to confirm immunohistochemical results at that time.

Although the frequency of *RB* mutation in EOC is of debate, more concrete evidence exists demonstrating that the RB pathway is frequently altered. Mutations in either INK4 protein $p16^{INK4a}$ (*p16*), *RB*, or *cyclin D1/CDK4* are observed in almost 50% of EOC clinical samples in a very thorough piece of work (33, 35). To control Cdk-mediated inhibitory phosphorylation of RB, tumor suppressor p16 specifically antagonizes cyclin D-dependent kinases leading to continued RB-E2F binding and repressing activation of the E2F transcriptional program. Specifically analyzing *p16* expression and alteration, numerous studies reported that alteration in *p16* via either mutation, LOH, or promoter methylation occur in between 30% and 65% of EOCs, although a far lower percentage has also been reported (Table 2).

Of great interest is the observation that over 50% of EOC patients have mutations in both the p53 and RB pathways, including 40% of serous carcinomas (33). It is well known that extensive interaction exists between these two pathways (30). The *INK4a* locus encodes two proteins through use of an alternative reading frame: $p16^{INK4a}$ and a second tumor suppressor involved in activating p53 and $p14^{ARF}$ ($p19^{Arf}$ in mice). p14 represses HDM2, modulating the p53-HDM2 negative feedback pathway. In *p14*-null cell lines, E2F over expression enforces S phase entry (118), while deregulated E2F induces p14 expression. Together, these data provide several possibilities for p53-RB pathway interaction and indicate the significance of concomitant deregulation in both pathways.

3.3 Mouse Models to Analyze p53 and RB Function in EOC

Given the aforementioned data from clinical samples, several groups have attempted to model the roles of p53 and RB using the mouse as a model system. The first

Table 2 Defects of *p16* and *Rb* in human ovarian carcinomas

Gene	Defect (average %)	Defective/total cases (%)	Reference
p16	Homozygous mutation (7%)	2/7 (29)	(100)
		1/50 (2)	(101)
		2/27 (7)	(102)
		2/70 (3)	(32)
		2/88 (2)	(28)
		5/30 (17)	(103)
		0/22 (0)	(104)
		1/94 (1)	(105)
		0/23 (0)	(106)
		0/49 (0)	(107)
		1/35 (3)	(108)
		8/45 (18)	(109)
	Methylation (15%)	8/43 (17)	(32)
		16/44 (36)	(105)
		0/23 (0)	(110)
		6/23 (26)	(106)
		2/49 (4)	(111)
		2/37 (5)	(112)
		0/35 (0)	(108)
		6/46 (13)	(33)
		100/249 (40)	(34)
		5/50 (10)	(113)
	Loss of expression (37%)	22/60 (37)	(32)
		19/94 (20)	(105)
		6/22 (27)	(106)
		20/59 (34)	(35)
		22/29 (76)	(112)
		28/47 (60)	(36)
		10/46 (22)	(33)
		70/117 (60)	(108)
		28/82 (34)	(107)
		9/73 (12)	(37)
		23/107 (21)	(33)
		60/134 (45)	(31)
RB	Homozygous mutation (9%)	1/24 (4)	(114)
		2/15 (13)	(96)
	Loss of or aberrant expression (19%)	2/25 (8)	(98)
		2/26 (8)	(99)
		3/22 (14)	(115)
		7/34 (21)	(116)
		2/59 (3)	(35)
		5/46 (11)	(33)
		7/9 (78)	(97)
		10/84 (12)	(107)
		1/78 (1)	(117)
		28/134 (21)	(31)
		12/107 (37)	(33)

Only experiments on freshly collected surgical material are included approach taken was to direct expression of the transforming region of SV40 large T antigen (SV40 Tag) in the mouse OSE by using the *Mullerian inhibitory substance type II receptor (MISIIR)* promoter. SV40 Tag binds and inactivates both p53 and Rb proteins. Necropsy of *MISIIR-SV40-Tag* transgenic mice revealed bilateral ovarian masses in 50% of cases and bloody ascites were frequently present in the abdominal cavity (119). Pathological analysis classified the tumors as poorly differentiated carcinomas.

Even though a clearly important breakthrough in EOC modeling, this approach has several shortcomings. First, although the *MISIIR* promoter directs expression to the OSE, neoplastic lesions were also observed at other sites demonstrating a degree of promoter leakiness. Second, expression of MISIIR is also evident during early embryonic development; tumors therefore arise during early adult life, which is unlike than that observed in humans. Third, and more importantly, through alternative splicing, *SV40* early region encodes several viral proteins including small t and 17kT antigens in addition to large T. All three proteins directly bind Hsc70 through a J domain at the N terminus, while large T and 19kT share a LXCXE-binding motif allowing inactivation of all known members of the RB family. RB family members *p107* and *p130* are rarely mutated in human neoplasms (120). Furthermore, small t antigen has been implicated in cell transformation (121).

To test that p53 and Rb are directly involved in epithelial ovarian carcinogenesis, we established a more defined and controlled approach to inactivate *p53* and/or *Rb* in the mouse OSE through Cre-*loxP* technology (122). By taking advantage of the enclosed anatomical location of the mouse ovary within the ovarian bursa, selective exposure of OSE to any agent can be achieved. To inactivate *p53* and/or *Rb*, adenovirus expressing Cre recombinase under control of the *immediate early cytomegalovirus* promoter (Ad*Cre*) is injected through the oviductal infundibulum into the bursa of transgenic mice carrying conditional alleles of each gene. $Rb^{loxP/loxP}$ mice do not have any ovarian tumors and only 6% of $p53^{loxP/loxP}$ mice develop neoplasia, while 97% of $p53^{loxP/loxP}$ $Rb^{loxP/loxP}$ mice develop ovarian tumor after single exposure to Ad*Cre*. Following a similar clinical course to that seen in humans, tumors spread intraperitoneally (27%), form hemorrhagic or serous ascites (24%), and frequently metastasize to the contralateral ovary (15%), lung (18%), and liver (6%). Pathological evaluation of the early stages of carcinogenesis combined with cytokeratin 8 (CK8) immunostaining demonstrated an epithelial origin of induced neoplasms in 84% of cases. Consistent with a proposed role of p53 in the initiation of high grade serous adenocarcinomas, induced tumors were most comparable to this subset of human EOC tumors.

This approach has several advantages over other methods to model EOC in the mouse. First, intrabursal administration of Ad*Cre* removes the requirement for an OSE-specific promoter, of which none are currently known. Though OSE-specific infection was performed previously (83), our approach involves no cell culture stage and all tumor development is accomplished in adult immunocompetent mice. The approach also allows conditional and temporal control of the initiating events, which is particularly useful for modeling the early stages of EOC initiation. As such, an identical approach was recently used to demonstrate the role of *K-ras* and

Pten in the initiation of endometrioid ovarian cancer (123), and *Brca1* in preneoplastic changes (82). Taken together, these results clearly demonstrate that different genetic alterations lead to distinct subsets and stages of EOC.

4 Applications of Genetically Defined Models

Although the primary goal of generating genetically-engineered mouse models is to attain a better understanding of the molecular pathways behind EOC carcinogenesis, other significant goals are to allow rational drug design and testing in a defined and reproducible environment and to allow development of improved imaging techniques. In this section, we describe recent novel applications of mouse models of EOC.

4.1 Rational Drug Design

Treatment options for patients with advanced stages of ovarian cancer are almost nonexistent and severely limited in efficacy. Because of the high percentage of patients succumbing to the disease, ovarian cancer is a good candidate for chemoprevention.

A large body of work, largely in colorectal cancer studies, has indicated that nonsteroidal antiinflammatory drugs (NSAIDs), such as aspirin or sulindac, reduce the number and size of colonic polyps in patients with familial adenomatous polyposis (FAP) (124–126). Chronic administration of aspirin over a 10–15 year period has been reported to reduce risk of developing colon cancer by up to 50% (127), indicating a protective effect of NSAIDs. Nobel Prize winner John Vane proposed that the effects of NSAIDs are mediated by inhibiting the enzymatic activity of cyclooxygenase (COX) (128). COX is responsible for catalyzing arachadonic acid into PGG_2, which is then converted into PGH_2, and is subsequently converted into one of many prostaglandins: hormone-like, lipid soluble molecules involved in a wide range of physiological processes, including platelet aggregation, muscular contraction/relaxation, and immunity. Two isoforms of COX protein exist – COX-1 and COX-2, the later having received the most attention since COX-1 appears to be constitutively expressed, while COX-2 is not normally expressed unless induced by proinflammatory cytokines.

Work in $Apc^{\Delta 716}$ mice, which spontaneously develop numerous polyps in the intestinal tract similar to FAP in humans, confirmed a link between NSAIDs and COX-2. $Apc^{\Delta 716}$ mice on a *Cox-2*-null background develop significantly fewer polyps compared with a *Cox-2*-wild type background. Treatment of $Apc^{\Delta 716}$ mice on a *Cox-2*-wild type background with either sulindac or a novel Cox-2 inhibitor MF-tricyclic similarly reduced the polyp number (129). Since this initial report, there has been much interest in developing COX-2 isoform-specific inhibitors over

NSAIDs because of fewer adverse effects. In contrast, COX-1 has received little attention, despite having been purified and cloned prior to COX-2.

Although NSAIDs appear to reduce risk of cancers at sites such as esophagus and stomach (130), their effect in cancers of the ovary remain inconclusive. Although some groups have reported high levels of COX-2 in ovarian cancer (131, 132), others have reported elevated COX-1, but not COX-2, in ovarian cancer tissue samples (133, 134) or cell lines (135, 136), suggesting tissue-specific roles for each isoform. Furthermore, Cox-1 over expression was previously demonstrated in tumors arising from *p53*-null mouse OSE cells also over expressing either *c-myc* and *K-ras* or *c-myc* and *Akt* (137), while Cox-2 was either not expressed or expressed at very low levels. Therefore, in a large collaborative effort, Daikoku and coworkers investigated Cox-1/2 expression status in a defined and controlled manner, using three genetically-engineered mouse models to gain a better understanding of the roles of this class of protein in EOC, and whether Cox over expression is unique to specific genetic alterations or is widespread (138). The previously characterized models used were based on intrabursal Ad*Cre* administration to inactivate *p53* and *Rb* (122), or inactivate *Pten* and activate *K-ras* (123) or based on *MISIIR*-directed expression of *SV40 Tag* (119), as outlined earlier. In all three models, Cox-1, but not Cox-2, was over expressed in the mouse EOCs as judged by RT-PCR, in situ hybridization, Western blotting, and immunohistochemistry with Cox-1/2 isoform-specific primers, probes, and antibodies. The observation that Cox-1 is over expressed in an identical pattern in four different mouse models on the basis of different genetic lesions suggests that Cox-1 over expression may be widespread and a conserved aspect of EOC.

The investigation by Daikoku and colleagues has opened a new avenue for the rational design of preventive and therapeutic agents against ovarian cancer and may lead to a fundamental shift in approach toward COX inhibitors. Perhaps most significantly, in a microarray study comparing global gene expression between $p53^{loxP/loxP}$ $Rb^{loxP/loxP}$ OSE cells treated with either Ad*Cre* or control virus in culture, Cox-1 over expression was detected at the earliest passages (138), indicating the potential usefulness of Cox-1 as a screening marker.

4.2 Development of New Imaging Techniques

Although identification of screening markers associated with EOC are undoubtedly of critical importance to allow early and accurate diagnosis, it is extremely difficult to find markers that are flawless, as both a high degree of specificity and sensitivity is essential. The most widely used biomarker for ovarian tumors is the serum tumor marker CA125 (139). Unfortunately, though 80% of patients with advanced EOC have high CA125 serum levels, only half of them are positive at the early stage of disease (140, 141), whereas, conversely, CA125 concentration may be elevated in individuals free of disease, resulting in false-positive tests. For this reason, CA125 has limited diagnostic value, and positive results must be substantiated by exploratory surgery or laparoscopy, which, like all surgical procedures, carries a certain degree of risk. Consequently, adequate monitoring of patients, especially those at elevated risk of developing EOC, such as women carrying germline mutations in *BRCA* genes, is difficult and prophylactic oophorectomy is recommended, which is not a viable option for nulliparous women who wish to raise a family. For this reason, minimally invasive imaging techniques need to be developed to allow improved patient monitoring.

Multiphoton microscopy (MPM) (142) offers one possible means to improve diagnostic imaging. Two-photon MPM is based on the theory that two low-energy infrared photons may arrive simultaneously at a fluorophore and result in electronic transition normally observed upon absorption of a single photon. Several endogenous molecules, such as NAD(P)H and flavins, emit photons upon two-photon excitation, while fluorescent proteins such as green fluorescent protein is also detectable via MPM. In addition, second harmonic generation (SHG) allows direct imaging of anisotropic biological molecules such as collagen (143) with no requirement for exogenously added fluorophores and may be imaged at the same time as two-photon microscopy. MPM has several advantages over traditional fluorescence imaging because of its low phototoxicity and lack of out-of-focal plane excitation (143). Together with our collaborators Drs. Warren Zipfel, Rebecca Williams, and Watt Webb, we have demonstrated the utility of two-photon microscopy to image deep into the mouse ovary (144). In contrast to transvaginal ultrasonography and traditional laparoscopy, which provide either low resolution images or images only of the ovary surface, respectively, MPM is able to image at high resolution (cellular level) deep (~200–300 μm) into the ovary, allowing one to rapidly acquire images of quality comparable to that of traditional hematoxylin and eosin-stained histological sections.

MPM has been used to help answer diverse biological questions such as how gene expression correlates with metastasis, whether senile plaques change size in a mouse model of Alzheimer's disease, and the role of sensory deprivation in cortical plasticity (101, 145–147). In addition to low phototoxicity, MPM can allow analysis of individual cell migration and motility in a time-lapse manner (148), while long-term, repeated imaging procedures may be carried out by performing MPM during several rounds of survival surgery (145), allowing one to closely follow development of EOC from the very earliest stages of carcinogenesis. The construction of an endoscopic MPM device should facilitate translation of this imaging method into clinical practice. Such a device is currently under development.

5 Concluding Remarks

Because of asymptomatic development, the initiating events of ovarian cancer remain obscure and much of our current understanding is based on circumstantial and correlative evidences. To this end, the development of accurate mouse models of ovarian cancer is of utmost importance in expanding our knowledge of ovarian carcinogenesis.

On the basis of the observations that *p53* and *Rb* pathways are commonly altered in human EOC, we have inactivated both tumor suppressors in the mouse OSE and demonstrated formation of neoplasms that are most comparable to human high grade serous carcinomas of the ovary. Importantly, the approach for conditional induction of OSE-specific genetic alterations described in our work is well applicable to other studies seeking to test roles of specific genetic alterations in the OSE in a time-, location-, and lineage-dependant manner. This study, and others in the field, has given significant weight to the hypothesis that *p53* and *Rb* mutations play critical roles in ovarian carcinogenesis, in particular at the very earliest stages. We have gone on to demonstrate the usefulness of genetically-engineered mouse models in identifying proteins for therapeutic targeting and development of improved imaging techniques, and it is our hope that these approaches will lead to a more complete picture of ovarian carcinogenesis, as well as facilitate its detection, treatment, and prevention.

Acknowledgments This work was supported by NIH grants CA 112354 and RR 17595 to AYN and a Cornell Vertebrate Genomics Scholar Award to DCC.

References

1. Jemal, A., Siegel, R., Ward, E., Murray, T., Xu, J., Smigal, C., and Thun, M. J. (2006). Cancer statistics, 2006. CA Cancer J Clin *56*, 106–130.
2. Scully, R. E. a. S., L.H. (1999). Histological typing of ovarian tumours, Vol 9, (Berlin Hiedelberg New York: Springer).
3. Auersperg, N., Wong, A. S., Choi, K. C., Kang, S. K., and Leung, P. C. (2001). Ovarian surface epithelium: biology, endocrinology, and pathology. Endocr Rev *22*, 255–288.
4. Nikitin, A. Y., Connolly, D. C., and Hamilton, T. C. (2004). Pathology of ovarian neoplasms in genetically modified mice. Comp Med *54*, 26–28.
5. Scully, R. E. (1977). Ovarian tumors. A review. Am J Pathol *87*, 686–720.
6. Vanderhyden, B. C., Shaw, T. J., and Ethier, J. F. (2003). Animal models of ovarian cancer. Reprod Biol Endocrinol *1*, 67.
7. Szotek, P. P., Pieretti-Vanmarcke, R., Masiakos, P. T., Dinulescu, D. M., Connolly, D., Foster, R., Dombkowski, D., Preffer, F., Maclaughlin, D. T., and Donahoe, P. K. (2006). Ovarian cancer side population defines cells with stem cell-like characteristics and Mullerian Inhibiting Substance responsiveness. Proc Natl Acad Sci USA *103*, 11154–11159.
8. Chiba, T., Kita, K., Zheng, Y. W., Yokosuka, O., Saisho, H., Iwama, A., Nakauchi, H., and Taniguchi, H. (2006). Side population purified from hepatocellular carcinoma cells harbors cancer stem cell-like properties. Hepatology *44*, 240–251.
9. Haraguchi, N., Utsunomiya, T., Inoue, H., Tanaka, F., Mimori, K., Barnard, G. F., and Mori, M. (2006). Characterization of a side population of cancer cells from human gastrointestinal system. Stem Cells *24*, 506–513.
10. Hirschmann-Jax, C., Foster, A. E., Wulf, G. G., Nuchtern, J. G., Jax, T. W., Gobel, U., Goodell, M. A., and Brenner, M. K. (2004). A distinct "side population" of cells with high drug efflux capacity in human tumor cells. Proc Natl Acad Sci USA *101*, 14228–14233.
11. Kruger, J. A., Kaplan, C. D., Luo, Y., Zhou, H., Markowitz, D., Xiang, R., and Reisfeld, R. A. (2006). Characterization of stem cell-like cancer cells in immune competent mice. Blood *108*, 3906–3912.
12. Fathalla, M. F. (1971). Incessant ovulation–a factor in ovarian neoplasia? Lancet *2*, 163.
13. Fathalla, M. F. (1972). Factors in the causation and incidence of ovarian cancer. Obstet Gynecol Surv *27*, 751–768.

14. Riman, T., Dickman, P. W., Nilsson, S., Correia, N., Nordlinder, H., Magnusson, C. M., and Persson, I. R. (2002). Risk factors for invasive epithelial ovarian cancer: results from a Swedish case-control study. Am J Epidemiol *156*, 363–373.
15. Risch, H. A., Marrett, L. D., and Howe, G. R. (1994). Parity, contraception, infertility, and the risk of epithelial ovarian cancer. Am J Epidemiol *140*, 585–597.
16. Risch, H. A., Weiss, N. S., Lyon, J. L., Daling, J. R., and Liff, J. M. (1983). Events of reproductive life and the incidence of epithelial ovarian cancer. Am J Epidemiol *117*, 128–139.
17. Titus-Ernstoff, L., Perez, K., Cramer, D. W., Harlow, B. L., Baron, J. A., and Greenberg, E. R. (2001). Menstrual and reproductive factors in relation to ovarian cancer risk. Br J Cancer *84*, 714–721.
18. Whittemore, A. S., Harris, R., and Itnyre, J. (1992). Characteristics relating to ovarian cancer risk: collaborative analysis of 12 US case-control studies. IV. The pathogenesis of epithelial ovarian cancer. Collaborative Ovarian Cancer Group. Am J Epidemiol *136*, 1212–1220.
19. Bose, C. K. (2005). Does hormone replacement therapy prevent epithelial ovarian cancer? Reprod Biomed Online *11*, 86–92.
20. Bukulmez, O., and Arici, A. (2000). Leukocytes in ovarian function. Hum Reprod Update *6*, 1–15.
21. Cramer, D. W., and Welch, W. R. (1983). Determinants of ovarian cancer risk. II. Inferences regarding pathogenesis. J Natl Cancer Inst *71*, 717–721.
22. Fleming, J. S., Beaugie, C. R., Haviv, I., Chenevix-Trench, G., and Tan, O. L. (2006). Incessant ovulation, inflammation and epithelial ovarian carcinogenesis: revisiting old hypotheses. Mol Cell Endocrinol *247*, 4–21.
23. Konishi, I. (2006). Gonadotropins and ovarian carcinogenesis: a new era of basic research and its clinical implications. Int J Gynecol Cancer *16*, 16–22.
24. Mohle, J., Whittemore, A., Pike, M., and Darby, S. (1985). Gonadotrophins and ovarian cancer risk. J Natl Cancer Inst *75*, 178–180.
25. Ness, R. B., and Cottreau, C. (1999). Possible role of ovarian epithelial inflammation in ovarian cancer. J Natl Cancer Inst *91*, 1459–1467.
26. Nikitin, A. Y., and Hamilton, T.C. (2005). Modeling ovarian cancer in the mouse. Res Adv in Cancer *5*, 49–59.
27. Horiuchi, A., Itoh, K., Shimizu, M., Nakai, I., Yamazaki, T., Kimura, K., Suzuki, A., Shiozawa, I., Ueda, N., and Konishi, I. (2003). Toward understanding the natural history of ovarian carcinoma development: a clinicopathological approach. Gynecol Oncol *88*, 309–317.
28. Shih Ie, M., and Kurman, R. J. (2004). Ovarian tumorigenesis: a proposed model based on morphological and molecular genetic analysis. Am J Pathol *164*, 1511–1518.
29. Hahn, W. C., and Weinberg, R. A. (2002). Modelling the molecular circuitry of cancer. Nat Rev Cancer *2*, 331–341.
30. Sherr, C. J., and McCormick, F. (2002). The RB and p53 pathways in cancer. Cancer Cell *2*, 103–112.
31. Bali, A., O'Brien, P. M., Edwards, L. S., Sutherland, R. L., Hacker, N. F., and Henshall, S. M. (2004). Cyclin D1, p53, and p21Waf1/Cip1 expression is predictive of poor clinical outcome in serous epithelial ovarian cancer. Clin Cancer Res *10*, 5168–5177.
32. Fujita, M., Enomoto, T., Haba, T., Nakashima, R., Sasaki, M., Yoshino, K., Wada, H., Buzard, G. S., Matsuzaki, N., Wakasa, K., and Murata, Y. (1997). Alteration of *p16* and *p15* genes in common epithelial ovarian tumors. Int J Cancer *74*, 148–155.
33. Hashiguchi, Y., Tsuda, H., Yamamoto, K., Inoue, T., Ishiko, O., and Ogita, S. (2001). Combined analysis of p53 and RB pathways in epithelial ovarian cancer. Hum Pathol *32*, 988–996.
34. Katsaros, D., Cho, W., Singal, R., Fracchioli, S., Rigault De La Longrais, I. A., Arisio, R., Massobrio, M., Smith, M., Zheng, W., Glass, J., and Yu, H. (2004). Methylation of tumor suppressor gene p16 and prognosis of epithelial ovarian cancer. Gynecol Oncol *94*, 685–692.
35. Kusume, T., Tsuda, H., Kawabata, M., Inoue, T., Umesaki, N., Suzuki, T., and Yamamoto, K. (1999). The p16-cyclin D1/CDK4-pRb pathway and clinical outcome in epithelial ovarian cancer. Clin Cancer Res *5*, 4152–4157.

36. Sui, L., Dong, Y., Ohno, M., Goto, M., Inohara, T., Sugimoto, K., Tai, Y., Hando, T., and Tokuda, M. (2000). Inverse expression of Cdk4 and p16 in epithelial ovarian tumors. Gynecol Oncol 79, 230–237.
37. Tachibana, M., Watanabe, J., Matsushima, Y., Nishida, K., Kobayashi, Y., Fujimura, M., and Shiromizu, K. (2003). Independence of the prognostic value of tumor suppressor protein expression in ovarian adenocarcinomas: A multivariate analysis of expression of p53, retinoblastoma, and related proteins. Int J Gynecol Cancer 13, 598–606.
38. Curtin, J. C., and Spinella, M. J. (2005). p53 in human embryonal carcinoma: identification of a transferable, transcriptional repression domain in the N-terminal region of p53. Oncogene 24, 1481–1490.
39. D'Souza, S., Xin, H., Walter, S., and Choubey, D. (2001). The gene encoding p202, an interferon-inducible negative regulator of the p53 tumor suppressor, is a target of p53-mediated transcriptional repression. J Biol Chem 276, 298–305.
40. Hammond, E. M., and Giaccia, A. J. (2005). The role of p53 in hypoxia-induced apoptosis. Biochem Biophys Res Commun 331, 718–725.
41. Hoffman, W. H., Biade, S., Zilfou, J. T., Chen, J., and Murphy, M. (2002). Transcriptional repression of the anti-apoptotic survivin gene by wild type p53. J Biol Chem 277, 3247–3257.
42. Imbriano, C., Gurtner, A., Cocchiarella, F., Di Agostino, S., Basile, V., Gostissa, M., Dobbelstein, M., Del Sal, G., Piaggio, G., and Mantovani, R. (2005). Direct p53 transcriptional repression: in vivo analysis of CCAAT-containing G2/M promoters. Mol Cell Biol 25, 3737–3751.
43. Wang, L., Wu, Q., Qiu, P., Mirza, A., McGuirk, M., Kirschmeier, P., Greene, J. R., Wang, Y., Pickett, C. B., and Liu, S. (2001). Analyses of p53 target genes in the human genome by bioinformatic and microarray approaches. J Biol Chem 276, 43604–43610.
44. el-Deiry, W. S., Tokino, T., Velculescu, V. E., Levy, D. B., Parsons, R., Trent, J. M., Lin, D., Mercer, W. E., Kinzler, K. W., and Vogelstein, B. (1993). WAF1, a potential mediator of p53 tumor suppression. Cell 75, 817–825.
45. Miyashita, T., and Reed, J. C. (1995). Tumor suppressor p53 is a direct transcriptional activator of the human bax gene. Cell 80, 293–299.
46. Oda, E., Ohki, R., Murasawa, H., Nemoto, J., Shibue, T., Yamashita, T., Tokino, T., Taniguchi, T., and Tanaka, N. (2000). Noxa, a BH3-only member of the Bcl-2 family and candidate mediator of p53-induced apoptosis. Science 288, 1053–1058.
47. Owen-Schaub, L. B., Zhang, W., Cusack, J. C., Angelo, L. S., Santee, S. M., Fujiwara, T., Roth, J. A., Deisseroth, A. B., Zhang, W. W., Kruzel, E., and et al. (1995). Wild-type human p53 and a temperature-sensitive mutant induce Fas/APO-1 expression. Mol Cell Biol 15, 3032–3040.
48. Sigal, A., and Rotter, V. (2000). Oncogenic mutations of the p53 tumor suppressor: the demons of the guardian of the genome. Cancer Res 60, 6788–6793.
49. Chene, P. (1998). In vitro analysis of the dominant negative effect of p53 mutants. J Mol Biol 281, 205–209.
50. Kern, S. E., Pietenpol, J. A., Thiagalingam, S., Seymour, A., Kinzler, K. W., and Vogelstein, B. (1992). Oncogenic forms of p53 inhibit p53-regulated gene expression. Science 256, 827–830.
51. Shaulian, E., Zauberman, A., Ginsberg, D., and Oren, M. (1992). Identification of a minimal transforming domain of p53: negative dominance through abrogation of sequence-specific DNA binding. Mol Cell Biol 12, 5581–5592.
52. Unger, T., Mietz, J. A., Scheffner, M., Yee, C. L., and Howley, P. M. (1993). Functional domains of wild-type and mutant p53 proteins involved in transcriptional regulation, transdominant inhibition, and transformation suppression. Mol Cell Biol 13, 5186–5194.
53. Blagosklonny, M. V. (2000). p53 from complexity to simplicity: mutant p53 stabilization, gain-of-function, and dominant-negative effect. Faseb J 14, 1901–1907.
54. O'Neill, C. J., Deavers, M. T., Malpica, A., Foster, H., and McCluggage, W. G. (2005). An immunohistochemical comparison between low-grade and high-grade ovarian serous

carcinomas: significantly higher expression of p53, MIB1, BCL2, HER-2/neu, and C-KIT in high-grade neoplasms. Am J Surg Pathol 29, 1034–1041.
55. Singer, G., Stohr, R., Cope, L., Dehari, R., Hartmann, A., Cao, D. F., Wang, T. L., Kurman, R. J., and Shih Ie, M. (2005). Patterns of p53 mutations separate ovarian serous borderline tumors and low- and high-grade carcinomas and provide support for a new model of ovarian carcinogenesis: a mutational analysis with immunohistochemical correlation. Am J Surg Pathol 29, 218–224.
56. Lassus, H., Leminen, A., Lundin, J., Lehtovirta, P., and Butzow, R. (2003). Distinct subtypes of serous ovarian carcinoma identified by p53 determination. Gynecol Oncol 91, 504–512.
57. Chan, W. Y., Cheung, K. K., Schorge, J. O., Huang, L. W., Welch, W. R., Bell, D. A., Berkowitz, R. S., and Mok, S. C. (2000). Bcl-2 and p53 protein expression, apoptosis, and p53 mutation in human epithelial ovarian cancers. Am J Pathol 156, 409–417.
58. Gadducci, A., Di Cristofano, C., Zavaglia, M., Giusti, L., Menicagli, M., Cosio, S., Naccarato, A. G., Genazzani, A. R., Bevilacqua, G., and Cavazzana, A. O. (2006). P53 gene status in patients with advanced serous epithelial ovarian cancer in relation to response to paclitaxel- plus platinum-based chemotherapy and long-term clinical outcome. Anticancer Res 26, 687–693.
59. Havrilesky, L., Darcy, M., Hamdan, H., Priore, R. L., Leon, J., Bell, J., and Berchuck, A. (2003). Prognostic significance of p53 mutation and p53 overexpression in advanced epithelial ovarian cancer: a Gynecologic Oncology Group Study. J Clin Oncol 21, 3814–3825.
60. Caduff, R. F., Svoboda-Newman, S. M., Ferguson, A. W., Johnston, C. M., and Frank, T. S. (1999). Comparison of mutations of Ki-RAS and p53 immunoreactivity in borderline and malignant epithelial ovarian tumors. Am J Surg Pathol 23, 323–328.
61. Fujita, M., Enomoto, T., Inoue, M., Tanizawa, O., Ozaki, M., Rice, J. M., and Nomura, T. (1994). Alteration of the p53 tumor suppressor gene occurs independently of K-ras activation and more frequently in serous adenocarcinomas than in other common epithelial tumors of the human ovary. Jpn J Cancer Res 85, 1247–1256.
62. Henriksen, R., Strang, P., Wilander, E., Backstrom, T., Tribukait, B., and Oberg, K. (1994). p53 expression in epithelial ovarian neoplasms: relationship to clinical and pathological parameters, Ki-67 expression and flow cytometry. Gynecol Oncol 53, 301–306.
63. Renninson, J., Baker, B. W., McGown, A. T., Murphy, D., Norton, J. D., Fox, B. W., and Crowther, D. (1994). Immunohistochemical detection of mutant p53 protein in epithelial ovarian cancer using polyclonal antibody CMI: correlation with histopathology and clinical features. Br J Cancer 69, 609–612.
64. Dogan, E., Saygili, U., Tuna, B., Gol, M., Gurel, D., Acar, B., and Koyuncuoglu, M. (2005). p53 and mdm2 as prognostic indicators in patients with epithelial ovarian cancer: a multivariate analysis. Gynecol Oncol 97, 46–52.
65. Eltabbakh, G. H., Belinson, J. L., Kennedy, A. W., Biscotti, C. V., Casey, G., Tubbs, R. R., and Blumenson, L. E. (1997). p53 overexpression is not an independent prognostic factor for patients with primary ovarian epithelial cancer. Cancer 80, 892–898.
66. Ho, E. S., Lai, C. R., Hsieh, Y. T., Chen, J. T., Lin, A. J., Hung, M. H., and Liu, F. S. (2001). p53 mutation is infrequent in clear cell carcinoma of the ovary. Gynecol Oncol 80, 189–193.
67. Otis, C. N., Krebs, P. A., Quezado, M. M., Albuquerque, A., Bryant, E., San Juan, X., Kleiner, D., Sobel, M. E., and Merino, M. J. (2000). Loss of heterozygosity in *P53, BRCA1*, and estrogen receptor genes and correlation to expression of p53 protein in ovarian epithelial tumors of different cell types and biological behavior. Hum Pathol 31, 233–238.
68. Kupryjanczyk, J., Bell, D. A., Dimeo, D., Beauchamp, R., Thor, A. D., and Yandell, D. W. (1995). *p53* gene analysis of ovarian borderline tumors and stage I carcinomas. Hum Pathol 26, 387–392.
69. Kupryjanczyk, J., Thor, A. D., Beauchamp, R., Merritt, V., Edgerton, S. M., Bell, D. A., and Yandell, D. W. (1993). *p53* gene mutations and protein accumulation in human ovarian cancer. Proc Natl Acad Sci USA 90, 4961–4965.
70. Skomedal, H., Kristensen, G. B., Abeler, V. M., Borresen-Dale, A. L., Trope, C., and Holm, R. (1997). TP53 protein accumulation and gene mutation in relation to overexpression of MDM2 protein in ovarian borderline tumours and stage I carcinomas. J Pathol 181, 158–165.

71. Zheng, J., Benedict, W. F., Xu, H. J., Hu, S. X., Kim, T. M., Velicescu, M., Wan, M., Cofer, K. F., and Dubeau, L. (1995). Genetic disparity between morphologically benign cysts contiguous to ovarian carcinomas and solitary cystadenomas. J Natl Cancer Inst *87*, 1146–1153.
72. Cuatrecasas, M., Villanueva, A., Matias-Guiu, X., and Prat, J. (1997). K-ras mutations in mucinous ovarian tumors: a clinicopathologic and molecular study of 95 cases. Cancer *79*, 1581–1586.
73. Diebold, J., Seemuller, F., and Lohrs, U. (2003). K-RAS mutations in ovarian and extraovarian lesions of serous tumors of borderline malignancy. Lab Invest *83*, 251–258.
74. Singer, G., Kurman, R. J., Chang, H. W., Cho, S. K., and Shih Ie, M. (2002). Diverse tumorigenic pathways in ovarian serous carcinoma. Am J Pathol *160*, 1223–1228.
75. Singer, G., Oldt, R., 3rd, Cohen, Y., Wang, B. G., Sidransky, D., Kurman, R. J., and Shih Ie, M. (2003a). Mutations in BRAF and KRAS characterize the development of low-grade ovarian serous carcinoma. J Natl Cancer Inst *95*, 484–486.
76. Singer, G., Shih Ie, M., Truskinovsky, A., Umudum, H., and Kurman, R. J. (2003b). Mutational analysis of K-ras segregates ovarian serous carcinomas into two types: invasive MPSC (low-grade tumor) and conventional serous carcinoma (high-grade tumor). Int J Gynecol Pathol *22*, 37–41.
77. Pothuir, B., Leitao, M., Barakat, R., Akram, M., Bogomolniy, F., Olvera, N. and Lin, O. (2001). Genetic analysis of ovarian carcinoma histogenesis. In Society of Gynecologic Oncologists 32nd Annual Meeting.
78. Werness, B. A., Parvatiyar, P., Ramus, S. J., Whittemore, A. S., Garlinghouse-Jones, K., Oakley-Girvan, I., DiCioccio, R. A., Wiest, J., Tsukada, Y., Ponder, B. A., and Piver, M. S. (2000). Ovarian carcinoma in situ with germline BRCA1 mutation and loss of heterozygosity at BRCA1 and TP53. J Natl Cancer Inst *92*, 1088–1091.
79. Scully, R., and Livingston, D. M. (2000). In search of the tumour-suppressor functions of BRCA1 and BRCA2. Nature *408*, 429–432.
80. Villeneuve, J. B., Silverman, M. B., Alderete, B., Cliby, W. A., Li, H., Croghan, G. A., Podratz, K. C., and Jenkins, R. B. (1999). Loss of markers linked to BRCA1 precedes loss at important cell cycle regulatory genes in epithelial ovarian cancer. Genes Chromosomes Cancer *25*, 65–69.
81. Xing, D., and Orsulic, S. (2006). A mouse model for the molecular characterization of brca1-associated ovarian carcinoma. Cancer Res *66*, 8949–8953.
82. Clark-Knowles, K. V., Garson, K., and Vanderhyden, B. C. (2006). Conditional inactivation of Brca1 in the mouse ovarian surface epithelium results in an increase in preneoplastic changes. Exp Cell Res *313*, 133–145.
83. Orsulic, S., Li, Y., Soslow, R. A., Vitale-Cross, L. A., Gutkind, J. S., and Varmus, H. E. (2002). Induction of ovarian cancer by defined multiple genetic changes in a mouse model system. Cancer Cell *1*, 53–62.
84. Friend, S. H., Bernards, R., Rogelj, S., Weinberg, R. A., Rapaport, J. M., Albert, D. M., and Dryja, T. P. (1986). A human DNA segment with properties of the gene that predisposes to retinoblastoma and osteosarcoma. Nature *323*, 643–646.
85. Fung, Y. K., Murphree, A. L., T'Ang, A., Qian, J., Hinrichs, S. H., and Benedict, W. F. (1987). Structural evidence for the authenticity of the human retinoblastoma gene. Science *236*, 1657–1661.
86. Knudson, A. G., Jr. (1971). Mutation and cancer: statistical study of retinoblastoma. Proc Natl Acad Sci USA *68*, 820–823.
87. Lee, W. H., Bookstein, R., Hong, F., Young, L. J., Shew, J. Y., and Lee, E. Y. (1987). Human retinoblastoma susceptibility gene: cloning, identification, and sequence. Science *235*, 1394–1399.
88. Weissman, B. E., Saxon, P. J., Pasquale, S. R., Jones, G. R., Geiser, A. G., and Stanbridge, E. J. (1987). Introduction of a normal human chromosome 11 into a Wilms' tumor cell line controls its tumorigenic expression. Science *236*, 175–180.
89. Huh, M. S., Parker, M. H., Scime, A., Parks, R., and Rudnicki, M. A. (2004). Rb is required for progression through myogenic differentiation but not maintenance of terminal differentiation. J Cell Biol *166*, 865–876.

90. Iavarone, A., King, E. R., Dai, X. M., Leone, G., Stanley, E. R., and Lasorella, A. (2004). Retinoblastoma promotes definitive erythropoiesis by repressing Id2 in fetal liver macrophages. Nature *432*, 1040–1045.
91. Macleod, K. F., Hu, Y., and Jacks, T. (1996). Loss of Rb activates both p53-dependent and independent cell death pathways in the developing mouse nervous system. Embo J *15*, 6178–6188.
92. Brehm, A., Miska, E. A., McCance, D. J., Reid, J. L., Bannister, A. J., and Kouzarides, T. (1998). Retinoblastoma protein recruits histone deacetylase to repress transcription. Nature *391*, 597–601.
93. Luo, R. X., Postigo, A. A., and Dean, D. C. (1998). Rb interacts with histone deacetylase to repress transcription. Cell *92*, 463–473.
94. Magnaghi-Jaulin, L., Groisman, R., Naguibneva, I., Robin, P., Lorain, S., Le Villain, J. P., Troalen, F., Trouche, D., and Harel-Bellan, A. (1998). Retinoblastoma protein represses transcription by recruiting a histone deacetylase. Nature *391*, 601–605.
95. Wang, D., Kennedy, S., Conte, D., Jr., Kim, J. K., Gabel, H. W., Kamath, R. S., Mello, C. C., and Ruvkun, G. (2005). Somatic misexpression of germline P granules and enhanced RNA interference in retinoblastoma pathway mutants. Nature *436*, 593–597.
96. Liu, Y., Heyman, M., Wang, Y., Falkmer, U., Hising, C., Szekely, L., and Einhorn, S. (1994). Molecular analysis of the retinoblastoma gene in primary ovarian cancer cells. Int J Cancer *58*, 663–667.
97. Gras, E., Pons, C., Machin, P., Matias-Guiu, X., and Prat, J. (2001). Loss of heterozygosity at the RB-1 locus and pRB immunostaining in epithelial ovarian tumors: a molecular, immunohistochemical, and clinicopathologic study. Int J Gynecol Pathol *20*, 335–340.
98. Dodson, M. K., Cliby, W. A., Xu, H. J., DeLacey, K. A., Hu, S. X., Keeney, G. L., Li, J., Podratz, K. C., Jenkins, R. B., and Benedict, W. F. (1994). Evidence of functional RB protein in epithelial ovarian carcinomas despite loss of heterozygosity at the RB locus. Cancer Res *54*, 610–613.
99. Kim, T. M., Benedict, W. F., Xu, H. J., Hu, S. X., Gosewehr, J., Velicescu, M., Yin, E., Zheng, J., D'Ablaing, G., and Dubeau, L. (1994). Loss of heterozygosity on chromosome 13 is common only in the biologically more aggressive subtypes of ovarian epithelial tumors and is associated with normal retinoblastoma gene expression. Cancer Res *54*, 605–609.
100. Kamb, A., Gruis, N. A., Weaver-Feldhaus, J., Liu, Q., Harshman, K., Tavtigian, S. V., Stockert, E., Day, R. S. III, Johnson, B. E., and Skolnick, M. H. (1994). A cell cycle regulator potentially involved in genesis of many tumor types. Science *264*, 436–440.
101. Brown, E. B., Campbell, R. B., Tsuzuki, Y., Xu, L., Carmeliet, P., Fukumura, D., and Jain, R. K. (2001). In vivo measurement of gene expression, angiogenesis and physiological function in tumors using multiphoton laser scanning microscopy. Nat Med *7*, 864–868.
102. Wong, Y. F., Chung, T. K., Cheung, T. H., Nobori, T., Yim, S. F., Lai, K. W., Phil, M., Yu, A. L., Diccianni, M. B., Li, T. Z., and Chang, A. M. (1997). p16INK4 and p15INK4B alterations in primary gynecologic malignancy. Gynecol Oncol *65*, 319–324.
103. Kanuma, T., Nishida, J., Gima, T., Barrett, J. C., and Wake, N. (1997). Alterations of the *p16INK4A* gene in human ovarian cancers. Mol Carcinog *18*, 134–141.
104. Shigemasa, K., Hu, C., West, C. M., Clarke, J., Parham, G. P., Parmley, T. H., Korourian, S., Baker, V. V., and O'Brien, T. J. (1997). p16 overexpression: a potential early indicator of transformation in ovarian carcinoma. J Soc Gynecol Investig *4*, 95–102.
105. Milde-Langosch, K., Ocon, E., Becker, G., and Loning, T. (1998). p16/MTS1 inactivation in ovarian carcinomas: high frequency of reduced protein expression associated with hypermethylation or mutation in endometrioid and mucinous tumors. Int J Cancer *79*, 61–65.
106. Niederacher, D., Yan, H. Y., An, H. X., Bender, H. G., and Beckmann, M. W. (1999). *CDKN2A* gene inactivation in epithelial sporadic ovarian cancer. Br J Cancer *80*, 1920–1926.
107. Havrilesky, L. J., Alvarez, A. A., Whitaker, R. S., Marks, J. R., and Berchuck, A. (2001). Loss of expression of the p16 tumor suppressor gene is more frequent in advanced ovarian cancers lacking p53 mutations. Gynecol Oncol *83*, 491–500.

108. Saegusa, M., Machida, B. D., and Okayasu, I. (2001). Possible associations among expression of p14(ARF), p16(INK4a), p21(WAF1/CIP1), p27(KIP1), and p53 accumulation and the balance of apoptosis and cell proliferation in ovarian carcinomas. Cancer 92, 1177–1189.
109. Kudoh, K., Ichikawa, Y., Yoshida, S., Hirai, M., Kikuchi, Y., Nagata, I., Miwa, M., and Uchida, K. (2002). Inactivation of p16/CDKN2 and p15/MTS2 is associated with prognosis and response to chemotherapy in ovarian cancer. Int J Cancer 99, 579–582.
110. Ryan, A., Al-Jehani, R. M., Mulligan, K. T., and Jacobs, I. J. (1998). No evidence exists for methylation inactivation of the p16 tumor suppressor gene in ovarian carcinogenesis. Gynecol Oncol 68, 14–17.
111. Wong, Y. F., Chung, T. K., Cheung, T. H., Nobori, T., Yu, A. L., Yu, J., Batova, A., Lai, K. W., and Chang, A. M. (1999). Methylation of p16INK4A in primary gynecologic malignancy. Cancer Lett 136, 231–235.
112. McCluskey, L. L., Chen, C., Delgadillo, E., Felix, J. C., Muderspach, L. I., and Dubeau, L. (1999). Differences in *p16* gene methylation and expression in benign and malignant ovarian tumors. Gynecol Oncol 72, 87–92.
113. Ibanez de Caceres, I., Battagli, C., Esteller, M., Herman, J. G., Dulaimi, E., Edelson, M. I., Bergman, C., Ehya, H., Eisenberg, B. L., and Cairns, P. (2004). Tumor cell-specific BRCA1 and RASSF1A hypermethylation in serum, plasma, and peritoneal fluid from ovarian cancer patients. Cancer Res 64, 6476–6481.
114. Sasano, H., Comerford, J., Silverberg, S. G., and Garrett, C. T. (1990). An analysis of abnormalities of the retinoblastoma gene in human ovarian and endometrial carcinoma. Cancer 66, 2150–2154.
115. Taylor, R. R., Linnoila, R. I., Gerardts, J., Teneriello, M. G., Nash, J. D., Park, R. C., and Birrer, M. J. (1995). Abnormal expression of the retinoblastoma gene in ovarian neoplasms and correlation to p53 and K-ras mutations. Gynecol Oncol 58, 307–311.
116. Niemann, T. H., Trgovac, T. L., McGaughy, V. R., Lewandowski, G. S., and Copeland, L. J. (1998). Retinoblastoma protein expression in ovarian epithelial neoplasms. Gynecol Oncol 69, 214–219.
117. Konstantinidou, A. E., Korkolopoulou, P., Vassilopoulos, I., Tsenga, A., Thymara, I., Agapitos, E., Patsouris, E., and Davaris, P. (2003). Reduced retinoblastoma gene protein to Ki-67 ratio is an adverse prognostic indicator for ovarian adenocarcinoma patients. Gynecol Oncol 88, 369–378.
118. Qin, X. Q., Livingston, D. M., Kaelin, W. G., Jr., and Adams, P. D. (1994). Deregulated transcription factor E2F-1 expression leads to S-phase entry and p53-mediated apoptosis. Proc Natl Acad Sci USA 91, 10918–10922.
119. Connolly, D. C., Bao, R., Nikitin, A. Y., Stephens, K. C., Poole, T. W., Hua, X., Harris, S. S., Vanderhyden, B. C., and Hamilton, T. C. (2003). Female mice chimeric for expression of the simian virus 40 TAg under control of the MISIIR promoter develop epithelial ovarian cancer. Cancer Res 63, 1389–1397.
120. Weinberg, R. A. (1991). Tumor suppressor genes. Science 254, 1138–1146.
121. Hahn, W. C., Dessain, S. K., Brooks, M. W., King, J. E., Elenbaas, B., Sabatini, D. M., DeCaprio, J. A., and Weinberg, R. A. (2002). Enumeration of the simian virus 40 early region elements necessary for human cell transformation. Mol Cell Biol 22, 2111–2123.
122. Flesken-Nikitin, A., Choi, K. C., Eng, J. P., Shmidt, E. N., and Nikitin, A. Y. (2003). Induction of carcinogenesis by concurrent inactivation of p53 and Rb1 in the mouse ovarian surface epithelium. Cancer Res 63, 3459–3463.
123. Dinulescu, D. M., Ince, T. A., Quade, B. J., Shafer, S. A., Crowley, D., and Jacks, T. (2005). Role of K-ras and Pten in the development of mouse models of endometriosis and endometrioid ovarian cancer. Nat Med 11, 63–70.
124. Giardiello, F. M., Hamilton, S. R., Krush, A. J., Piantadosi, S., Hylind, L. M., Celano, P., Booker, S. V., Robinson, C. R., and Offerhaus, G. J. (1993). Treatment of colonic and rectal adenomas with sulindac in familial adenomatous polyposis. N Engl J Med 328, 1313–1316.

125. Labayle, D., Fischer, D., Vielh, P., Drouhin, F., Pariente, A., Bories, C., Duhamel, O., Trousset, M., and Attali, P. (1991). Sulindac causes regression of rectal polyps in familial adenomatous polyposis. Gastroenterology *101*, 635–639.
126. Nugent, K. P., Farmer, K. C., Spigelman, A. D., Williams, C. B., and Phillips, R. K. (1993). Randomized controlled trial of the effect of sulindac on duodenal and rectal polyposis and cell proliferation in patients with familial adenomatous polyposis. Br J Surg *80*, 1618–1619.
127. Thun, M. J., Namboodiri, M. M., and Heath, C. W., Jr. (1991). Aspirin use and reduced risk of fatal colon cancer. N Engl J Med *325*, 1593–1596.
128. Vane, J. R. (1971). Inhibition of prostaglandin synthesis as a mechanism of action for aspirin-like drugs. Nat New Biol *231*, 232–235.
129. Oshima, M., Dinchuk, J. E., Kargman, S. L., Oshima, H., Hancock, B., Kwong, E., Trzaskos, J. M., Evans, J. F., and Taketo, M. M. (1996). Suppression of intestinal polyposis in Apc delta716 knockout mice by inhibition of cyclooxygenase 2 (COX-2). Cell *87*, 803–809.
130. Farrow, D. C., Vaughan, T. L., Hansten, P. D., Stanford, J. L., Risch, H. A., Gammon, M. D., Chow, W. H., Dubrow, R., Ahsan, H., Mayne, S. T., et al. (1998). Use of aspirin and other nonsteroidal anti-inflammatory drugs and risk of esophageal and gastric cancer. Cancer Epidemiol Biomarkers Prev *7*, 97–102.
131. Klimp, A. H., Hollema, H., Kempinga, C., van der Zee, A. G., de Vries, E. G., and Daemen, T. (2001). Expression of cyclooxygenase-2 and inducible nitric oxide synthase in human ovarian tumors and tumor-associated macrophages. Cancer Res *61*, 7305–7309.
132. Matsumoto, Y., Ishiko, O., Deguchi, M., Nakagawa, E., and Ogita, S. (2001). Cyclooxygenase-2 expression in normal ovaries and epithelial ovarian neoplasms. Int J Mol Med *8*, 31–36.
133. Dore, M., Cote, L. C., Mitchell, A., and Sirois, J. (1998). Expression of prostaglandin G/H synthase type 1, but not type 2, in human ovarian adenocarcinomas. J Histochem Cytochem *46*, 77–84.
134. Gupta, R. A., Tejada, L. V., Tong, B. J., Das, S. K., Morrow, J. D., Dey, S. K., and DuBois, R. N. (2003). Cyclooxygenase-1 is overexpressed and promotes angiogenic growth factor production in ovarian cancer. Cancer Res *63*, 906–911.
135. Kino, Y., Kojima, F., Kiguchi, K., Igarashi, R., Ishizuka, B., and Kawai, S. (2005). Prostaglandin E2 production in ovarian cancer cell lines is regulated by cyclooxygenase-1, not cyclooxygenase-2. Prostaglandins Leukot Essent Fatty Acids *73*, 103–111.
136. Yang, W. L., Roland, I. H., Godwin, A. K., and Xu, X. X. (2005). Loss of TNF-alpha-regulated COX-2 expression in ovarian cancer cells. Oncogene *24*, 7991–8002.
137. Daikoku, T., Wang, D., Tranguch, S., Morrow, J. D., Orsulic, S., DuBois, R. N., and Dey, S. K. (2005). Cyclooxygenase-1 is a potential target for prevention and treatment of ovarian epithelial cancer. Cancer Res *65*, 3735–3744.
138. Daikoku, T., Tranguch, S., Trofimova, I. N., Dinulescu, D. M., Jacks, T., Nikitin, A. Y., Connolly, D. C., and Dey, S. K. (2006). Cyclooxygenase-1 is overexpressed in multiple genetically engineered mouse models of epithelial ovarian cancer. Cancer Res *66*, 2527–2531.
139. Verheijen, R. H., von Mensdorff-Pouilly, S., van Kamp, G. J., and Kenemans, P. (1999). CA 125: fundamental and clinical aspects. Semin Cancer Biol *9*, 117–124.
140. Nagele, F., Petru, E., Medl, M., Kainz, C., Graf, A. H., and Sevelda, P. (1995). Preoperative CA 125: an independent prognostic factor in patients with stage I epithelial ovarian cancer. Obstet Gynecol *86*, 259–264.
141. Zurawski, V. R., Jr., Knapp, R. C., Einhorn, N., Kenemans, P., Mortel, R., Ohmi, K., Bast, R. C., Jr., Ritts, R. E., Jr., and Malkasian, G. (1988). An initial analysis of preoperative serum CA 125 levels in patients with early stage ovarian carcinoma. Gynecol Oncol *30*, 7–14.
142. Denk, W., Strickler, J. H., and Webb, W. W. (1990). Two-photon laser scanning fluorescence microscopy. Science *248*, 73–76.
143. Williams, R. M., Zipfel, W. R., and Webb, W. W. (2001). Multiphoton microscopy in biological research. Curr Opin Chem Biol *5*, 603–608.
144. Zipfel, W. R., Williams, R. M., Christie, R., Nikitin, A. Y., Hyman, B. T., and Webb, W. W. (2003). Live tissue intrinsic emission microscopy using multiphoton-excited native fluorescence and second harmonic generation. Proc Natl Acad Sci USA *100*, 7075–7080.

145. Christie, R. H., Bacskai, B. J., Zipfel, W. R., Williams, R. M., Kajdasz, S. T., Webb, W. W., and Hyman, B. T. (2001). Growth arrest of individual senile plaques in a model of Alzheimer's disease observed by in vivo multiphoton microscopy. J Neurosci 21, 858–864.
146. Lendvai, B., Stern, E. A., Chen, B., and Svoboda, K. (2000). Experience-dependent plasticity of dendritic spines in the developing rat barrel cortex in vivo. Nature 404, 876–881.
147. Wang, W., Wyckoff, J. B., Frohlich, V. C., Oleynikov, Y., Huttelmaier, S., Zavadil, J., Cermak, L., Bottinger, E. P., Singer, R. H., White, J. G., et al. (2002). Single cell behavior in metastatic primary mammary tumors correlated with gene expression patterns revealed by molecular profiling. Cancer Res 62, 6278–6288.
148. Flesken-Nikitin, A., Williams, R. M., Zipfel, W. R., Webb, W. W., and Nikitin, A. Y. (2005). Use of multiphoton imaging for studying cell migration in the mouse. Methods Mol Biol 294, 335–345.

Ovulatory Factor in Ovarian Carcinogenesis

William J. Murdoch

1 Introduction

Ovaries of mammals are covered by a simple layer of epithelial cells that originate from the coelomic mesothelium during embryonic development. The surface epithelium is supported over the ovarian cortical interstitium by a basement membrane, and is adjoined by desmosomes and gap or tight junctional complexes. Although the surface epithelium represents only a small fraction of the diverse cell types that populate the ovary, it is thought to account for approximately 90% of its malignancies (1).

Common cancer of the ovary is an ovulation-related disease. It has been known for decades that circumstances that avert ovulation, namely oral contraceptive use and pregnancy/lactation, protect against ovarian cancer (2–5). Until recently, it was unclear how the processes of ovulation and carcinogenesis might be linked.

A follicle selected to ovulate emerges from the ovarian cortex and comes into apposition with the surface epithelium. A complex interplay of proteolytic enzymes and inflammatory mediators liberated within the formative site of ovulation (i.e., at the follicular–ovarian surface interface) degrade collagen matrices and provoke cellular death. Surface epithelial cells within a limited diffusion radius become committed to apoptosis and are sloughed. In the finale, a physical force sustained by contractile elements within the basal wall of the follicle ruptures the devitalized fabric at the apex and expels the ovum (6).

DNA-damaging reactive oxygen species are generated by leukocytes, which are attracted into the vicinity of the ovulatory stigma and undergo a respiratory burst. Another contributing determinant of genotoxicity is the ischemia-reperfusion flux coincident with ovulation and wound reparation. Bystander epithelial cells, which survive the trauma of ovulation, are subjected to oxidative DNA perturbations, which accrue into the postovulatory period. These cells proliferate and migrate to mend the void along the ovarian surface created by ovulation. It is conceivable that clonal expansion of a cell with unrepaired DNA is an initiating factor in the etiology of ovarian cancer (7).

Epithelial ovarian cancer is a deadly insidious disease in women because it typically remains asymptomatic until it has metastasized; it carries an 1-in-70 lifetime risk (8). Early-stage of the disease is characterized by the formation of an inclusion cyst (9), which contains surface epithelial cells that have invaded the ovarian cortex via processes secreted at ovulation (10) or by entrapment during remodeling (11). Apparently, the microenvironment of an inclusion cyst is conducive to metaplastic and dysplastic changes that precede tumorigenesis (1). Malignant cells seed the abdominal cavity when a cyst ruptures. A mutant cell exfoliated during the mechanics of ovulation may account for cases of diffuse intraperitoneal disease in which the ovaries remain relatively uninvolved (12).

2 Epidemiological Evidence for an Ovulation–Cancer Connection

An epidemiology-based hypothesis of ovarian neoplasia involving "incessant" ovulations was proposed by Fathalla in 1971 (13). It was surmised that repeated ovulations, without intervening dormant periods afforded by pregnancy, caused transformation of the ovarian epithelium. Exposures to injury and estrogen-rich follicular fluid were suspect.

Positive correlations clearly exist between increasing numbers of lifetime ovulations, ovarian precursor lesions, and carcinoma in women (14–21). In one study (20), there was an overall 6% increase in cancer risk with each ovulatory year; the most aggressive and damaging ovulations evidently occur in the third decade of life (i.e., during the peak reproductive years). Indeed, ovulatory factor may be more significant in premenopausal than postmenopausal onset ovarian cancer (22).

It follows that assisted reproductive programs that implement ovulation-inducing strategies would increase the risk for development of ovarian cancer. Yet results of surveys relating use of fertility drugs to ovarian cancer have been inconclusive. Some have deduced that women who do not become pregnant and those who are subjected to multiple treatments are at an elevated risk (23–25), while others suggest weak or no significant correlations (26–31). Among nulligravid women, it appears that exposure to ovulation-stimulating hormones is associated with borderline serous tumors, but not with metastatic histologic subtypes (32, 33). Furthermore, rates of ovarian cancer have remained relatively constant despite the widespread application of ovulatory stimulants (34). Nevertheless, because the prospective latency between initiation (i.e., at ovulation) and manifestation of established disease can be quite long (30–40 years or more), it will be important to continue to monitor recipients of superovulation protocols.

Support for the ovulation–cancer concept comes from histopathological studies of intensive egg-laying hens. These animals ovulate nearly every day and develop intraperitoneal carcinomas at a relatively high frequency (4–40% depending on reproductive history and age) (35–37). Moreover, inhibition of ovulation (with a progestin) protected hens from ovarian cancer (38).

There is essentially no published data on spontaneous rates of ovarian cancer among nonhuman mammals. One would expect incidences to be very low because females of most species are either pregnant, lactating or seasonally anovulatory for the bulk of their reproductive lives. Inclusion bodies of surface epithelium have been detected in ovaries of ewes (11). There is also evidence in rodents that surface epithelial stratification and ovarian invaginations/cysts are related to total lifetime ovulations (39, 40) and cycles of ovulation induction or estrogen administration (41, 42). Progression to cancer occurred in superovulated rats whose ovaries were exposed locally to a mutagen (43).

3 Carcinogenic Implication of Ovulatory Genotoxicity

Base damages of DNA caused by reactive oxygen species are an inevitable byproduct of physiological metabolism. To combat this predicament, animals have evolved elaborate enzymatic antioxidant defense mechanisms (superoxide dismutase, glutathione perioxidase, catalase); however, these are less than perfect, and toxic oxidants find their way to DNA targets (44). Oxidative damage products in DNA are a significant contributor to the risk of cancer development (45–47).

The N7–C8 bond of guanine is particularly susceptible to attack by the unpaired electron of hydroxyradical. 8-Oxoguanine is arguably the most important mutagenic lesion in DNA; mispairing with adenine during replications can yield GC-to-TA transversions often detected in tumor cells (48–50). Ovarian surface epithelial cells isolated from the perimeters of ovulated sheep, human, and hen follicles contained levels of 8-oxoguanine that exceeded those of cells obtained from extrinsic areas not affected by ovulation (51–53). Challenges to the genetic integrity of the ovarian surface epithelium were negated by pharmacological ovulation blockade (51).

A defective tumor suppressor gene, such as those that overexpress competitive mutant forms of the growth-inhibitory BRCA1/BRCA2, TP53, DAB2, or DIRAS3, is a probable basis for developing ovarian neoplasia as a result of ovulation (54, 55). Mutations in BRCA1/BRCA2 appear to be responsible for aggressive early-onset hereditary disease (56). Oxidative damages to guanine persisted in ovine ovarian surface epithelial cells that were affected by ovulation in vivo and in which synthesis of TP53 was then negated in culture by an antisense oligonucleotide; this was related to discordant cellular growth rates and expression of the cancer antigen CA-125 (57). More than one-half of human ovarian adenocarcinomas have discernible mutations in TP53 (54). Chromosomal anomalies and metaplasia have been detected in repetitive subcultures (to mimic recurrent ovulation-wound repair) of ovarian surface epithelial cells of rodents (58, 59).

Fortunately, dilemmas of DNA corruptions instigated by ovulation are normally reconciled by housekeeping cell-cycle arrest and base-excision repair mechanisms. TP53 and polymerase β were upregulated in response to the oxidative stress of ovulation imposed upon the ovarian surface epithelium of sheep (51). Production of TP53 and polymerase β was enhanced by progesterone (60). TP53 allots the time required for repair and proof-reading (61). Polymerase β performs the penultimate gap-filling function in the short-patch pathway (62, 63). Progesterone also stimulated poly(ADP-ribose) polymerase (PARP) in ovine cells (64). PARP serves as an adjunct in DNA repair. Binding of PARP and the synthesis of branched polymers of ADP-ribose in areas adjacent to a single-strand interruption functions as an antirecombinogenic element (65). Progesterone inhibited proliferation (66) and induced apoptosis (67, 68) in cultures of ovarian surface epithelial cells of macaques. The ovarian epithelium bordering postovulatory follicles of hens (which do not form a corpus luteum) undergo apoptosis, and are resorbed during follicular atresia (53). Ovarian inclusion bodies of surface epithelium can evidently be eliminated via the Fas apoptotic system (69). It is a unifocal escape from these reactions that could be problematic.

4 Prospective Use of Antioxidants in Ovarian Cancer Prevention

Since the prognosis for ovarian cancer patients with metastatic disease is so poor, and early detection has proven elusive, it is imperative that methods of chemoprevention be explored. DNA of ovarian surface epithelial cells associated with the ovulation stigma of ewes was protected from oxidative base damage by pretreatment with D-α-tocopherol (natural-source vitamin E). Programmed death within the surface epithelium at the apex of the preovulatory follicle (mediated by tumor necrosis factor) and correspondingly ovulation (and pregnancy outcome) were not altered by α-tocopherol (52). Ischemia-reperfusion injury to grafts of ovarian tissues was reduced by vitamin E (70).

As far as is known, vitamin E is the most effective chain-breaking antioxidant in cellular membranes, and thereby contributes to membrane phospholipid stability and safeguards intracellular molecules against damage imposed by free radicals (71, 72). Vitamin E also can act via mechanisms beyond its oxidant-quenching properties. Nitric oxide production by endothelial cells and superoxide release by leukocytes is suppressed by vitamin E (73). Nonredox modes of α-tocopherol action include inhibitory and stimulatory effects on rates of mitosis and removal of damaged DNA, respectively (74–77). Therefore, vitamin E could act during the immediate postovulatory period to impede untoward proliferative responses of ovarian surface epithelial cells until repairs to DNA can be accomplished.

Supplemental vitamin E could be of particular value in women at risk for the development of ovarian cancer (e.g., those with a genetic predisposition that are not using a contraceptive that inhibits ovulation). There is epidemiological evidence suggesting an inverse relationship between consumption of vitamin E and risk of ovarian carcinoma (78, 79). Similar reports have advocated protective effects of vitamin E against cancers of the lung, colorectum, cervix, and prostate gland (80). It appears that in general, incidences of oxidative DNA lesions and susceptibility to cancer are potentiated by micronutrient (e.g., antioxidant vitamin) deficiencies (81). Interestingly, the circulatory antioxidant status of ovarian cancer patients was reduced compared with age-matched controls (82).

5 Conclusions

The ovulatory event is extraordinary in that it involves a self-inflicted, surge gonadotropic-induced injury. Ovulation is a rate-limiting event for the perpetuation of a species; unfortunately, it also imparts a cancer risk to the ovarian surface epithelium.

OVULATION
↓
OXIDATIVE DAMAGE TO DNA OF THE OVARIAN SURFACE EPITHELIUM
malfunctional tumor suppressor genes, steroid hormonal imbalance
↓
DEFECTIVE DNA REPAIR/APOPTOTIC MECHANISMS
metaplasia, mutagenesis
↓
MALIGNANT TRANSFORMATION
growth stimulation (hormones, growth factors, cytokines)
↓
CLONAL EXPANSION
proteolysis, angiogenesis
↓
METASTASIS

Fig. 1 Proposed role of ovulation in the chronology of epithelial ovarian cancer

The DNA of ovarian surface epithelial cells contiguous with the site of ovulation is compromised by oxyradicals. It is proposed that this constitutes a first step in the etiology of ovarian tumorigenesis (Fig. 1). To avoid accumulations of potentially harmful mutations, it is essential that accurate restoration or proficient removal of anomalous cells comes to fruition. The level of danger hence escalates when a cell (as a prelude to mutation) escapes (e.g., because of a malfunctional tumor suppressor mechanism) repair or death. Perhaps the ovarian epithelium is vulnerable to genetic damages that are not reconciled because it has not been under a strong evolutionary pressure to respond to superfluous ovulations (83). The number of lifetime ovulations in most animals are kept to a minimum by pregnancy and season.

It remains uncertain why, in particular, the ovarian surface epithelium is so prone to neoplastic transformation; after all, it represents only a small fraction of the diverse cell-types that populate the ovary. Susceptibility may hinge on the fact that normal ovarian surface epithelial cells are of an uncommitted phenotype. Unlike the Mullerian epithelia of the female reproductive tract, development of ovarian surface cells is arrested at an immature pluripotent (stem) stage (1).

The sequences of events that lead to common ovarian cancer are multifactorial (Fig. 1). Several aberrant phases are undoubtedly required to yield a malignant phenotype with distinct growth and metastatic advantages. Ovarian cancer is generally considered to have some level of hormonal involvement; progestins are protective, and gonadotropins, androgens, and estrogens are facilitative (84–88). Paracrine-autocrine modulators (e.g., growth factors and cytokines) can also influence ovarian cancer cell behaviors (1, 89, 90). Metastatic spread is protease-dependent; urokinase and downstream matrix metalloproteinases, which digest basement membranes and interstitial connective tissues, are of particular importance (91). Vascular endothelial growth/permeability factor is secreted by ovarian cancer cells, and has been related to ascites formation and metastasis (92).

It is important in closing to emphasize that a circumstantial association between ovulation and the initiation of common ovarian cancer does not prove causal effect, and that an "ovulation model" is not absolute and does not explain the genesis of all epithelial ovarian tumors. For example: protection is conferred by tubal ligation or hysterectomy in spite of uninterrupted ovulation; protection provided by one gestation with breast feeding or short-term oral contraceptive use is superior to the predicted benefits of those missed ovulations that would have occurred otherwise; reduced numbers of ovulatory cycles due to menstrual irregularities and infertility (e.g., polycystic ovarian syndrome) are independent risk factors for ovarian cancer; and in addition to ovulation, other inflammatory responses (endometriosis and exposure of the ovarian surface to exogenous irritants such as talc or viruses) have been linked to ovarian cancer (33, 93–98)

Acknowledgment Supported by NIH RR-016474.

References

1. Auersperg N, Wong AST, Choi KC, Kang SK, Leung PCK. (2001) Ovarian surface epithelium: biology, endocrinology, and pathology. Endo Rev 22, 255–288.
2. Risch HA. (2000) Oral-contraceptive use, anovulatory action, and risk of epithelial ovarian cancer. Epidemiology 11, 614–615.
3. Schildkraut JM, Calingaert B, Marchbanks PA, Moorman PG, Rodriguez GC. (2002) Impact of progestin and estrogen potency in oral contraceptives on ovarian cancer risk. J Natl Cancer Inst 94, 32–38.
4. Riman T, Nilsson S, Persson IR. (2004) Review of epidemiological evidence for reproductive and hormonal factors in relation to the risk of epithelial ovarian malignancies. Acta Obstet Gynecol Scand 83, 783–795.
5. Bandera CA. (2005) Advances in understanding of risk factors for ovarian cancer. J Reprod Med 50, 399–406.
6. Murdoch WJ, McDonnel AC. (2002) Roles of the ovarian surface epithelium in ovulation and carcinogenesis. Reproduction 123, 743–750.
7. Murdoch WJ. (2005) Carcinogenic potential of ovulatory genotoxicity. Biol Reprod 73, 586–590.
8. Runnebaum IB, Stickeler E. (2001) Epidemiological and molecular aspects of ovarian cancer risk. J Cancer Res Clin Oncol 127, 73–79.
9. Feeley KM, Wells M. (2001) Precursor lesions of ovarian epithelial malignancy. Histopathology 38, 87–95.
10. Yang WL, Godwin AK, Xu XX. (2004) Tumor necrosis factor-α-induced matrix proteolytic enzyme production and basement membrane remodeling by human ovarian surface epithelial cells: molecular basis linking ovulation and cancer risk. Cancer Res 64, 1534–1540.
11. Murdoch WJ. (1994) Ovarian surface epithelium during ovulatory and anovulatory ovine estrous cycles. Anat Rec 240, 322–326.
12. Hamilton TC. (1992) Ovarian cancer, biology. Curr Prob Cancer 16, 5–57.
13. Fathalla MF. (1971) Incessant ovulation – a factor in ovarian neoplasia? Lancet 2, 163.
14. Casagrande JT, Louie EW, Pike MC, Roy S, Ross RK, Henderson BE. (1979) "Incessant ovulation" and ovarian cancer. Lancet 2, 170–173.
15. Hildreth NG, Kelsey JL, LiVolsi VA, Fischer DB, Holford TR, Mostow ED, Schwartz PE, White C. (1981) An epidemiologic study of epithelial carcinoma of the ovary. Am J Epidemiol 114, 398–405.
16. Risch HA, Weiss NS, Lyon JL, Daling JR, Liff JM. (1983) Events of reproductive life and the incidence of epithelial ovarian cancer. Am J Epidemiol 117, 128–139.

17. La Vecchia C, Franceschi S, Gallus G, Decarli A, Libertati A, Tognoni G. (1983) Incessant ovulation and ovarian cancer: a critical approach. Int J Epidemiol 12, 161–164.
18. Mori M, Harabuchi I, Miyake H, Casagrande JT, Henderson BE, Ross RK. (1988) Reproductive, genetic, and dietary risk factors for ovarian cancer. Am J Epidemiol 128, 771–777.
19. Whittemore AS, Harris R, Itnyre J. (1992) Characteristics relating to ovarian cancer risk: collaborative analysis of 12 US case-control studies. IV. The pathogenesis of epithelial ovarian cancer. Am J Epidemiol 136, 1212–1220.
20. Purdie DM, Bain CJ, Siskind V, Webb PM, Green AC. (2003) Ovulation and risk of epithelial ovarian cancer. Int J Cancer 104, 228–232.
21. Heller DS, Murphy P, Westhoff C. (2005) Are germinal inclusion cysts markers of ovulation? Gynecol Oncol 96, 496–499.
22. Tung KH, Wilkens LR, Wu AH, McDuffie K, Nomura AM, Kolonel LN, Terada KY, Goodman MT. (2005) Effect of anovulation factors on pre- and postmenopausal ovarian cancer risk: revisiting the incessant ovulation hypothesis. Am J Epidemiol 161, 321–329.
23. Whittemore AS, Harris R, Itnyre J. (1992) Characteristics relating to ovarian cancer risk: collaborative analysis of 12 US case-control studies. II. Invasive epithelial ovarian cancer in white women. Am J Epidemiol 136, 1184–1203.
24. Rossing MA, Daling JR, Weiss NS, Moore DE, Self SG. (1994) Ovarian tumors in a cohort of infertile women. New Engl J Med 331, 771–776.
25. Nieto JJ, Crow J, Sundaresan M, Constantinovici N, Perrett CW, MacLean AB, Hardiman PJ. (2001) Ovarian epithelial dysplasia in relation to ovulation induction and nulliparity. Gynecol Oncol 82, 344–349.
26. Venn A, Watson L, Lumley J, Giles G, King C, Healy D. (1995) Breast and ovarian cancer incidence after infertility and in vitro fertilisation. Lancet 346, 995–1000.
27. Mosgaard BJ, Lidegaard O, Kjaer SK, Schou G, Andersen AN. (1997) Infertility, fertility drugs, and invasive ovarian cancer: a case-control study. Fertil Steril 67, 1005–1012.
28. Modan B, Ron E, Lerner-Geva L, Blumstein T, Menczer J, Rabinovici J, Oelsner G, Freedman L, Mashiach S, Lunenfeld B. (1998) Cancer incidence in a cohort of infertile women. Am J Epidemiol 147, 1038–1042.
29. Potashnik G, Lerner-Geva L, Genkin L, Chetrit A, Lunenfeld E, Porath A. (1999) Fertility drugs and the risk of breast and ovarian cancers: results of a long-term follow-up study. Fertil Steril 71, 853–859.
30. Brinton LA, Lamb EJ, Moghissi KS, Scoccia B, Althuis MD, Mabie JE, Westhoff CL. (2004) Cancer risk after the use of ovulation-stimulating drugs. Obstet Gynecol Surv 59, 657–659.
31. Mahdavi A, Pejovic T, Nezhat F. (2006) Induction of ovulation and ovarian cancer: a critical review of the literature. Fertil Steril 85, 819–826.
32. Shoham Z. (1994) Epidemiology, etiology, and fertility drugs in ovarian epithelial carcinoma: where are we today? Fertil Steril 62, 433–448.
33. Ness RB, Cramer DW, Goodman MT, Kjaer SK, Mallin K, Mosgaard BJ, Purdie DM, Risch HA, Vergona R, Wu AH. (2002) Infertility, fertility drugs, and ovarian cancer: a pooled analysis of case-control studies. Am J Epidemiol 155, 217–224.
34. Glud E, Kjaer SK, Troisi R, Brinton LA. (1998) Fertility drugs and ovarian cancer. Epidemiol Rev 20, 237–257.
35. Fredrickson TN. (1978) Ovarian tumors of the hen. Environ Health Perspect 73, 35–51.
36. Damjanov I. (1989) Ovarian tumours in laboratory and domestic animals. Curr Top Pathol 78, 1–10.
37. Rodriguez-Burford C, Barnes MN, Berry W, Partridge EE, Grizzle WE. (2001) Immunohistochemical expression of molecular markers in an avian model: a potential model for preclinical evaluation of agents for ovarian cancer chemoprevention. Gynecol Oncol 81, 373–379.
38. Barnes MN, Berry WD, Straughn JM, Kirby TO, Leath CA, Huh WK, Grizzle WE, Partridge EE. (2002) A pilot study of ovarian cancer chemoprevention using medroxyprogesterone acetate in an avian model of spontaneous ovarian carcinogenesis. Gynecol Oncol 87, 57–63.

39. Clow OL, Hurst PR, Fleming JS. (2002) Changes in the mouse ovarian surface epithelium with age and ovulation number. Mol Cell Endocrinol 191, 105–111.
40. Tan OL, Hurst PR, Fleming JS. (2005) Location of inclusion cysts in mouse ovaries in relation to age, pregnancy, and total ovulation number: implications for ovarian cancer? J Pathol 205, 483–490.
41. Celik C, Gezginc K, Aktan M, Acar A, Yaman ST, Gungor S, Akyurek C. (2004) Effects of ovulation induction on ovarian morphology: an animal study. Int J Gynecol Cancer 14, 600–606.
42. Burdette JE, Kurley SJ, Kilen SM, Mayo KE, Woodruff TK. (2006) Gonadotropin-induced superovulation drives ovarian surface epithelia proliferation in CD1 mice. Endocrinology 147, 2338–2345.
43. Stewart SL, Querec TD, Ochman AR, Gruver BN, Bao R, Babb JS, Wong TS, Koutroukides T, Pinnola AD, Klein-Szanto A, Hamilton TC, Patriotis C. (2004) Characterization of a carcinogenesis rat model of ovarian preoplasia and neoplasia. Cancer Res 64, 8177–8183.
44. Collins AR. (1999) Oxidative DNA damage, antioxidants, and cancer. Bioessays 21, 238–246.
45. Marnett LJ. (2000) Oxyradicals and DNA damage. Carcinogenesis 21, 361–370.
46. Cooke MS, Evans MD, Dizdaroglu M, Lunec J. (2003) Oxidative DNA damage: mechanisms, mutation, and disease. FASEB J 17, 1195–1214.
47. Valko M, Izakovic M, Mazur M, Rhodes CJ, Telser J. (2004) Role of oxygen radicals in DNA damage and cancer incidence. Molec Cell Biochem 266, 37–56.
48. Grollman AP, Moriya M. (1993) Mutagenesis by 8-oxoguanine: an enemy within. Trends Genet 9, 246–249.
49. Cunningham RP. (1997) DNA repair: caretakers of the genome. Curr Biol 7, R576–R579.
50. Fortini P, Pascucci B, Parlanti E, D'Errico M, Simonelli V, Dogliotti E. (2003) 8-Oxoguanine DNA damage: at the crossroad of alternative repair pathways. Mutat Res 531, 127–139.
51. Murdoch WJ, Townsend RS, McDonnel AC. (2001) Ovulation-induced DNA damage in ovarian surface epithelial cells of ewes: prospective regulatory mechanisms of repair/survival and apoptosis. Biol Reprod 65, 1417–1424.
52. Murdoch WJ, Martinchick JF. (2004) Oxidative damage to DNA of ovarian surface epithelial cells affected by ovulation: carcinogenic implication and chemoprevention. Exp Biol Med 229, 553–559.
53. Murdoch WJ, Van Kirk EA, Alexander BM. (2005) DNA damages in ovarian surface epithelial cells of ovulatory hens. Exp Biol Med 230, 429–433.
54. Aunoble B, Sanches R, Didier E, Bignon YJ. (2000) Major oncogenes and tumor suppressor genes involved in epithelial ovarian cancer. Int J Oncol 16, 567–576.
55. Cvetkovic D. (2003) Early events in ovarian oncogenesis. Reprod Biol Endocrinol 1, 68.
56. Holschneider CH, Berek JS. (2000) Ovarian cancer: epidemiology, biology, and prognostic factors. Semin Surg Oncol 19, 3–10.
57. Murdoch WJ. (2003) Metaplastic potential of p53 down-regulation in ovarian surface epithelial cells affected by ovulation. Cancer Lett 191, 75–81.
58. Godwin AK, Testa JR, Handel LM, Liu Z, Vanderveer LA, Tracey PA, Hamilton TC. (1992) Spontaneous transformation of rat ovarian surface epithelial cells: association with cytogenetic changes and implications of repeated ovulation in the etiology of ovarian cancer. J Natl Cancer Inst 84, 592–601.
59. Roby KF, Taylor CC, Sweetwood JS, Cheng Y, Pace JL, Tawfik O, Persons DL, Smith PG, Terranova PF. (2000) Development of a syngeneic mouse model for events related to ovarian cancer. Carcinogenesis 21, 585–591.
60. Murdoch WJ, Van Kirk EA. (2002) Steroid hormonal regulation of proliferative, p53 tumor suppressor, and apoptotic responses of sheep ovarian surface epithelial cells. Mol Cell Endocrinol 186, 61–67.
61. Vousden KH, Lu X. (2002) Live or let die: the cell's response to p53. Nat Rev Cancer 2, 595–604.
62. Fortini P, Pascucci B, Parlanti E, D'Errico M, Simonelli V, Dogliotti E. (2003) 8-Oxoguanine DNA damage: at the crossroad of alternative repair pathways. Mutat Res 531, 127–139.

63. Sung JS, Demple B. (2006) Roles of base excision repair subpathways in correcting oxidized abasic sites in DNA. FEBS J 273, 1620–1629.
64. Murdoch WJ. (1998) Perturbation of sheep ovarian surface epithelial cells by ovulation: evidence for roles of progesterone and poly(ADP-ribose) polymerase in the restoration of DNA integrity. J Endocrinol 156, 503–508.
65. Lindahl T, Wood RD. (1999) Quality control by DNA repair. Science 286, 1897–1905.
66. Wright JW, Toth-Fejel S, Stouffer RL, Rodland KD. (2002) Proliferation of rhesus ovarian surface epithelial cells in culture: lack of mitogenic response to steroid or gonadotropic hormones. Endocrinology 143, 2198–2207.
67. Rodriguez GC, Walmer DK, Cline M, Krigman H, Lessey BA, Whitaker RS, Dodge R, Hughes CL. (1998) Effect of progestin on the ovarian epithelium of macaques: cancer prevention through apoptosis? J Soc Gynecol Invest 5, 271–276.
68. Rodriguez G. (2003) New insights regarding pharmacologic approaches for ovarian cancer prevention. Hematol Oncol Clin N Am 17, 1007–1020.
69. Ghahremani M, Foghi A, Dorrington JH. (1999) Etiology of ovarian cancer: a proposed mechanism. Med Hypoth 52, 23–26.
70. Nugent D, Newton H, Gallivan L, Gosden RG. (1998) Protective effect of vitamin E on ischaemia-reperfusion injury in ovarian grafts. J Reprod Fertil 114, 341–346.
71. Morrissey PA, Sheehy PJA. (1999) Optimal nutrition: vitamin E. Proc Nutr Soc 58, 459–468.
72. Herrera E, Barbas C. (2001) Vitamin E: action, metabolism and perspectives. J Physiol Biochem 57, 43–56.
73. Azzi A, Ricciarelli R, Zingg JM. (2002) Non-antioxidant molecular functions of alpha-tocopherol (vitamin E). FEBS Lett 519, 8–10.
74. Claycombe KJ, Meydani SN. (2001) Vitamin E and genome stability. Mutat Res 475, 37–44.
75. Kline K, Yu W, Sanders BG. (2001) Vitamin E: mechanisms of action as tumor cell growth inhibitors. J Nutr 131, 161S–163S.
76. Ricciarelli R, Zingg JM, Azzi A. (2001) Vitamin E: protective role of a Janus molecule. FASEB J 15, 2314–2325.
77. Neuzil J, Kagedal K, Andera L, Weber C, Brunk UT. (2002) Vitamin E analogs: a new class of multiple action agents with anti-neoplastic and anti-atherogenic activity. Apoptosis 7, 179–187.
78. Fairfield KM, Hankinson SE, Rosner BA, Hunter DJ, Colditz GA, Willett WC. (2001) Risk of ovarian carcinoma and consumption of vitamins A, C, and E and specific carotenoids. Cancer 92, 2318–2326.
79. McCann SE, Moysich KB, Mettlin C. (2001) Intakes of selected nutrients and food groups and risk of ovarian cancer. Nutr Cancer 39, 19–28.
80. Tamini RM, Lagiou P, Adami HO, Trichopoulos D. (2002) Prospects for chemoprevention of cancer. J Int Med 251, 286–300.
81. Ames BN, Wakimoto P. (2002) Are vitamins and mineral deficiencies a major cancer risk? Nat Rev Cancer 2, 694–704.
82. Senthil K, Aranganathan S, Nalini N. (2004) Evidence of oxidative stress in the circulation of ovarian cancer patients. Clin Chim Acta 339, 27–32.
83. Auersperg N, Edelson MI, Mok SC, Johnson SW, Hamilton TC. (1998) The biology of ovarian cancer. Semin Oncol 25, 281–304.
84. Risch HA. (1998) Hormonal etiology of epithelial ovarian cancer, with a hypothesis concerning the role of androgens and progesterone. J Natl Cancer Inst 90, 1774–1786.
85. Konishi I, Kuroda H, Mandai M. (1999) Gonadotropins and development of ovarian cancer. Oncol Suppl 57(2), 45–48.
86. Ho SM. (2003) Estrogen, progesterone and epithelial ovarian cancer. Reprod Biol Endocrinol 1, 73.
87. Cunat S, Hoffman P, Pujol P. (2004) Estrogens and epithelial ovarian cancer. Gynecol Oncol 94, 25–32.

88. Wang PH, Chang C. (2004) Androgens and ovarian cancers. Eur J Gynaecol Oncol 25, 157–163.
89. Nash MA, Ferrandia G, Gordinier M, Loercher A, Freedman RS. (1999) The role of cytokines in both normal and malignant ovary. Endo Related Cancer 6, 93–107.
90. Grundker C, Emons G. (2003) Role of gonadotropin-releasing hormone (GnRH) in ovarian cancer. Reprod Biol Endocrinol 1, 65.
91. Stack MS, Ellerbroek SM, Fishman DA. (1998) The role of proteolytic enzymes in the pathology of epithelial ovarian carcinoma. Int J Oncol 12, 569–576.
92. Bamberger ES, Perrett CW. (2002) Angiogenesis in epithelian ovarian cancer. Mol Pathol 55, 348–359.
93. Ness RB, Grisso JA, Cottreau C, Klapper J, Vergona R, Wheeler JE, Morgan M, Schlesselman JJ. (2000) Factors related to inflammation of the ovarian epithelium and risk of ovarian cancer. Epidemiology 11, 111–117.
94. Scott JS. (1984) How to induce ovarian cancer: and how not to. Brit Med J 289, 781–782.
95. Bristow RE, Kaplan BY. (1996) Ovulation induction, infertility, and ovarian cancer risk. Fertil Steril 66, 499–507.
96. Holschneider CH, Berek JS. (2000) Ovarian cancer: epidemiology, biology, and prognostic factors. Semin Surg Oncol 19, 3–10.
97. Siskind V, Green A, Bain C, Purdie D. (2000) Beyond ovulation: oral contraceptives and epithelial ovarian cancer. Epidemiology 11, 106–110.
98. Moorman PG, Schildkraut JM, Calingaert B, Halabi S, Vine MF, Berchuck A. (2002) Ovulation and ovarian cancer: a comparison of two methods for calculating lifetime ovulatory cycles (United States). Cancer Causes Control 13, 807–811.

Part II
Ovarian Cancer Therapeutics

Gynecologic Oncology Group (GOG-USA) Trials in Ovarian Cancer

Robert F. Ozols

1 Introduction

In the United States, the Gynecologic Oncology Group (GOG) has been the leading clinical trial organization for women with gynecologic cancers. Recently, GOG has established collaborations with other clinical trials groups throughout the world to facilitate the rapid completion of large, randomized, controlled trials in women with ovarian cancer.

In ovarian cancer, GOG has performed a series of clinical trials in three distinct subsets of women with markedly different prognostic factors: early stage (FIGO I and II), advanced stage (FIGO III) with optimally resected disease (no residual tumor nodule greater than 1 cm), and advanced stage (FIGO III and IV) with suboptimal disease. GOG clinical trials in ovarian cancer have evaluated different strategies in these three distinct patient groups, including adjuvant therapy, maintenance therapy, consolidation approaches, interval debulking surgery, the role of second-look operations, new chemotherapeutic regimens, and the role of intraperitoneal (IP) chemotherapy. More recently, GOG has been evaluating the role of biological therapies (molecular-targeted treatment), either as single-agent treatment or in combination with chemotherapy.

2 Early-Stage Ovarian Cancer

Less than one-third of ovarian cancers are diagnosed when the disease is localized to the pelvis and can be completely resected. In an early GOG study (1), it was demonstrated that patients with stage IA or IB disease and favorable histology did not require adjuvant chemotherapy, as survival after surgical staging alone was greater than 90%. Another subset of patients with early-stage ovarian cancer has a markedly worse prognosis with recurrence rates of approximately 25–40% (2, 3). High-risk features include stage IA or IB and unfavorable histology, including grade 3 or clear cell, stage IC, or stage II. The GOG has performed three randomized trials in this unfavorable group of patients with early-stage ovarian cancer.

The first study (1) demonstrated that IP phosphorus-32 or oral melphalan resulted in overall survival rates of approximately 80% at 6 years, and the radioisotope was selected for further evaluation because of less toxicity. The next study (4) compared IP phosphorus-32 to three cycles of chemotherapy with intravenous cisplatin plus cyclophosphamide. The recurrence rate with chemotherapy was 31% lower than that observed with the phosphorus-32 regimen, although this difference was not statistically significant. These results, however, were in accordance with the results of a multicenter trial in Italy comparing cisplatin to IP phosphorus-32 in stage IC patients, which reported that cisplatin significantly reduced the relapse rate by 61% (5). Consequently, chemotherapy became the standard treatment for patients with early-stage, high-risk, ovarian cancer when compared with treatments using radioisotopes. Studies in advanced-stage disease (described later) demonstrated that carboplatin plus paclitaxel was the optimum regimen for patients with advanced-stage disease. On the basis of those trials, the same two-drug combination has been evaluated in the last GOG trials in patients with high-risk, early, ovarian cancer.

GOG 157 compared three cycles with six cycles of adjuvant carboplatin plus paclitaxel in this same group of patients with high-risk features (6). Paclitaxel was administered at a dose of 175 mg m^{-2} over 3 h and carboplatin was dosed to an area-under-the-curve (AUC) of 7.5. Both drugs were administered every 21 days. Four hundred fifty-seven patients were included in this trial, and 93% were histologically and medically eligible. Despite the fact that thorough surgical staging was a protocol requirement, there was incomplete or inadequate documentation in 29% of patients. The recurrence rate for six cycles was 24% lower (hazard ratio [HR]: 0.76; 95% confidence interval [CI] 0.51–1.13, $p = 0.18$), and the estimated probability of recurrence within 5 years was 20.1% (six cycles) vs. 25.4% (three cycles). The death rate was similar for three or six cycles of treatment (HR: 1.02; 95% CI 0.66–1.57). Figure 1 shows the overall survival of randomized patients.

The estimated probability of survival at 5 years is 81% (three cycles) vs. 83% (six cycles). Six cycles of treatment caused significantly more neurotoxicity with grade 3 or 4 neurotoxicity developing in 11% compared with 2% of patients treated with three cycles. In addition, six cycles also caused significantly more severe anemia and granulocytopenia.

In the current GOG trial, patients with high-risk, early-stage disease are treated with three cycles of adjuvant carboplatin plus paclitaxel followed by randomization to observation or to weekly paclitaxel at a low dose for six months. It is hypothesized that weekly low-dose paclitaxel may have an antiangiogenic effect.

There appears to have been only modest, if any, improvement in survival for patients treated on the last two clinical trials with different chemotherapy regimens. The small number of patients who are diagnosed with early-stage disease coupled with the favorable natural history combine to make clinical trials in this group of patients difficult (7). It has been recommended that patients with high-risk features in the future perhaps be included in advanced-stage trials that are evaluating molecular-targeted therapy, either alone or in combination with chemotherapy (8). It has already been accepted by most cooperative groups throughout the world that patients with stage II disease should be included in advanced-stage trials.

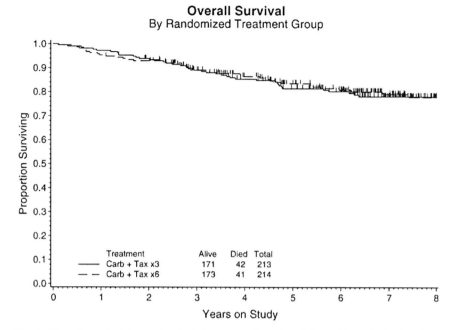

Fig. 1 Overall survival by randomized treatment. Reproduced from Gynecologic Oncology, September 2006, with kind permission from Elsevier

3 Evaluation of New Chemotherapy Regimens

In the late 1880s and early 1990s, it was established that platinum compounds were the most effective agents in patients with advanced ovarian cancer. Phase II trials demonstrated the activity of paclitaxel in patients with platinum-resistant disease, and GOG performed a pivotal trial comparing what had been the standard therapy of cisplatin plus cyclophosphamide vs. cisplatin plus paclitaxel in patients with suboptimal stage III and IV disease (9). This landmark study, which was subsequently confirmed in a separate trial (10), demonstrated that combination therapy with paclitaxel plus cisplatin prolonged both progression-free and overall survival compared with cisplatin plus cyclophosphamide. Median survival for patients treated with paclitaxel plus cisplatin was 37 months compared with 25 months for patients treated with cyclophosphamide and cisplatin.

GOG subsequently performed a three-arm study in patients with suboptimal stage III and IV disease, who were randomized to single-agent cisplatin, single-agent paclitaxel, or the combination of cisplatin plus paclitaxel (11). Patients who were randomized to single-agent therapy were frequently crossed over even before progression to the other agent, and many have interpreted this trial as a comparison of six cycles of cisplatin plus paclitaxel vs. sequential single-agent therapy with the same drugs. There was no difference in overall survival in any of the three arms of the study, and cisplatin plus paclitaxel was still considered to be the standard because of less toxicity compared with multiple cycles of single-agent sequential therapy.

Carboplatin was developed as a less toxic analogue to cisplatin, and phase I/II trials of carboplatin plus paclitaxel have demonstrated that this two-drug combination could be safely administered and resulted in a high response rate (75%) (12) in these uncontrolled trials. However, there was concern that carboplatin may be less effective than cisplatin, and survival may be compromised in patients with optimal stage III disease in whom a substantial proportion of patients could be expected to survive at least 5 years. GOG consequently performed a noninferiority study of cisplatin (75 mg m^{-2}) plus paclitaxel (135 mg m^{-2} in a 24-h infusion) with carboplatin (AUC 7.5) plus paclitaxel (175 mg m^{-2} in a 3-h infusion) in 800 patients with optimal stage III disease (13). Median progression-free survival and overall survival were 19.4 and 48.7 months, respectively, for patients treated with cisplatin plus paclitaxel compared with 20.7 and 57.4 months for carboplatin plus paclitaxel-treated patients (Fig. 2).

The relative risk (RR) of progression for the carboplatin plus paclitaxel group was 0.88 (95% CI 0.75–1.03), and the RR of death was 0.84 (95% CI 0.7–1.02). The carboplatin plus paclitaxel-treated group of patients experienced less toxicity, and it was concluded that carboplatin plus paclitaxel was less toxic, easier to administer, and clearly not inferior when compared with cisplatin plus paclitaxel.

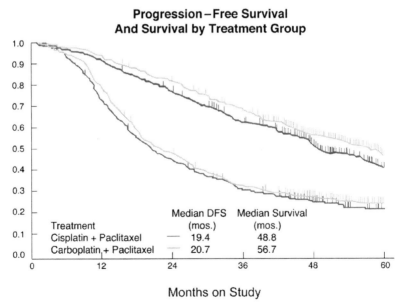

Fig. 2 Progression-free survival (*lower curves*) and survival (*upper curves*) for optimal stage III patients treated with cisplatin plus paclitaxel or carboplatin plus paclitaxel

In fact, the 16% reduction in the risk of death and an 8-month improvement in median survival compared with treatment with cisplatin plus paclitaxel helped to establish carboplatin plus paclitaxel as the regimen of choice against which all new chemotherapy treatments were to be evaluated (14). A similar trial by the Arbeitsgemeinschaft Gynäkologische Onkologie (AGO) Group, which included both optimal and suboptimal patients, also concluded that carboplatin plus paclitaxel was the preferred regimen for patients with advanced disease (15).

Figure 3 depicts the survival of patients on GOG 158 following relapse (13). Although 90% of patients did achieve a clinical complete remission following six cycles of either carboplatin plus paclitaxel or cisplatin plus paclitaxel, median time to progression was still less than 2 years. Following progressive disease, median survival was approximately 2 years, and it can be seen that there is little likelihood, if any, for cure for patients with recurrent ovarian cancer despite the fact that some patients can survive with recurrent disease for as long as 4 or 5 years.

Although the two-drug combination of carboplatin plus paclitaxel has been accepted as the worldwide standard for patients with advanced ovarian cancer, it is also clear that overall survival has been minimally improved. Treatment with carboplatin plus paclitaxel resulted in a clinical complete remission rate of approximately 75% in patients with advanced ovarian cancer (50% for patients with suboptimal disease and 90% for patients with optimal disease). However, median progression-free survival ranges from 16 to 22 months, also depending upon the volume of disease at the time chemotherapy was initiated, and median overall survival ranges from 24 to approximately 60 months, and likewise is dependent upon the volume of disease at the time of diagnosis. Numerous agents have been shown

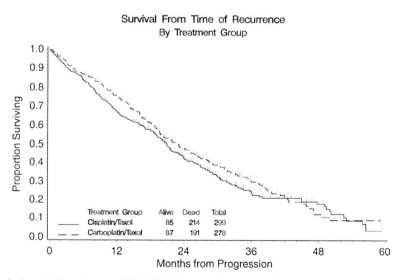

Fig. 3 Survival for patients on GOG 158 following relapse from initial therapy with either cisplatin/Taxol or carboplatin/Taxol

to have activity in patients with recurrent ovarian cancer, and these agents have been subsequently incorporated into a large number of randomized trials comparing novel combinations vs. standard therapy with paclitaxel plus carboplatin.

GOG in collaboration with the Gynecologic Cancer Intergroup (GCIG) has recently reported the results of the largest trial ever performed in patients with advanced ovarian cancer. Over 4,000 patients with stage III or IV ovarian cancer, optimal and suboptimal disease, or with primary peritoneal cancer were randomized to one of five different chemotherapy regimens (Fig. 4). Bookman et al. reported that there was no significant difference in progression-free survival or in overall survival in these five different regimens, and consequently carboplatin plus paclitaxel remains the standard of care (16).

4 Surgical Issues in Advanced Disease

An attempt at maximum debulking surgery at the time of diagnosis is generally accepted to be the standard of care for patients with advanced-stage ovarian cancer. However, there has never been a trial in which previously untreated patients were randomly assigned to debulking surgery or no debulking surgery. The reason that

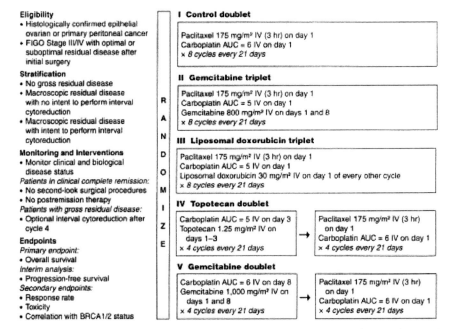

Fig. 4 Overview of eligibility criteria, stratification, monitoring, treatment interventions, endpoints, and schema of GOG 182-ICON5

surgical debulking has been accepted to be the standard of care is primarily based on uncontrolled trials. Two large metaanalyses suggested that the presence of small-volume residual disease was associated with improved survival (17, 18). In a study by Bristow et al. (17), each 10% decrease in residual tumor volume produced a 5.5% increase in median survival.

There have been, however, two large randomized trials of surgical debulking, but neither study evaluated initial surgical debulking (19, 20). The European Organisation for Research and Treatment of Cancer (EORTC) study randomly assigned patients in whom the initial debulking attempt was not considered to be successful to six cycles of chemotherapy (cisplatin plus cyclophosphamide) alone, or three cycles of the same chemotherapy followed by an attempt at surgical cytoreduction, and then three more cycles of chemotherapy (19). In this study, patients assigned to interval surgical cytoreduction had improved rates of progression-free and overall survival. A GOG study was designed to confirm the results of the EORTC study for secondary surgical cytoreduction. The GOG study, however, showed no significant differences between patients who underwent secondary surgical cytoreduction compared with those who did not (20). It appears that the major difference between these two studies was in the extent of the initial surgery (21). In the GOG study, it was required that all patients undergo initial maximum surgery to debulk the tumor whereas in the EORTC study it appears that most patients did not have what would be considered to be a maximal effort at debulking, according to the GOG standards. There is no doubt that patients who have the least amount of disease after initial surgery have a better prognosis. However, it is also important to recognize the impact of the biology of the tumor on prognosis. A retrospective GOG trial demonstrated that patients whose tumor burden was ≤1 cm at the time of diagnosis and consequently did not require debulking surgery had better survival than those patients who were surgically debulked and were left with the same amount of residual disease as those patients who did not require debulking surgery (22).

5 Second-Look Laparotomy

Second-look laparotomy (SLL) was incorporated into standard ovarian cancer management in the 1970s in an effort to minimize the unnecessary exposure to alkylating agents that were known to result in second malignancies with prolonged use. It was also recognized that approximately one-half of patients in a clinical complete remission would be found to have residual disease at a second-look operation. It was hypothesized that in patients who were in a clinical remission if residual disease was found at SLL, then further therapy would be beneficial and in contrast if the results of the SLL showed no evidence of disease, patients could stop treatment and be spared from unnecessary toxicity. Two randomized trials of SLL that failed to demonstrate any survival benefit have been criticized

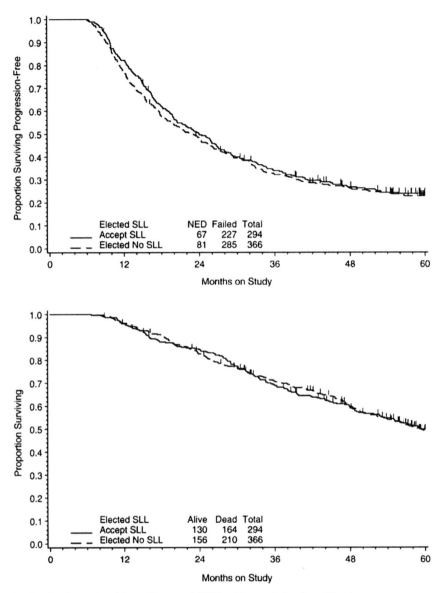

Fig. 5(a, b) Reproduced from Greer et al. 2005, with permission from Elsevier

because of methodologic flaws, including the type of chemotherapy used (23, 24). GOG 158 was designed to prospectively evaluate the results of SLL in patients with optimal stage III disease (25). A SLL was not required, but was allowed as an option to assess response to treatment with carboplatin plus paclitaxel or cisplatin plus paclitaxel. Prior to randomization, patients were informed of the controversial issues surrounding SLL and they chose whether or not to undergo an SLL at the completion of chemotherapy. In this study, there were 393 patients who elected SLL and 399 patients who elected no SLL. Residual ovarian cancer was found in 46% of 294 (75%) patients undergoing SLL. Patients with residual disease at SLL underwent a wide variety of different treatments selected by the individual physician and the patient. Figure 5a and b depicts the progression-free survival and the overall survival of patients who accepted an SLL vs. those who elected not to have an SLL. There is clearly no difference in overall survival, and the results suggest that there is no benefit from administering second-line chemotherapy in patients who have a positive SLL compared with administering that same therapy when there is clinical progression. Although this was not a randomized trial and the results suggest that SLL remains the best available technique for determining the posttreatment status of ovarian cancer, there is no evidence that this operation results in improvement in survival. It is now generally accepted that an SLL be limited to research protocols, which would be dependent on the findings at SLL.

6 Consolidation and Maintenance Studies

As previously noted, most patients with advanced ovarian cancer will obtain a clinical complete remission following surgery and chemotherapy. However, most patients who do achieve a clinical complete remission will ultimately relapse. Numerous strategies have been evaluated in an effort to prevent recurrences in patients who achieve a clinical complete remission. Maintenance therapy usually refers to administration of cytotoxic drugs and biological agents for extended periods of time or until the time of relapse. Consolidation usually refers to a short course of treatment, such as high-dose chemotherapy with a stem cell transplant, IP therapy, or radiation (external beam, IP radioisotopes, or IP radioimmunoconjugates). There is no evidence that any form of maintenance therapy or consolidation therapy improves survival in this group of patients. Maintenance therapy and consolidation therapy were recently addressed in two GOG trials.

In one trial, patients in a clinical complete remission were randomized to 3 vs. 12 cycles of monthly paclitaxel (175 mg m^{-2} in a 3-h infusion) (26). This study was closed by the Data Safety Monitoring Board after scheduled interim analysis demonstrated that patients who received the extended treatment with paclitaxel had a statistically significant improvement in progression-free survival (28 months vs. 21 months). Patients who received three cycles of chemotherapy were given the option to receive an additional nine cycles of chemotherapy. This crossover, coupled with the early closure of the study, does not permit an adequate assessment of survival. It should be noted that maintenance paclitaxel was also associated with significant toxicity, particularly neurotoxicity. The GOG has initiated another randomized trial in this group of patients in which survival is the endpoint. Patients who achieve a clinical complete remission are randomized to receive no further therapy, treatment with 12 months of maintenance paclitaxel, or treatment with a novel pegylated paclitaxel compound.

IP p32 was also evaluated prospectively in a GOG trial in patients who achieved a surgically confirmed complete remission (27). Neither progression-free nor overall survival was statistically improved by IP p32.

7 IP Chemotherapy

IP chemotherapy has been studied for over two decades in patients with small-volume ovarian cancer. IP chemotherapy is based on the rationale that ovarian cancer remains primarily an IP disease, and the administration of drugs directly into the peritoneal cavity will lead to a pharmacologic advantage with higher intratumoral drug concentrations when compared with intravenous administration. The GOG has performed three large randomized trials in this group of patients (28–30). In the most recent study, patients were randomized to receive either intravenous chemotherapy with paclitaxel plus cisplatin or to an IP regimen, which consisted of paclitaxel in a 24-h infusion on day 1 followed by IP cisplatin (100 mg m^{-2}) on day 2, and IP chemotherapy with paclitaxel (60 mg m^{-2} on day 8). The patients treated with the IP regimen received significant improvement in progression-free survival and a 16-month improvement in overall survival compared with patients treated with intravenous cisplatin plus paclitaxel (Fig. 6). The results of this study

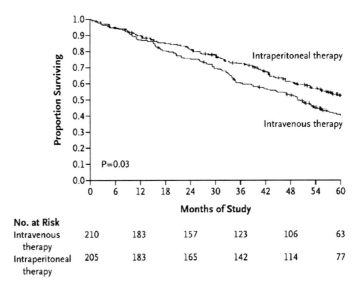

Fig. 6 Reproduced from Armstrong et al. 2006, with permission from the New England Journal of Medicine

prompted the National Cancer Institute to issue a Clinical Announcement that patients and physicians should strongly consider IP therapy in women who have optimal disease after cytoreductive surgery (31). However, controversy remains whether IP therapy should be accepted as standard treatment (32).

There has been no randomized comparison of IP therapy with IV carboplatin plus paclitaxel in patients with optimal stage III disease. A cross-trial comparison of IV carboplatin plus paclitaxel to IP regimens suggests very similar efficacy (32). Furthermore, IP therapy has for-midable toxicity, and in the most recently study of the GOG, only 42% of patients could complete the planned six courses. The GOG is exploring less toxic IP regimens in phase II trials, and it is likely that IP therapy will have limited acceptance until an IP regimen is developed with acceptable toxicity and superior efficacy to IV carboplatin plus paclitaxel.

8 Molecular-Targeted Therapy in Ovarian Cancer

The GOG has undertaken an extensive effort to evaluate new biological agents that target specific molecular pathways that may be involved in the pathogenesis of ovarian cancer (33). Molecular pathways that are under evaluation as targets include: angiogenesis, the erb family of receptors and tyrosine kinases, proteosomes, and the mitogen-activated protein kinase pathway. The GOG phase II trial of bevacizumab resulted in such notable activity (a 17.7% objective response and almost 40% of patients without disease progression at 6 months) that a placebo-controlled trial of carboplatin plus paclitaxel plus or minus bevacizumab in previously untreated patients with suboptimal stage III and IV disease has recently been initiated.

References

1. R.C. Young, L.A. Walton, S.S. Ellenberg, H.D. Homesley, G.D. Wilbanks, D.G. Decker, A. Miller, R. Park, and F. Major Jr., Adjuvant therapy in stage I and stage II epithelial ovarian cancer. *N Engl J Med* 322(15), 1021–1027 (1990).
2. I. Vergote, J. De Brabanter, A. Fyles, K. Bertelsen, N. Einhorn, P. Sevelda, M.E. Gore, J. Kaern, H. Verreist, K. Sjovall, D. Timmerman, J. Vandewalle, M. Van Gramberen, and C.G. Trope, Prognostic importance of degree of differentiation and cyst rupture in stage I invasive epithelial ovarian carcinoma, *Lancet* 357(9251), 176–182 (2001).
3. I.B. Vergote, J. Kaern, V.M. Abeler, E.O. Pettersen, L.N. De Vos, and G. Trope, Analysis of prognostic factors in stage I epithelial ovarian carcinoma: importance of degree of differentiation and deoxyribonucleic acid ploidy in predicting relapse, *Am J Obstet Gynecol* 169(1), 40–52 (1993).
4. R.C. Young, M.F. Brady, R.K. Nieberg, H.J. Long, A.R. Mayer, S.S. Lentz, J. Hurteau, and D.S. Alberts, Adjuvant treatment for early ovarian cancer: a randomized phase III trial of intraperitoneal ^{32}P or intravenous cyclophosphamide and cisplatin – a Gynecologic Oncology Group study, *J Clin Oncol* 21(23), 4350–4355 (2003).
5. G. Bolis, N. Colombo, S. Pecorelli, V. Torri, S. Marsoni, C. Bonazzi, S. Chiari, G. Favalli, G. Mangili, and M. Presti, Adjuvant treatment for early epithelial ovarian cancer: results of two randomised clinical trials comparing cisplatin to no further treatment or chromic phosphate

(32P). GICOG: Gruppo Interregionale Collaborativo in Ginecologia Oncologica, *Ann Oncol* 6(9), 887–893 (1995).
6. J. Bell, M.F. Brady, R.C. Young, J. Lage, J.L. Walker, K.Y. Look, G.S.Rose, and N.M. Spirtos, Randomized phase III trial of three versus six cycles of adjuvant carboplatin and paclitaxel in early stage epithelial ovarian carcinoma: A Gynecologic Oncology Group study, *Gynecol Oncol* 102 432–439 (2006).
7. R.C. Young, Early-stage ovarian cancer: to treat or not to treat, *J Natl Cancer Inst* 95(2), 94–95 (2003).
8. I. Vergote and F. Amant, Time to include high-risk early ovarian cancer in randomized phase III trials of advanced ovarian cancer, *Gynecol Oncol* 102(3), 415–417 (2006).
9. W.P. McGuire, W.J. Hoskins, M.F. Brady, P.R. Kucera, E.E. Partridge, K.Y. Look, D.L. Clarke-Pearson, and M. Davidson, Cyclophosphamide and cisplatin compared with paclitaxel and cisplatin in patients with stage III and stage IV ovarian cancer, *N Engl J Med* 334(1), 1–6 (1996).
10. M.J. Piccart, K. Bertelsen, K. James, J. Cassidy, C. Mangioni, E. Simonsen, G. Stuart, S. Kaye, I. Vergote, R. Blom, R. Grimshaw, R.J. Atkinson, K.D. Swenerton, C. Trope, M. Nardi, J. Kaern, S. Tumolo, P. Timmers, J.A. Roy, F. Lhoas, B. Lindvall, M. Bacon, A. Birt, J.E. Andersen, B. Zee, J. Paul, B. Baron, and S. Pecorelli, Randomized intergroup trial of cisplatin-paclitaxel versus cisplatin-cyclophosphamide in women with advanced epithelial ovarian cancer: three-year results, *J Natl Cancer Inst* 92(9), 699–708 (2000).
11. F.M. Muggia, P.S. Braly, M.F. Brady, G. Sutton, T.H. Niemann, S.L. Lentz, R.D. Alvarez, P.R. Kucera, and J.M. Small, Phase III randomized study of cisplatin versus paclitaxel versus cisplatin and paclitaxel in patients with suboptimal stage III or IV ovarian cancer: a Gynecologic Oncology Group study, *J Clin Oncol* 18(1), 106–115 (2000).
12. M.A. Bookman, W.P. McGuire III, D. Kilpatrick, E. Keenan, W.M. Hogan, S.W. Johnson, P. O'Dwyer, E. Rowinsky, H.H. Gallion, and R.F. Ozols, Carboplatin and paclitaxel in ovarian carcinoma: a phase I study of the Gynecologic Oncology Group, *J Clin Oncol* 14 1895–1902 (1996).
13. R. Ozols, B.N. Bundy, B.E. Greer, J.M. Fowler, D. Clarke-Pearson, R.A. Burger, R.S. Mannel, K. DeGeest, E.M. Hartenbach, and R. Baergen, Phase III trial of cisplatin and paclitaxel compared to carboplatin and paclitaxel in patients with "optimally" resected stage III ovarian cancer: a GOG study, *J Clin Oncol* 21(17), 3194–3200 (2003).
14. A. du Bois, M. Quinn, T. Thigpen, J. Vermorken, E. Avall-Lundquist, M. Bookman, D. Bowtell, M. Brady, A. Casado, A. Cervantes, E. Eisenhauer, M. Friedlaender, K. Fujiwara, S. Grenman, J.P. Guastalla, P. Harper, T. Hogberg, S. Kaye, H. Kitchener, G. Kristensen, R. Mannel, W. Meier, B. Miller, J.P. Neijt, A. Oza, R. Ozols, M. Parmar, S. Pecorelli, J. Pfisterer, A. Poveda, D. Provencher, E. Pujade-Lauraine, M. Randall, J. Rochon, G. Rustin, S. Sagae, F. Stehman, G. Stuart, E. Trimble, P. Vasey, I. Vergote, R. Verheijen, U. Wagner, 2004 consensus statements on the management of ovarian cancer: final document of the 3rd International Gynecologic Cancer Intergroup Ovarian Cancer Consensus Conference (GCIG OCCC 2004), *Ann Oncol* 16(suppl 8), viii7–viii12 (2005).
15. A. du Bois, H.J. Luck, W. Meier, H.P. Adams, V. Mobus, S. Costa, T. Bauknecht, B. Richter, M. Warm, W. Schroder, S. Olbricht, U. Nitz, C. Jackisch, G. Emons, U. Wagner, W. Kuhn, J. Pfisterer, A randomized clinical trial of cisplatin/paclitaxel versus carboplatin/paclitaxel as first-line treatment of ovarian cancer, *J Natl Cancer Inst* 95(17), 1320–1329 (2003).
16. M.A. Bookman, GOG0182-ICON5: 5-arm phase III randomized trial of paclitaxel (P) and carboplatin (C) vs combinations with gemcitabine (G), PEG-liposomal doxorubicin (D), or topotecan (T) in patients (pts) with advanced-stage epithelial ovarian (EOC) or primary peritoneal (PPC) carcinoma, *Proc Am Soc Clin Oncol* 24(18S), abstract 5002 (2006).
17. R.E. Bristow, R.S. Tomacruz, D.K. Armstrong, E.L. Trimble, F.J. Montz, Survival effect of maximal cytoreductive surgery for advanced ovarian carcinoma during the platinum era: a meta-analysis, *J Clin Oncol* 20 1248–1259 (2002).
18. D.G. Allen, A.P.M. Heintz, F.W.M.M. Touw, A meta-analysis of residual disease and survival in stage III and IV carcinoma of the ovary, *Eur J Gynaecol Oncol* 16 349–356 (1995).

19. M.E.L. van der Burg, M. van Lent, M. Buyse The effect of debulking surgery after induction chemotherapy on the prognosis in advanced epithelial ovarian cancer, *N Engl J Med* 332 629–634 (1995).
20. P.G. Rose, S. Nerenstone, M.F. Brady Secondary surgical cytoreduction for advanced ovarian carcinoma, *N Engl J Med* 351 2489–2497 (2004).
21. T. Thigpen, The if and when of surgical debulking for ovarian carcinoma, *N Engl J Med* 351 2544–2546 (2004).
22. W.J. Hoskins, W.P. McGuire, M.F. Brady The effect of diameter of largest residual disease on survival after primary cytoreductive surgery in patients with suboptimal residual epithelial ovarian carcinoma, *Am J Obstet Gynecol* 170 974–980 (1994).
23. D. Luesley, F. Lawton, G. Blackledge Failure of second-look laparotomy to influence survival in epithelial ovarian cancer, *Lancet* 2(8611), 599–603 (1988).
24. M.O. Nicoletto, S. Tumolo, R. Talamini Surgical second look in ovarian cancer: a randomized study in patients with laparoscopic complete remission – a Northeastern Oncology Cooperative Group – Ovarian Cancer Cooperative Group study, *J Clin Oncol* 15(3), 994–999 (1997).
25. B.E. Greer, B.N. Bundy, R.F. Ozols, J.M. Fowler, D. Clarke-Pearson, R.A. Burger, R. Mannel, K. DeGeest, E.M. Hartenbach, R.N. Baergen, L.J. Copeland, Implications of second-look laparotomy in the context of optimally resected stage III ovarian cancer: a non-randomized comparison using an explanatory analysis: a Gynecologic Oncology Group study, *Gynecol Oncol* 99 71–79 (2005).
26. M. Markman, P.Y. Liu, S. Wilczynski, B. Monk, L.J. Copeland, R.D. Alvarez, C. Jiang, and D. Alberts, Phase III randomized trial of 12 versus 3 months of maintenance paclitaxel in patients with advanced ovarian cancer after complete response to platinum and paclitaxel-based chemotherapy: a Southwest Oncology Group and Gynecologic Oncology Group trial, *J Clin Oncol* 21(13), 2460–2465 (2003).
27. M.A. Varia, F.B. Stehman, B.N. Bundy, J.A. Benda, D.L. Clarke-Pearson, R.D. Alvarez, and H.J. Long, *J Clin Oncol* 21(15), 2849–2855 (2003).
28. P. Benedetti-Panici, S. Greggi, F. Maneschi Anatomical and pathological study of retroperitoneal nodes in epithelial ovarian cancer, *Gynecol Oncol* 51 150–154 (1993).
29. J.L. Walker, D. Armstrong, H.Q. Huang Intraperitoneal catheter outcomes in a phase III trial of intravenous versus intraperitoneal chemotherapy in optimal stage III ovarian and primary peritoneal cancer: a Gynecologic Oncology Group study, *Gynecol Oncol* 100 27–32 (2006).
30. D.K. Armstrong, B. Bundy, L. Wenzel, H.Q. Huang, R. Baergen, S. Lele, L.J. Copeland, J.L. Walker, and R.A. Burger, Intraperitoneal cisplatin and paclitaxel in ovarian cancer, *N Engl J Med* 354 34–43 (2006).
31. Clinical advisory: NCI issues clinical announcement for preferred method of treatment for advanced ovarian cancer, Available at www.nlm.nih.gov/databases/alerts/ovarian_ip_chemo.html (2006).
32. R.F. Ozols, Intraperitoneal cisplatin therapy in ovarian cancer: comparison with standard intravenous carboplatin and paclitaxel, *Gynecol Oncol* 103 1–6 (2006).
33. L. Martin and R.J. Schilder, Novel non-cytotoxic therapy in ovarian cancer: current status and future prospects, *JNCCN* 4 955–966 (2006).
34. R.A. Burger, M. Sill, B.J. Monk Phase II trial of bevacizumab in persistent or recurrent epithelial ovarian cancer (EOC) or primary peritoneal cancer (PPC): a Gynecologic Oncology Group (GOG) study, *J Clin Oncol* 23 (Suppl 1), 457s, Abstract 5009.

Intraperitoneal Chemotherapy for Ovarian Cancer

Mark A. Morgan

1 Introduction

Cytoreductive surgery followed by chemotherapy with a platinum and taxane combination constitutes the accepted standard treatment for patients with advanced epithelial ovarian cancer. Intravenous (IV) carboplatin and a 3-h infusion of paclitaxel were considered to be the preferable regimen at a recent international consensus conference. This was due to issues related to ease of administration, toxicity, and quality of life, with no evidence of inferiority to other regimens (1). Despite this, the long-term outlook for patients with advanced ovarian cancer remains poor, although patients with optimally debulked stage III disease have median survivals approaching 5 years. It is in this group of patients (generally with less than 2 cm residual nodules) that the intraperitoneal (IP) administration of chemotherapy has been studied for over 20 years. Since ovarian cancer remains clinically in the peritoneal cavity for much of its natural history, this approach seemed rational. Despite encouraging early results, the IP approach has not been incorporated into the mainstream management of advanced ovarian cancer.

On January 4, 2006, the National Cancer Institute (NCI) of the United States issued a clinical advisory regarding the "Preferred method of treatment for advanced ovarian cancer. … The new NCI clinical announcement recommends that women with advanced ovarian cancer who undergo effective surgical debulking receive a combination of IV and IP chemotherapy. IP chemotherapy allows higher doses and more frequent administration of drugs, and it appears to be more effective in killing cancer cells in the peritoneal cavity, where ovarian cancer is likely to spread or recur first." Despite this announcement, IP chemotherapy remains controversial. Toxicity and administration concerns persist and because of previous trial design concerns, it is felt that the ideal IP chemotherapy regimen has yet to be defined.

2 Theory of IP Chemotherapy

IP chemotherapy using nitrogen mustard was first used in the 1950s, primarily to control ascites (2). In 1978, Dedrick et al. (3) demonstrated a pharmacokinetic rationale for IP drug administration in ovarian cancer. They suggested that a severalfold log increase in dose-intensity could be obtained by IP treatment compared with systemic administration of the same drugs. Clinical and preclinical studies have demonstrated that larger and water insoluble drugs have a larger peritoneal cavity/plasma ratio and a prolonged dwell time in the peritoneal cavity (Table 1) (4, 5). However, studies also revealed that most chemotherapy drugs only penetrate the tumor surface for a few millimeters (6–9). It is for this reason that IP chemotherapy has been largely restricted to patients with small volume residual disease. Agents such as cisplatin and carboplatin, with a smaller pharmacologic advantage intraperitoneally, obtain significant systemic levels, which may permit better treatment of the inner core of tumor nodules.

3 Clinical Trials

Early clinical trials, primarily in the 1980s and 1990s tested multiple agents, typically in a salvage setting. These trials helped to refine administration techniques, define toxicities related to the IP treatment, and identify drugs and regimens worthy of further study in randomized trials in first-line treatment.

3.1 Phase I and Phase II Trials

Phase I and II trials, primarily in the salvage setting with small volume disease evaluated drugs such as thiotepa, mitoxantrone, etoposide, 5-flourouracil, cisplatin, carboplatin and paclitaxel alone or in combination. Despite IP administration, systemic effects such as neurotoxicity with cisplatin and paclitaxel and myelotoxicity with carboplatin, etoposide, thiotepa, and mitoxantrone can be seen and be dose-limiting. Local toxicity, manifested by abdominal pain, seems to be more pronounced in more slowly absorbed drugs such as paclitaxel and mitoxantrone, but can be seen with any agent.

Complete surgical response rates have been demonstrated in approximately one third of patients with small volume residual disease at second look laparotomy

Table 1 Physical characteristics and peritoneal/plasma ratios of selected drugs

Drug	Molecular weight	H_2O solubility	Peak	AUC
Cisplatin	300.05	+	20	12
Carboplatin	371.25	+	24	18
Melphalan	305.20	−	93	65
Doxorubicin	543.53	±	474	—
5-FU	130.08	±	298	367
Methotrexate	454.44	−	92	100
Paclitaxel	853.92	−	—	1,000
Mitoxantrone	517.40	−	—	1,400

when treated with an IP cisplatin-based regimen (10). In addition, improved or prolonged survival has been suggested for a t least some subsets of patients treated with IP chemotherapy in the consolidation or salvage setting (11, 12).

3.2 Phase III Trials

One phase III randomized trial, conducted by the European Organization for Research and Treatment of Cancer (EORTC), examined the role of IP cisplatin chemotherapy for consolidation therapy (13). Unfortunately, the study was closed prematurely because of a low accrual rate. After a median follow-up of 8 years, there is no difference in overall survival (OS) or progression-free survival (PFS), although there was nonsignificant improvement in median survival of 13 months with IP consolidation (78 months vs. 91 months).

Six randomized trials reporting survival have been published assessing IP chemotherapy in the front-line treatment of ovarian cancer. Three studies compared IV and IP regimens with similar dosing (Table 2) (14–16), and three studies compared different regimens and dosing (Table 3) (17–19). The estimated relative death rates for six studies (including the EORTC consolidation trial) are shown in Fig. 1. On an average, IP therapy was associated with a 21.6% decrease in the risk of death (hazard ratio = 0.79; 95% CI: 0.70–0.89). For ovarian cancer patients who are optimally debulked and have an expected median survival of 4 years, this should translate into a gain in median survival of about 12 months. This analysis led to the NCI clinical announcement regarding the potential benefit of IP chemotherapy.

Three cooperative group trials performed in the United States were particularly important in prompting the NCI announcement. Only one (14) compared identical regimens with the only difference being the route of administration of cisplatin. The median survival for the IP arm was significantly better than the IV arm (49 months vs. 41 months). Although abdominal pain was more common in the IP arm, myelotoxicity, ototoxicity, and neurotoxicity were significantly more frequent in the IV arm. The IP regimen was not widely accepted, because in the same year, the

Table 2   Three trials with equivalent doses IV and IP

Reference	IP regimen	IV regimen	Survival IP (months)	Survival IV (months)	p value
(14)	Cisplatin 100 Cyclo 600	Cisplatin 100 Cyclo 600	49	41	0.02
(15)	Cisplatin 50 Epidox 50 Cyclo 600	Cisplatin 50 Epidox 50 Cyclo 600	67	51	0.14
(16)	Carbo 350 Cyclo 600	Carbo 350 Cyclo 600	26	25	NS

All doses are given in mg m^{-2}.
Cyclo, cyclophosphamide; *Epidox*, epidoxorubicin.

Table 3 Three trials with different regimens IV and IP

Reference	IP regimen	IV regimen	Survival IP (months)	Survival IV (months)	p value
(17)	IP cisplatin 100 (IV) AC or EC	Cisplatin 50 AC or EC	43	48	0.47
(18)	IV carbo AUC 9 IP cisplatin 100 IV paclitaxel 135	Cisplatin 75 Paclitaxel 135	67	51	0.05
(19)	IV paclitaxel 135 IP cisplatin 100 IP paclitaxel 60	Cisplatin 75 Paclitaxel 135	66	50	0.03

All doses are given in mg m^{-2}
Carbo, carboplatin; *AC*, doxorubicin and cyclophosphamide; *EC*, epirubicin and cyclophosphamide

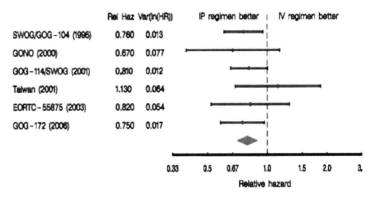

Fig. 1 NCI clinical announcement, 2006. The diamond is summary hazard ratio (0.79; 95% CI: 0.70–0.89)

Gynecologic Oncology Group (GOG) reported a 12 month improvement in median survival of patients with suboptimally debulked ovarian cancer by substituting IV paclitaxel for IV cyclophosphamide. This was felt to be more important than administering cisplatin by the IP route. The second (18) and third (19) trials compared markedly different regimens. The control arm of both trials employed IV cisplatin (75 mg m^{-2}) and IV paclitaxel (135 mg m^{-2}) over 24 h for six cycles. However, one trial used carboplatin IV, given at an AUC of 9 for 2 courses followed by six cycles IV paclitaxel and IP cisplatin (100 mg m^{-2}) and the other added IP paclitaxel (60 mg m^{-2}) on day 8 to IV paclitaxel and IP cisplatin. As expected, both IP regimens were significantly more toxic, but the median survival was increased by 11 and 16 months, respectively. Because the IV and IP regimens in these two trials were so different, it is difficult to conclude that the route of administration was the primary reason for survival difference. Regardless, the median survival of 66 months in the Armstrong trial is the longest yet reported for patients with advanced ovarian cancer, and makes a strong argument for considering this regimen to be the preferred one for optimally debulked patients.

3.3 Toxicity and Complications

When identical doses and regimens were used with the only difference being the IP delivery of cisplatin or carboplatin (see Table 2), the major toxicities that could be attributed to the IP therapy were abdominal pain or catheter complications (failure or infection). Systemic toxicity such as myelosuppression or neurotoxicity was generally lower with the IP regimen.

When different doses and regimens were used (see Table 3), systemic toxicities including myelosuppression, neurotoxicity, and metabolic toxicity were worse with the IP regimens. This was primarily because the dose intensity and total doses received were greater with the IP regimens.

Most studies have shown a lower completion of therapy rate for the IP arm vs. the IV arm. This was especially notable for GOG 172 (19), the trial that produced the longest median survival yet reported in advanced ovarian cancer and prompted the NCI announcement. In that trial, only 42% of patients completed all six cycles and 42% completed less than three cycles. Catheter-related issues (infection, blockage, leakage, and access problems) were responsible for at least one-third of the cases that failed to complete the proscribed therapy. There was no association between when a catheter was placed (at the time of debulking surgery or later) and completion of therapy, but patients who had left colon or rectosigmoid resections were less likely to initiate IP treatment (20).

4 Future Directions

Although IP regimens have been shown to convey a survival advantage in trials of advanced, optimally debulked ovarian cancer, many questions remain. The use of different and more aggressive IP regimens, the relatively high failure to complete therapy rate, and the comparison to control regimens of paclitaxel and cisplatin have been cited as reasons to doubt the role of IP therapy as the primary treatment variable responsible for improved survival. The regimen of carboplatin and paclitaxel, which has a lower toxicity profile and can be given to an outpatient, has been advocated by some investigators as the standard regimen to which IP therapy should be compared (21).

To develop more tolerable IP regimens, future research will focus on several approaches. One is to modify the GOG 172 regimen, which used a 24-h infusion of paclitaxel (175 mg m^{-2}) day 1, IP cisplatin (100 mg m^{-2}) day 2, and IP paclitaxel (60 mg m^{-2}) day 8. Approaches include shortening the IV paclitaxel infusion to 3 h, reducing the IP cisplatin dose to 75 mg m^{-2}, and eliminating the IP paclitaxel or substituing IP or IV paclitaxel with docetaxel. Another approach is to substitute IP carboplatin for IP cisplatin. Although in the past carboplatin has been considered inferior to cisplatin when given by the IP route, recent reanalysis of dosage and pharmacokinetic issues as well as clinical evidence from Japan suggests that IP carboplatin may be a less toxic alternative to IP cisplatin (22). This is being evaluated in a current GOG phase I trial. The potential role of biologic agents and targeted therapies given IP or in conjunction with IP chemotherapy is the area that needs to be studied further.

Finally, technical issues such as the ideal type and timing of IP catheter placement need to be clarified. The optimal number of IP cycles is not known at this time. It is also not known for sure whether IP chemotherapy should be employed in the presence of bulky disease, retroperitoneal nodal disease, or after rectosigmoid resection.

5 Conclusions

Over 20 years of research has demonstrated that it is possible to achieve prolonged survival in advanced ovarian cancer in the front-line and salvage setting, using regimens that include IP chemotherapy. Results from recent randomized clinical trials have done a great deal to raise the awareness of the potential of this route of therapy. However, much work is still needed to educate the oncology community in the safe administration of IP chemotherapy, and research will continue to try to refine current regimens to improve tolerability and at the same time maintain or increase efficacy.

References

1. Ozols RF, Bundy BN, Greer BE, Fowler JM, Clarke-Pearson D, Burger RA, Mannel RS, DeGeest K, Hartenbach EM, Baergen R. Phase III Trial of carboplatin and paclitaxel compared with cisplatin and paclitaxel in patients with optimally resected stage III ovarian cancer: a Gynecologic Oncology Group Study. J Clin Oncol 21, 3194–3200 (2003).
2. Weisberger AS, Levine B, Storaasli JP. Use of nitrogen mustard in treatment of serous effusions of neoplastic origin. J Am Med Assoc 159, 1704–1707 (1955).
3. Dedrick RL, Myers CE, Bungay PM, Devita VT. Pharmacokinetic rationale for peritoneal drug administration in the treatment of ovarian cancer. Cancer Treat Rep 62, 1–9 (1978).
4. Markman M. Intraperitoneal chemotherapy. Semin Oncol 18, 248–254 (1991).
5. Markman M, Powinsky E, Hakes T, Reichman B, Jones W, Lewis JL, Rubis S, Curtin J, Barakat R, Phillips M. Phase I trial of intraperitoneal taxol: a Gynecologic Oncology Group study. J Clin Oncol 10, 1485–1491 (1992).
6. Ozols RF, Locker GY, Doroshow JH, Grotzinger KR, Myers CE, Young RC. Pharmacokinetics of adriamycin and tissue penetration in murine ovarian cancer. Cancer Res 39, 3209–3214 (1979).

7. West GW, Weichselbaum R, Little JB. Limited penetration of methotrexate into human osteosarcoma spheroids as a proposed model for solid tumor resistance to adjuvant chemotherapy. Cancer Res 40, 3665–3668 (1980).
8. Nederman T, Carlsson J. Penetration and binding of vinblastine and 5-fluorouracil in cellular spheroids. Cancer Chemother Pharmacol 13,131–135 (1984).
9. Los G, Mutsaers PH, van der Vijgh WJ, Baldew GS, de Graaf PW, McVie JG. Direct diffusion of cis-diamminedichloroplatinum (II) in intraperitoneal rat tumors after intraperitoneal chemotherapy: a comparison with systemic chemotherapy. Cancer Res 49, 3380–3384 (1989).
10. Markman M. Intraperitoneal chemotherapy of ovarian cancer. Semin Oncol 25, 356–60 (1998).
11. Markman M, Reichman B, Hakes T, et al. Impact on survival of surgically defined favorable responses to salvage intraperitoneal chemotherapy in small-volume residual ovarian cancer. J Clin Oncol 10, 1479–1484 (1992).
12. Barakat RR, Sabbatini P, Bhaskaran D, Reuzin M, Smith A, Venkatraman E, Aghajanian L, Hensley M, Sorgnet S, Brown C, Soslow R, Markman M, Hoskins WJ, Spriggs D. Intraperitoneal chemotherapy for ovarian carcinoma: results of long-term follow-up. J Clin Oncol 20, 694–698 (2002).
13. Piccart MJ, Floquet A, Scarfone G, et al. Intraperitoneal cisplatin versus no further treatment: 8-year results of EORTC 55875, a randomized phase III study in ovarian cancer patients with a pathologically complete remission after platinum-based intravenous chemotherapy. Int J Gynecol Cancer 13 (Suppl 2), 196–203 (2003).
14. Alberts DS, Green S, Hannigan EV, O'Toole R, Stock-Novack D, Anderson P, Surwit EA, Malvlya VK, Nahhas WA, Jolles CJ. Improved therapeutic index of carboplatin plus cyclophosphamide versus cisplatin plus cyclophosphamide: final report by the Southwest Oncology Group of a phase III randomized trial in stages III and IV ovarian cancer. J Clin Oncol 10, 706–717 (1996).
15. Gadducci A, Carnino F, Chiara S, Brunetti I, Tanganelli L, Romanini A, Bruzzone M, Conte PF. Intraperitoneal versus intravenous cisplatin in combination with intravenous cyclophosphamide and epidoxorubicin in optimally cytoreduced advanced epithelial ovarian cancer: a randomized trial of the Gruppo Oncologico Nord-Ovest. Gynecol Oncol 76, 157–162 (2000).
16. Polyzos A, Tsavaris N, Kosmas C, Giannikos L, Katsikas M, Kalahanis N, Karatzas G, Christodoulou K, Giannakopoulos K, Stamatiadis D, Katsilambros N. A comparative study of intraperitoneal carboplatin versus intravenous carboplatin with intravenous cyclophosphamide in both arms as initial chemotherapy for stage III ovarian cancer. Oncology 56, 291–296 (1999).
17. Yen MS, Juang CM, Lai CR, Chao GC, Ng HT, Yuan CC. Intraperitoneal cisplatin-based chemotherapy vs. intravenous cisplatin-based chemotherapy for stage III optimally cytoreduced epithelial ovarian cancer. Int J Gynaecol Obstet 72, 55–60 (2001).
18. Markman M, Bundy BN, Alberts DS, Fowler JM, Clark-Pearson DL, Carson LF, Wadler S, Sickel J. Phase III trial of standard-dose intravenous cisplatin plus paclitaxel versus moderately high-dose carboplatin followed by intravenous paclitaxel and intraperitoneal cisplatin in small-volume stage III ovarian carcinoma: an intergroup study of the Gynecologic Oncology Group, Southwestern Oncology Group, and Eastern Cooperative Oncology Group. J Clin Oncol 19, 1001–1007 (2001).
19. Armstrong DK, Bundy B, Wenzel L, Huang HQ, Baergen R, Lele S, Copeland LJ, Walker Jl, Burger RA. Intraperitoneal cisplatin and paclitaxel in ovarian cancer. N Engl J Med 354, 34–43 (2006).
20. Walker JL, Huang H, Armstrong D, et al. Intraperitoneal catheter outcomes in a phase III trial of intravenous versus intraperitoneal chemotherapy in optimal stage III ovarian and primary peritoneal cancer: a Gynecologic Oncology Group study. Bynecol Oncol 100, 27–32 (2006).
21. Ozols RF, Bookman MA, Young RC. Intraperitoneal chemotherapy for ovarian cancer. N Engl J Med 354, 1641–1643 (2006).
22. Fujiwara K, Markman M, Morgan M, Coleman RL. Intraperitoneal carboplatin-based chemotherapy for epithelial ovarian cancer. Gynecol Oncol 97, 10–15 (2005).

Ovarian Cancer: Can We Reverse Drug Resistance?

David S.P. Tan, Joo Ern Ang, and Stan B. Kaye

1 Introduction

The treatment of ovarian cancer improved substantially with the introduction of platinum-based chemotherapy in the 1980s. In the 1990s, the results of randomised-controlled trials established paclitaxel in combination with a platinum agent as a standard initial chemotherapy for advanced ovarian cancer (1–3). However, over 90% of patients with advanced ovarian cancer will still subsequently die because of clinical failure of chemotherapy, i.e. drug resistance, resulting in an overall 5-year survival of only 30–40%. Clearly, there is a need for more progress in addressing this issue.

2 What Causes Drug Resistance?

In clinical practice, drug resistance refers to progressive disease that occurs at doses of drug treatment associated with manageable toxicity (4). As most chemotherapeutic agents have a low therapeutic index, even small-fold changes in the sensitivity of tumour cells in patients can render them clinically resistant. Overcoming drug resistance will therefore require a three-pronged approach: first, and crucially, to delineate the mechanisms underlying clinical drug resistance; second, based on this better understanding, to introduce rational circumvention strategies together with conventional cytotoxics to optimise current treatment methods; and third, to develop novel cytotoxic agents, which may reverse drug resistance mechanisms and/or possess cytotoxic activity as well.

Although the underlying mechanisms for clinical drug resistance remain to be elucidated, it is reasonable to suppose that they will fall into three categories: pharmacokinetic, tumour micro-environmental, and cancer-cell specific (4). Pharmacokinetic treatment failure is based on the hypothesis that tumour cells are exposed to insufficient doses of chemotherapeutic agents. Tumour microenviromental factors include tumour hypoxia, angiogenesis, and stromal and intercellular adhesion, all of which may affect drug penetration and cell survival. Cancer-cell specific factors refer to the intrinsic and somatically acquired genetic, epigenetic, and gene expression changes that alter the cellular response pathways leading to cell death following exposure to cytotoxic agents. The drug resistant phenotype is often the result of an interaction between all three factors and strategies designed to overcome drug resistance may need to target one or more of them.

3 Targeting Pharmacokinetic Factors

3.1 Increasing Dose Exposure

To achieve maximum tumour cell death, chemotherapy should be given at the maximum tolerated dose (MTD). Retrospective studies suggest that efficacy is proportional to the degree of myelotoxicity encountered. In the high dose chemotherapy Scottish Gynaecological Cancer Trials Group trial of cisplatin dose intensity in advanced ovarian cancer (50 vs. 100 mg m^{-2}) in combination with cyclphosphamide (5), a twofold dose increase (cisplatin) led to an initial improvement in median survival but limited long-term benefit at cost of increased toxicity. These results suggest that at a higher intravenous dose, the emergence of a drug resistant cell population is delayed but not prevented. Phase II trials of very high-dose chemotherapy with autologous bone marrow transplantation (ABMT) or peripheral blood stem cell transplantation (PBSCT) in previously treated patients with recurrent ovarian cancer have produced higher response rates than achieved with conventional doses (6) but recently, a Phase III randomised trial of this approach in previously untreated patients have reported no improvement in outcome (7).

An alternative way of maximising drug exposure in ovarian cancer is by intraperitoneal (IP) administration of chemotherapy. The rationale for IP therapy is to facilitate the local delivery of a higher concentration of cytotoxic agents to the peritoneum, which is the predominant site of tumour in ovarian cancer, while normal tissues, such as the bone marrow, are relatively spared. In the Gynecologic Oncology Group (GOG) 172 study (8), IP chemotherapy with cisplatin and taxol was shown to be of benefit in patients with post-debulking residual disease of <1 cm, with a median survival difference in the IV vs. IP arms of 15.9 months ($p < 0.03$). However, patients in the IP arm experienced higher rates of grade III/IV haematological toxicity and poorer quality of life 6 weeks after chemotherapy. Moreover, IP patients received 100 mg m^{-2} of cisplatin compared with only 75 mg m^{-2} in the IV arm, thus the overall benefit may ultimately be related to a higher systemic dose of cisplatin received by the IP patients. In addition, concern has been expressed that the control arm (paclitaxel and cisplatin) in this study was less effective than can now be achieved with paclitaxel and carboplatin at optimal doses (9).

3.2 Improving Drug Delivery

Inadequate drug exposure can also be caused by poor drug delivery to the tumour. This can occur because of low bioavailability, extensive first-pass metabolism, high plasma protein binding and low tissue binding. One approach is to use novel drug delivery systems, which include liposomes, pegylation (10) and polymer-drug carriers (11). Liposomes exploit the increased permeability of tumour vasculature relative to normal capillaries to achieve increased tumour-cell-specific drug delivery. However, there is no evidence in randomised trials to date of clear superiority for this approach in the context of drug resistance. Alternative means of improving drug delivery include lipid-, polymer- or poly(L)glutamic-acid conjugated paclitaxel (such as CT-2103), which has recently demonstrated response rates of 10% in heavily pretreated patients with recurrent ovarian cancer in a Phase II study (12). Antibody-directed enzyme prodrug therapy (ADEPT) (13) and gene-directed enzyme prodrug therapy (GDEPT) (14) approaches also aim to increase tumour-cell-specific drug exposure to overcome drug resistance with minimal toxicity to normal tissues.

3.3 Optimising Chemotherapeutic Schedules

Whenever drugs are given in combination for the treatment of a disease, the potential for synergistic, additive or antagonistic effects exists. Although the combination of carboplatin and taxol is now considered as the gold standard first line therapy for ovarian cancer on the basis of the two large randomised studies in the 1990s (1, 3), two subsequent studies by the International Collaborative Ovarian Neoplasm (ICON) group (15) and the GOG 132 (16) failed to demonstrate the benefit of adding paclitaxel to platinum. Whilst the reasons for this are unclear, the possibility of a negative interaction between paclitaxel and platinum cannot be excluded. In addition, carboplatin-associated thrombocytopenia is reduced in patients treated with combination carboplatin/paclitaxel, which further suggests some degree of antagonism in the interaction between both drugs (17). Furthermore, Phase III trials comparing the combination of carboplatin plus paclitaxel with various triple-agent and sequential doublet combinations (where carboplatin and paclitaxel were given in combination following carboplatin plus gemcitabine/topotecan) in patients with advanced-stage epithelial ovarian cancer and primary peritoneal carcinoma have proven negative (18, 19). Hence, there may be scope to consider alternative scheduling of chemotherapeutic drugs to maximise their effectiveness, particularly with the separation of paclitaxel and carboplatin by sequential delivery in the first-line treatment of ovarian cancer. Such an approach has been adopted in a series of feasibility studies by the Scottish group, including initial therapy with full doses of carboplatin followed by four courses of paclitaxel-based treatment, with initial results that are encouraging (20). In support of this approach, in vitro data suggest that the additive cytotoxic effects of paclitaxel and platinum are significantly superior when tumour cells are sequentially exposed to paclitaxel followed by carboplatin or cisplatin (21, 22). Furthermore, the sensitivity of tumour cells to paclitaxel and platinum compounds differs in that cells with mutant p53 are more sensitive to paclitaxel, whereas cells with wild-type p53 and/or non-functional BRCA1/BRCA2 are more sensitive to platinum (23, 24). Separating the two drugs may thus maximise these differing sensitivities in a heterogenous tumour cell population.

An alternative way of using existing chemotherapeutic drugs is to administer them on a weekly schedule. Dr. Van der burg in Rotterdam has pioneered this approach in the context of drug-sensitive disease (25). Her most recent data using weekly carboplatin (AUC 4) with paclitaxel (90 mg m^{-2}) are most interesting, as they show a response rate of 53% in patients with platinum-resistant disease (26). Randomised trials of this approach are about to commence.

4 Targeting the Tumour Microenvironment

Tumour progression is accompanied by angiogenesis in response to hypoxic stimuli (27). However, because of the haphazard growth of abnormal, leaky blood and lymph vessels, interstitial fibrosis and a contraction of the interstitial space by stromal fibroblasts, the resultant effect is a raised intratumoral interstitial fluid pressure, which may serve as a physical barrier to systemic drug delivery (28). Combining conventional chemotherapeutic agents with an angiogenesis inhibitor may, therefore, enhance intra-tumoural drug penetration by inhibiting tumour neovascularisation, thereby reducing the intra-tumoural pressure. Two recent retrospective analyses of bevacizumab, a vascular-endothelial growth factor (VEGF) inhibitor, therapy in platinum-refractory ovarian cancer have demonstrated overall response rates of 35% when used in combination with other chemotherapeutic agents, and 16% when used alone (29, 30). Although theoretically there could be a negative impact on the efficacy of chemotherapy through the induction of tumour hypoxia from this approach (31), clinical data in breast, colorectal and lung cancer point to a consistent enhancement of chemotherapy (32). Randomised trials in which bevacizumab is added to front line chemotherapy in ovarian cancer are now underway.

The extracellular matrix (ECM) has been suggested to provide protection against chemotherapy-induced apoptosis in various cancers (33, 34). Integrin-mediated interactions between tumour and stromal cells can also influence chemoresistance via anti-apoptotic signalling pathways (35). In addition, many *ECM* genes are elevated in cisplatin-resistant cells (36). In particular, *COL6A3*, which encodes for collagen VI, has been observed to be one of the most highly upregulated genes, and cultivation of cisplatin-sensitive cells in the presence of collagen VI protein promotes resistance in vitro (36). Collagen VI expression has been demonstrated in ovarian cancer in vivo, and is associated with tumour grade and prognosis. More recently, Choi and colleagues demonstrated increased antibody penetration into xenotransplanted ovarian cancer cells following exposure to collagenase (37).

These studies suggest that tumour cells may directly remodel their microenvironment to increase their survival in the presence of chemotherapeutic drugs, and point to new potential targets in modulating drug resistance.

5 Targeting Cancer Cell Factors

5.1 Drug Efflux and Detoxification

Experimental models have implicated various cell-specific mechanisms in the development of drug resistance in ovarian cancer cells. These include increased drug efflux from chemoresistant cells mediated by transport proteins encoded by the ATP-binding cassette (*ABC*) family of drug resistance genes, which include permeability glycoprotein (PGP) and the multi-drug resistance protein 1 (MRP1), and drug detoxification mechanisms such as increased inactivation of platinum by cytoplasmic thiols (38).

In taxane-resistant cells, the two mechanisms most commonly associated with the development of drug resistance are the overexpression of drug efflux proteins, like PGP, and alterations of tubulin, which is the cellular target of the taxanes (39). Biricodar (INCEL, VX-710) has been shown to restore drug sensitivity to PGP and MRP1 expressing cells in vitro, and the combination of Biricodar with paclitaxel in 45 paclitaxel-refractory advanced ovarian cancer patients has been examined in a Phase II trial, demonstrating three partial responses with stable disease observed in a further 12 cases (40). PSC-833 is a non-immunosuppressive cyclosporine analogue that inhibits the function of PGP in vitro. In parallel with the Biricodar studies, the combination of PSC-833 with paclitaxel in the setting of paclitaxel-refractory ovarian cancer has been examined in two Phase II trials with partial responses observed in up to 10% of subjects and stable disease in a further 25% (41, 42). However, there was no evidence of any improvement in outcome in a large randomized controlled Phase III study comparing concurrent PSC833 or placebo alongside first-line Carboplatin/Paclitaxel (43). This suggests that if modulators such as PSC-833 and Biricodar are likely have any prospect of success in reversing resistance in ovarian cancer, they will need to be applied in a pre-selected population of patients with established resistance to paclitaxel, based on PGP and MRP1 dysfunction.

Alkylating agents, including platinum, are potent electrophiles that are inactivated by intracellular electron-rich molecules glutathione and gluthatione-s-transferase-pi (GST) (44). Reductions in the levels of these intracellular molecules have been associated with reversal of cisplatin-resistance (44), and the glutathione pro-drug TLK286 was designed to capitalise on the fact that there is increased expression of GST in ovarian cancer cells compared with normal cells. TLK286 is activated by GST in experimental models, following which pro-apoptotic electrophilic fragments are released. Broad experimental activity has been demonstrated with responses observed not merely in patients with ovarian cancer (45). A response rate of 15% was seen in patients with platinum-refractory disease in Phase II trials (46), and a large randomized Phase III study of TLK286 vs. liposomal doxorubicin or topotecan is underway, in addition to an extensive combination programme. Buthionine sulphoximine (BSO) is a synthetic amino acid that inhibits the production of glutathione, whilst ethacrynic acid inhibits GST-mediated conjugation of glutathione to cisplatin (47, 48). Phase I studies have demonstrated good safety profiles in both agents. Tumour response and depletion of tumour intra-cellular glutathione levels was achieved with BSO (in combination with melphalan) in two patients with platinum refractory ovarian cancer (49, 50). The platinum analogue ZD0473 was designed to sterically hinder glutathione- and GST-mediated inactivation of alkylating agents. Sequence specificity of DNA-adduct formation is quite likely to differ from other platinum-analogues and may account for its activity in cisplatin-resistant cell lines. However, despite its in vitro potential, a response rate of only 8.3% was observed in a Phase II study in 59 platinum-refractory ovarian cancer patients (51).

5.2 DNA Repair

Experimental models have very usefully facilitated the identification of the cellular components responsible for the response to DNA/cellular damage from cytotoxic agents, the result of which is either DNA repair, or damage tolerance (resistance) or apoptosis (sensitivity). Erroneous DNA replication following platinum-adduct formation is recognised by the mismatch repair (MMR) pathway proteins, leading to cell death. Alternatively, platinum-adducts can be rectified by the nucleotide excision repair (NER) system. Therefore, down-regulation of MMR or up-regulation of NER could lead to platinum-resistance (52, 53).

Defects in *MMR* genes, which code for hMLH1 and hMSH2, result in failure of DNA damage recognition and tumours characterised by microsatellite instability (MSI) (38). Experimental data implicate MMR defects caused by loss of MLH1 expression in drug resistance to a wide variety of cytotoxic agents including cisplatin and doxorubicin (38). Sensitivity to the platinum analogue oxaliplatin is independent of cellular MMR status, and it was thought that oxaliplatin is quite likely to be active in platinum-resistant cases where MLH1 is frequently down-regulated (54). However, response rates of only 5–6% have been observed in Phase II studies of single-agent oxaliplatin involving platinum-resistant patients (55, 56). Thus, other methods for modulating MMR defects are currently being investigated. MLH1 promoter methylation has been noted in drug resistant, hMLH1 negative ovarian and colon tumour xenografts. The clinical relevance of methylation was borne out in a study that examined tumour DNA extracted from plasma in a large randomized clinical trial (SCOTROC 1). Patients who acquired methylation in plasma samples after disease relapse had a considerably worse outcome following further chemotherapy compared with those in whom methylation did not occur (57). Previously, the demethylating agent 2´-deoxy-5-azacytidine (Decitabine) had been shown to re-sensitise tumour cells to cisplatin in in vitro and in vivo experimental models (38, 58). In a recent Phase I clinical trial, the combination of carboplatin and decitabine has, in fact, been shown to be feasible with demethylation evident at well-tolerated doses (59). A randomised Phase II study in partially platinum-sensitive ovarian cancer patients (with treatment free intervals of 6–12months) is therefore about to open. Patients (30–40% of whom should have methylated DNA) will receive carboplatin with or without decitabine.

Up-regulation of genes involved in the NER pathway, e.g. *ERCC1* has been associated with resistance to platinum-based therapy in ovarian cancer (60, 61). Yondelis (ET-743) is a marine compound that binds to the minor groove of DNA (62), and in vitro models suggest that ET-743-mediated apoptosis is enhanced by NER and inhibited by MMR, making it an attractive prospect in platinum-resistant tumours. Moreover, ET-743 acts synergistically when used in combination with platinum (63). Initial Phase I studies have indicated that, as a single agent, hepatotoxicity was dose limiting (64), and that in combination with carboplatin, myelotoxicity was dose-limiting. Phase II studies have in fact demonstrated its efficacy in platinum-sensitive cases to be superior to that in platinum-resistant ones (65). A randomized Phase III trial is now exploring the combination of ET-743 with liposomal doxorubicin. Another compound that targets NER is the DNA polymerase alpha inhibitor, Aphidicolin, which has been shown to reverse platinum resistance in ovarian cancer because of enhanced NER. Its water-soluble analogue, aphidicolin glycinate, has been shown to be well tolerated in a Phase I study (66).

Cells with deficiencies in homologous recombination (HR) pathway genes, including *BRCA1, BRCA2* and *FANC*, are unable to repair DNA cross-links and DNA double-strand breaks by error-free HR, thus resulting in genomic instability and cancer predisposition (67). Hence, HR-deficient cells are particularly platinum-sensitive as they are unable to repair DNA-damage caused by inter- and intra-strand adducts, whereas platinum-resistance may occur, at least in part, through the function of intact HR DNA repair pathway genes, particularly *FANC* and *BRCA* (68). Recently, deficiencies in HR pathway genes have also been found to confer extreme sensitivity to inhibition of poly(ADP-ribose) polymerase (PARP) (69), an enzyme involved in base excision repair, which is a key pathway in the repair of DNA single-strand breaks. The prevalence of *BRCA1* mutations in ovarian cancer patients has been reported to be between 5% and 23% (70, 71), whilst inactivation of *BRCA1* by promoter methylation has been reported in up to 31% of sporadic ovarian cancer (72). In addition, BRCA1 and BRCA2 defects have been reported in 82% of mullerian tumours (ovarian, peritoneal and Fallopian tube tumours) (73). Hence, PARP inhibition may serve to enhance the efficacy of platinum in these patients or, indeed, induce cytotoxic effects by itself. Our unit is currently running a Phase I trial of an oral PARP inhibitor (Ku59436) and, interestingly, we have seen clear signs of anti-tumour efficacy in one patient with presumed BRCA mutations (74).

5.3 Apoptotic Evasion

One of the main mediators of tumour cell cytotoxicity following exposure to chemotherapy or radiotherapy is apoptosis. The response to chemotherapy may thus be attenuated by the up-regulation of anti-apoptotic proteins or down-regulation of pro-apoptotic proteins.

One of the most frequently mutated genes in human cancers is *p53* (38). Following DNA damage, active p53 induces the up-regulation of mitochondrial pro-apoptotic proteins like BAX, as well as those of cell death receptor pathways like TRAIL-R1 (38). Lack of functional p53 can therefore result in drug resistance, as these cells are unable to undergo apoptosis in response to DNA damage (75). Therefore, determining p53 status may prove useful in predicting therapeutic response to specific drugs. Thus, agents such as the triplatinum analogue BBR3464 – which compared with other platinum analogues, has demonstrated greater potency, a more rapid rate of DNA binding, the ability to form longer-term DNA cross-links and to induce apoptosis independent of p53 cellular status in vitro, as well as good in vivo activity against p53 mutant xenografts (76) – have been investigated in Phase I and II trials. Unfortunately, although responses were demonstrated in platinum-sensitive patients, little activity was seen in platinum-resistant patients (77).

Proteins involved in the anti-apoptotic phosphatase and tensin homolog/phosphoinositide-3-kinase (PTEN/PI3K) pathway have been found to be overexpressed in ovarian cancer (38). Down-regulation of PTEN, which negatively regulates PI3K, is observed in numerous tumour types including lung, melanoma and prostate cancer, and results in the activation of AKT and consequently mammalian target of rapamycin (mTOR) kinase signalling (78). Amplification of the gene coding for the p110 alpha subunit of PI3K has been observed in 40% of ovarian cancer (79). Lysophosphatidic acid (LPA), which is a growth factor found in ovarian cancer ascites, has been shown to promote cell survival by activating the PI3K/AKT pathway (78). In addition, amplification of AKT has been observed in undifferentiated ovarian cancer and may lead to resistance to various drugs, including taxanes, because of apoptotic failure (80).

In this context, another potentially important anti-apoptotic molecular target is the HSP90 molecular chaperone, which serves to stabilize a number of mutated and over-expressed signalling proteins that promote cell survival and proliferation (81), including those in the PI3K/AKT pathway. Hence, HSP-90 inhibition appears as an attractive target for modulating dug resistance. The HSP-90 inhibitor, 17-allyamino, 17-demethoxygeldanamycin (17-AAG), has been shown to sensitise ovarian tumour cells with constitutively active AKT to paclitaxel (82), and may also have additive or synergistic effects in combination with cisplatin, doxorubicin and paclitaxel (83).

5.4 Tumour Cell Alterations of Drug-Specific Targets

Specific point mutations in beta-tubulin, e.g. an alanine to threonine substitution at residue 364, have been shown to confer resistance to paclitaxel (84). Epothilones are cytotoxic macrolides with a similar mechanism to paclitaxel but with the advantages of increased potency as well as retaining activity in taxane-resistant settings in preclinical models, where overexpression of PGP or beta-tubulin mutations are present (85). Four epothilones are in early clinical trials for cancer treatment, and Phase I studies have shown that the dose-limiting toxicities are generally neurotoxicity and diarrhoea (86). Significantly, Ixabepithlone and Patupilone have shown promising efficacy in platinum and taxane-refractory ovarian cancer (86). Trials with these drugs are ongoing, including a randomized Phase III trial of Patupilone compared with liposomal doxorubicin in patients with platinum-resistant disease.

6 Conclusions: Understanding Resistance and Finding New Targets for the Future

The low response rates observed in many of the aforementioned clinical trials, based initially on promising results from experimental models, suggest that although certain molecular mechanisms may contribute to drug resistance, e.g. that of drug efflux and detoxification, they are unlikely to be the primary mechanisms in these patients. These results highlight the inherent difficulties in translating the results from studies using in vitro models, which are often too reductionistic and over-simplistic in their focus on specific pathways, to clinical use. Nonetheless, bearing in mind this caveat, in vitro studies on drug response and resistance remain a useful experimental tool in providing new clues for hypothesis testing and validation in the clinical setting. An example of such a "bottom-up" approach would begin with the identification of biologically important molecules in vitro, in vivo and clinical relevance of which may subsequently be validated using tissue microarrays (87) made up of tumour samples from patients with clinical data available, ideally including paired samples from the same patients before treatment and when resistance develops.

In recent years, the advent of high-throughput technologies using microarrays to facilitate molecular profiling of tumours has allowed detailed analysis of the genomic and gene expression changes in ovarian cancer (88). By comparing the molecular profiles of drug-sensitive and drug-resistant ovarian cancers, these techniques have the potential to provide us with a greater and more clinically relevant insight into the underlying mechanisms of drug resistance in these cancers. Recent studies using gene-expression arrays have derived a chemotherapy-response profile (CRP), which is predictive of pathological complete response to first line platinum/taxane chemotherapy in 60 patients with ovarian cancer (89), and reported 230 differentially expressed genes between primary chemosensitive and primary chemoresistant ovarian tumours (90).

However, microarray techniques are subject to considerable data variability, due in part to variability in methods of RNA/DNA extraction, probe labelling and hybridisation, the type of microarray platform used, the number and histological type of samples analysed, the methods used for microarray and statistical analysis, and methods used for results validation (91). Although no formal reliable method for power calculation has been devised for microarray-based studies, a minimum number of 50 subjects have been suggested (92). Unfortunately, most of the studies addressing the question of drug resistance in ovarian cancer have been hampered by insufficient sample sizes, a lack of stringency in cohort selection with regard to the uniformity of selected histological tumour subtype and type of treatment received by the subjects analysed, and the absence of matched pre- and post-chemotherapy/chemoresistant tumour samples. Indeed, given the clinical phenotypic diversity exhibited by different histological subtypes of ovarian cancer, it is very likely that more useful information will be gained from microrarray studies by analysing sufficient numbers of each ovarian cancer histological subtype before attempting cross-subtype comparisons. It is envisaged that future well-designed microarray studies involving sufficient numbers of histological subtype-specific ovarian tumour samples will lead to a more conclusive picture of the underlying mechanisms behind drug resistance, and the identification of new targets to reverse this phenomenon in the treatment of ovarian cancer.

References

1. McGuire WP, Hoskins WJ, Brady MF, Kucera PR, Partridge EE, Look KY, et al. Cyclophosphamide and cisplatin compared with paclitaxel and cisplatin in patients with stage III and stage IV ovarian cancer. New Engl J Med. 1996 Jan 4;334(1):1–6.
2. Ozols RF, Bundy BN, Greer BE, Fowler JM, Clarke-Pearson D, Burger RA, et al. Phase III trial of carboplatin and paclitaxel compared with cisplatin and paclitaxel in patients with optimally resected stage III ovarian cancer: a Gynecologic Oncology Group study. J Clin Oncol. 2003 Sep 1;21(17):3194–200.
3. Piccart MJ, Bertelsen K, James K, Cassidy J, Mangioni C, Simonsen E, et al. Randomized intergroup trial of cisplatin-paclitaxel versus cisplatin-cyclophosphamide in women with advanced epithelial ovarian cancer: three-year results. J Natl Cancer Inst. 2000 May 3;92(9):699–708.
4. Agarwal R, Kaye SB. Ovarian cancer: strategies for overcoming resistance to chemotherapy. Nat Rev Cancer. 2003 Jul;3(7):502–16.
5. Kaye SB, Paul J, Cassidy J, Lewis CR, Duncan ID, Gordon HK, et al. Mature results of a randomized trial of two doses of cisplatin for the treatment of ovarian cancer. Scottish Gynecology Cancer Trials Group. J Clin Oncol. 1996 Jul;14(7):2113–19.
6. Ozols RF, Gore M, Trope C, Grenman S. Intraperitoneal treatment and dose-intense therapy in ovarian cancer. Ann Oncol. 1999;10 (Suppl 1):59–64.
7. Pedrazzoli P, Ledermann J, Lotz JP, Leyvraz S, Aglietta M, Rosti G, et al. High dose chemotherapy with autologous hematopoietic stem cell support for solid tumors other than breast cancer in adults. Ann Oncol. 2006 Oct;17(10):1479–88.
8. Armstrong DK, Bundy B, Wenzel L, Huang HQ, Baergen R, Lele S, et al. Intraperitoneal cisplatin and paclitaxel in ovarian cancer. New Engl J Med. 2006 Jan 5;354(1):34–43.
9. Gore M, du Bois A, Vergote I. Intraperitoneal chemotherapy in ovarian cancer remains experimental. J Clin Oncol. 2006 Oct 1;24(28):4528–30.
10. Harris JM, Chess RB. Effect of pegylation on pharmaceuticals. Nat Rev. 2003 Mar;2(3):214–21.
11. Duncan R. The dawning era of polymer therapeutics. Nat Rev. 2003 May;2(5):347–60.
12. Sabbatini P, Aghajanian C, Dizon D, Anderson S, Dupont J, Brown JV, et al. Phase II study of CT-2103 in patients with recurrent epithelial ovarian, fallopian tube, or primary peritoneal carcinoma. J Clin Oncol. 2004 Nov 15;22(22):4523–31.

13. Houba PH, Boven E, van der Meulen-Muileman IH, Leenders RG, Scheeren JW, Pinedo HM, et al. Pronounced antitumor efficacy of doxorubicin when given as the prodrug DOX-GA3 in combination with a monoclonal antibody beta-glucuronidase conjugate. Int J Cancer. 2001 Feb 15;91(4):550–4.
14. McNeish IA, Green NK, Gilligan MG, Ford MJ, Mautner V, Young LS, et al. Virus directed enzyme prodrug therapy for ovarian and pancreatic cancer using retrovirally delivered *E. coli* nitroreductase and CB1954. Gene Ther. 1998 Aug;5(8):1061–9.
15. Colombo N. Randomised trial of paclitaxel (PTX) and carboplatin (CBDCA) versus a control arm of carboplatin or CAP (cyclophosphamide, doxorubicin & cisplatin: the Third International Collaborative Ovarian Neoplasm Study (ICON3). Proc ASCO 2000 Jun;19:2349–51.
16. Muggia FM, Braly PS, Brady MF, Sutton G, Niemann TH, Lentz SL, et al. Phase III randomized study of cisplatin versus paclitaxel versus cisplatin and paclitaxel in patients with suboptimal stage III or IV ovarian cancer: a gynecologic oncology group study. J Clin Oncol. 2000 Jan;18(1):106–15.
17. Moss C, Kaye SB. Ovarian cancer: progress and continuing controversies in management. Eur J Cancer 2002 Sep;38(13):1701–7.
18. Bookman MA. GOG0182-ICON5: 5-arm Phase III randomized trial of paclitaxel (P) and carboplatin (C) vs combinations with gemcitabine (G), PEG-liposososmal doxorubicin (D), or topotecan (T) in patients (pts) with advanced-stage epithelial ovarian (EOC) or primary peritoneal (PPC) carcinoma. ASCO Annual Meeting Proceedings. 2006;24(18S):A5002.
19. Scarfone G, Scambia G, Raspagliesi F, Mangili G, Danese S, Presti M, et al. A multicenter, randomized, Phase III study comparing paclitaxel/carboplatin (PC) versus topotecan/paclitaxel/carboplatin (TPC) in patients with stage III (residual tumor > 1 cm after primary surgery) and IV ovarian cancer (OC). ASCO Annual Meeting Proceedings 2006.
20. Harries M, Moss C, Perren T, Gore M, Hall G, Everard M, et al. A Phase II feasibility study of carboplatin followed by sequential weekly paclitaxel and gemcitabine as first-line treatment for ovarian cancer. Br J Cancer. 2004 Aug 16;91(4):627–32.
21. Rowinsky EK, Citardi MJ, Noe DA, Donehower RC. Sequence-dependent cytotoxic effects due to combinations of cisplatin and the antimicrotubule agents taxol and vincristine. J Cancer Res Clin Oncol. 1993;119(12):727–33.
22. Waltmire CN, Alberts DS, Dorr RT. Sequence-dependent cytotoxicity of combination chemotherapy using paclitaxel, carboplatin and bleomycin in human lung and ovarian cancer. Anti Cancer Drugs. 2001 Aug;12(7):595–602.
23. Gadducci A, Cosio S, Muraca S, Genazzani AR. Molecular mechanisms of apoptosis and chemosensitivity to platinum and paclitaxel in ovarian cancer: biological data and clinical implications. Eur J Gynaecol Oncol. 2002;23(5):390–6.
24. Lavarino C, Pilotti S, Oggionni M, Gatti L, Perego P, Bresciani G, et al. *p53* gene status and response to platinum/paclitaxel-based chemotherapy in advanced ovarian carcinoma. J Clin Oncol. 2000 Dec 1;18(23):3936–45.
25. de Jongh FE, de Wit R, Verweij J, Sparreboom A, van den Bent MJ, Stoter G, et al. Dose-dense cisplatin/paclitaxel. a well-tolerated and highly effective chemotherapeutic regimen in patients with advanced ovarian cancer. Eur J Cancer. 2002 Oct;38(15):2005–13.
26. Van der Burg ME, Vergote I, Burger CW, van der Gaast A. Phase II study of weekly paclitaxel carboplatin in the treatment of progressive ovarian cancer. ASCO Annual Meeting Proceedings (Post-Meeting Edition) 2004;22(14S):A5058.
27. Carmeliet P, Dor Y, Herbert JM, Fukumura D, Brusselmans K, Dewerchin M, et al. Role of HIF-1alpha in hypoxia-mediated apoptosis, cell proliferation and tumour angiogenesis. Nature. 1998 Jul 30;394(6692):485–90.
28. Heldin CH, Rubin K, Pietras K, Ostman A. High interstitial fluid pressure – an obstacle in cancer therapy. Nat Rev Cancer. 2004 Oct;4(10):806–13.
29. Monk BJ, Han E, Josephs-Cowan CA, Pugmire G, Burger RA. Salvage bevacizumab (rhuMAB VEGF)-based therapy after multiple prior cytotoxic regimens in advanced refractory epithelial ovarian cancer. Gynecologic Oncol. 2006 Aug;102(2):140–4.

30. Wright JD, Hagemann A, Rader JS, Viviano D, Gibb RK, Norris L, et al. Bevacizumab combination therapy in recurrent, platinum-refractory, epithelial ovarian carcinoma: a retrospective analysis. Cancer. 2006 Jul 1;107(1):83–9.
31. Teicher BA. Hypoxia and drug resistance. Cancer Metastasis Rev. 1994 Jun;13(2):139–68.
32. de Castro Junior G, Puglisi F, de Azambuja E, El Saghir NS, Awada A. Angiogenesis and cancer: a cross-talk between basic science and clinical trials (the "do ut des" paradigm). Crit Rev Oncol/Hematol. 2006 Jul;59(1):40–50.
33. Sethi T, Rintoul RC, Moore SM, MacKinnon AC, Salter D, Choo C, et al. Extracellular matrix proteins protect small cell lung cancer cells against apoptosis: a mechanism for small cell lung cancer growth and drug resistance *in vivo*. Nat Med. 1999 Jun;5(6):662–8.
34. Hazlehurst LA, Dalton WS. Mechanisms associated with cell adhesion mediated drug resistance (CAM-DR) in hematopoietic malignancies. Cancer Metastasis Rev. 2001;20(1–2): 43–50.
35. Elliott T, Sethi T. Integrins and extracellular matrix: a novel mechanism of multidrug resistance. Expert Rev Anticancer Ther. 2002 Aug;2(4):449–59.
36. Sherman-Baust CA, Weeraratna AT, Rangel LB, Pizer ES, Cho KR, Schwartz DR, et al. Remodeling of the extracellular matrix through overexpression of collagen VI contributes to cisplatin resistance in ovarian cancer cells. Cancer cell. 2003 Apr;3(4):377–86.
37. Choi J, Credit K, Henderson K, Deverkadra R, He Z, Wiig H, et al. Intraperitoneal immunotherapy for metastatic ovarian carcinoma: Resistance of intratumoral collagen to antibody penetration. Clin Cancer Res. 2006 Mar 15;12(6):1906–12.
38. Baird RD, Kaye SB. Drug resistance reversal – are we getting closer? Eur J Cancer. 2003 Nov;39(17):2450–61.
39. Fojo AT, Menefee M. Microtubule targeting agents: basic mechanisms of multidrug resistance (MDR). Semin Oncol. 2005 Dec;32(6 Suppl 7):S3–8.
40. Seiden MV, Swenerton KD, Matulonis U, Campos S, Rose P, Batist G, et al. A Phase II study of the MDR inhibitor biricodar (INCEL, VX-710) and paclitaxel in women with advanced ovarian cancer refractory to paclitaxel therapy. Gynecol Oncol. 2002 Sep;86(3):302–10.
41. Fracasso PM, Brady MF, Moore DH, Walker JL, Rose PG, Letvak L, et al. Phase II study of paclitaxel and valspodar (PSC 833) in refractory ovarian carcinoma: a gynecologic oncology group study. J Clin Oncol. 2001 Jun 15;19(12):2975–82.
42. Fields A, Hochster H, Runowicz C, Speyer J, Goldberg G, Cohen C, et al. PSC833: initial clinical results in refractory ovarian cancer patients. Curr Opin Oncol. 1998 Aug;10 (Suppl 1):S21.
43. Joly F, Mangioni C, Nicoletto Mea. A Phase 3 study of PSC 833 in combination with paclitaxel and carboplatin versus paclitaxel and carboplatin alone in patients with stage IV or suboptimally debulked stage III epithlial ovarian cancer or primary cancer of the peritoneum. Proc Am Soc Clin Oncol. 2002;21:202a.
44. Townsend DM, Shen H, Staros AL, Gate L, Tew KD. Efficacy of a glutathione *S*-transferase pi-activated prodrug in platinum-resistant ovarian cancer cells. Mol Cancer Therapeut. 2002 Oct;1(12):1089–95.
45. Tew KD. TLK-286: a novel glutathione *S*-transferase-activated prodrug. Expert Opin Investig Drugs. 2005 Aug;14(8):1047–54.
46. Kavanagh JJ, Gershenson DM, Choi H, Lewis L, Patel K, Brown GL, et al. Multi-institutional Phase 2 study of TLK286 (TELCYTA, a glutathione *S*-transferase P1-1 activated glutathione analog prodrug) in patients with platinum and paclitaxel refractory or resistant ovarian cancer. Int J Gynecol Cancer. 2005 Jul-Aug;15(4):593–600.
47. Renschler MF. The emerging role of reactive oxygen species in cancer therapy. Eur J Cancer. 2004 Sep;40(13):1934–40.
48. Mulder GJ, Ouwerkerk-Mahadevan S. Modulation of glutathione conjugation *in vivo*: how to decrease glutathione conjugation *in vivo* or in intact cellular systems *in vitro*. Chem Biol Interact. 1997 Jun 6;105(1):17–34.
49. Bailey HH. L-S,R-buthionine sulfoximine: historical development and clinical issues. Chem Biol Interact. 1998 Apr 24;111–112:239–54.

50. O'Dwyer PJ, Hamilton TC, LaCreta FP, Gallo JM, Kilpatrick D, Halbherr T, et al. Phase I trial of buthionine sulfoximine in combination with melphalan in patients with cancer. J Clin Oncol. 1996 Jan;14(1):249–56.
51. Gore ME, Atkinson RJ, Thomas H, Cure H, Rischin D, Beale P, et al. A Phase II trial of ZD0473 in platinum-pretreated ovarian cancer. Eur J Cancer. 2002 Dec;38(18):2416–20.
52. Dijt FJ, Fichtinger-Schepman AM, Berends F, Reedijk J. Formation and repair of cisplatin-induced adducts to DNA in cultured normal and repair-deficient human fibroblasts. Cancer Res. 1988 Nov 1;48(21):6058–62.
53. Brown R, Hirst GL, Gallagher WM, McIlwrath AJ, Margison GP, van der Zee AG, et al. hMLH1 expression and cellular responses of ovarian tumour cells to treatment with cytotoxic anticancer agents. Oncogene. 1997 Jul 3;15(1):45–52.
54. Raymond E, Faivre S, Woynarowski JM, Chaney SG. Oxaliplatin: mechanism of action and antineoplastic activity. Semin Oncol. 1998 Apr;25(2 Suppl 5):4–12.
55. Piccart MJ, Green JA, Lacave AJ, Reed N, Vergote I, Benedetti-Panici P, et al. Oxaliplatin or paclitaxel in patients with platinum-pretreated advanced ovarian cancer: A randomized Phase II study of the European Organization for Research and Treatment of Cancer Gynecology Group. J Clin Oncol. 2000 Mar;18(6):1193–202.
56. Dieras V, Bougnoux P, Petit T, Chollet P, Beuzeboc P, Borel C, et al. Multicentre Phase II study of oxaliplatin as a single-agent in cisplatin/carboplatin + /- taxane-pretreated ovarian cancer patients. Ann Oncol. 2002 Feb;13(2):258–66.
57. Gifford G, Paul J, Vasey PA, Kaye SB, Brown R. The acquisition of hMLH1 methylation in plasma DNA after chemotherapy predicts poor survival for ovarian cancer patients. Clin Cancer Res. 2004 Jul 1;10(13):4420–6.
58. Plumb JA, Strathdee G, Sludden J, Kaye SB, Brown R. Reversal of drug resistance in human tumor xenografts by 2´-deoxy-5-azacytidine-induced demethylation of the hMLH1 gene promoter. Cancer Res. 2000 Nov 1;60(21):6039–44.
59. Lee C, Appleton K, Plumb Jea. A Phase I trial of the DNA-hypomethylating agent 5-aza-2´-deoxycytidine in combination with carboplatin both given 4 weekly by intravenous injection in patients with advanced solid tumours. J Clin Oncol. 2004;22(14S):128.
60. Dabholkar M, Bostick-Bruton F, Weber C, Bohr VA, Egwuagu C, Reed E. ERCC1 and ERCC2 expression in malignant tissues from ovarian cancer patients. J Natl Cancer Inst. 1992 Oct 7;84(19):1512–7.
61. Dabholkar M, Vionnet J, Bostick-Bruton F, Yu JJ, Reed E. Messenger RNA levels of XPAC and ERCC1 in ovarian cancer tissue correlate with response to platinum-based chemotherapy. J Clin Investig. 1994 Aug;94(2):703–8.
62. Agarwal R, Linch M, Kaye SB. Novel therapeutic agents in ovarian cancer. Eur J Surg Oncol. 2006 Oct;32(8):875–86.
63. Scotto KW. ET-743: more than an innovative mechanism of action. Anticancer Drugs. 2002 May;13(Suppl 1):S3–6.
64. Ryan DP, Supko JG, Eder JP, Seiden MV, Demetri G, Lynch TJ, et al. Phase I and pharmacokinetic study of ecteinascidin 743 administered as a 72-hour continuous intravenous infusion in patients with solid malignancies. Clin Cancer Res. 2001 Feb;7(2):231–42.
65. Colombo N, al e. Phase II and pharmacokinetic study of 3-hr infusion of Et-743 in ovarian cancer patients filing platinum-taxanes. Proc Am Soc Clin Oncol. 2002 Oct;21(221A):(Abstr 880).
66. Sessa C, Capri G, Gianni L, Peccatori F, Grasselli G, Bauer J, et al. Clinical and pharmacological Phase I study with accelerated titration design of a daily times five schedule of BBR3464, a novel cationic triplatinum complex. Ann Oncol. 2000 Aug;11(8):977–83.
67. Tutt A, Ashworth A. The relationship between the roles of *BRCA* genes in DNA repair and cancer predisposition. Trends Mol Med. 2002 Dec;8(12):571–6.
68. Olopade OI, Wei M. FANCF methylation contributes to chemoselectivity in ovarian cancer. Cancer Cell. 2003 May;3(5):417–20.
69. Farmer H, McCabe N, Lord CJ, Tutt AN, Johnson DA, Richardson TB, et al. Targeting the DNA repair defect in BRCA mutant cells as a therapeutic strategy. Nature. 2005 Apr 14;434(7035):917–21.

70. Stratton JF, Gayther SA, Russell P, Dearden J, Gore M, Blake P, et al. Contribution of BRCA1 mutations to ovarian cancer. New Engl J Med. 1997 Apr 17;336(16):1125–30.
71. Geisler JP, Hatterman-Zogg MA, Rathe JA, Buller RE. Frequency of BRCA1 dysfunction in ovarian cancer. J Natl Cancer Inst. 2002 Jan 2;94(1):61–7.
72. Esteller M, Silva JM, Dominguez G, Bonilla F, Matias-Guiu X, Lerma E, et al. Promoter hypermethylation and BRCA1 inactivation in sporadic breast and ovarian tumors. J Natl Cancer Inst. 2000 Apr 5;92(7):564–9.
73. Hilton JL, Geisler JP, Rathe JA, Hattermann-Zogg MA, DeYoung B, Buller RE. Inactivation of BRCA1 and BRCA2 in ovarian cancer. J National Cancer Inst. 2002 Sep 18;94(18):1396–406.
74. Fong PC, Spicer J, Reade S, Reid A, Vidal L, Schellens JH, et al. Phase I pharmacokinetic (PK) and pharmacodynamic (PD) evaluation of a small molecule inhibitor of Poly ADP-Ribose Polymerase (PARP), KU-0059436 (Ku) in patients (p) with advanced tumours. ASCO Annual Meeting Proceedings Part I. 2006;24(18S):A3022.
75. Lowe SW, Bodis S, McClatchey A, Remington L, Ruley HE, Fisher DE, et al. p53 status and the efficacy of cancer therapy *in vivo*. Science. 1994 Nov 4;266(5186):807–10.
76. Manzotti C, Pratesi G, Menta E, Di Domenico R, Cavalletti E, Fiebig HH, et al. BBR 3464: a novel triplatinum complex, exhibiting a preclinical profile of antitumor efficacy different from cisplatin. Clin Cancer Res. 2000 Jul;6(7):2626–34.
77. Calvert A*ea*. Phase II clinical study of BBR3464, a novel bifunctional platinum analogue, in patients with ovarian cancer. Eur J Cancer. 2001;37(A965).
78. Easton JB, Houghton PJ. mTOR and cancer therapy. Oncogene. 2006 Oct 16;25(48):6436–46.
79. Shayesteh L, Lu Y, Kuo WL, Baldocchi R, Godfrey T, Collins C, et al. PIK3CA is implicated as an oncogene in ovarian cancer. Nat Genet. 1999 Jan;21(1):99–102.
80. Bellacosa A, de Feo D, Godwin AK, Bell DW, Cheng JQ, Altomare DA, et al. Molecular alterations of the *AKT2* oncogene in ovarian and breast carcinomas. Int J Cancer. 1995 Aug 22;64(4):280–5.
81. Banerji U, Walton M, Raynaud F, Grimshaw R, Kelland L, Valenti M, et al. Pharmacokinetic-pharmacodynamic relationships for the heat shock protein 90 molecular chaperone inhibitor 17-allylamino, 17-demethoxygeldanamycin in human ovarian cancer xenograft models. Clin Cancer Res. 2005 Oct 1;11(19 Pt 1):7023–32.
82. Sain N, Krishnan B, Ormerod MG, De Rienzo A, Liu WM, Kaye SB, et al. Potentiation of paclitaxel activity by the HSP90 inhibitor 17-allylamino-17-demethoxygeldanamycin in human ovarian carcinoma cell lines with high levels of activated AKT. Mol Cancer Therapeut. 2006 May;5(5):1197–208.
83. Nguyen DM, Lorang D, Chen GA, Stewart JHt, Tabibi E, Schrump DS. Enhancement of paclitaxel-mediated cytotoxicity in lung cancer cells by 17-allylamino geldanamycin: *in vitro* and *in vivo* analysis. Ann Thorac Surg. 2001 Aug;72(2):371–8; discussion 8–9.
84. Giannakakou P, Sackett DL, Kang YK, Zhan Z, Buters JT, Fojo T, et al. Paclitaxel-resistant human ovarian cancer cells have mutant beta-tubulins that exhibit impaired paclitaxel-driven polymerization. J Biol Chem. 1997 Jul 4;272(27):17118–25.
85. Goodin S, Kane MP, Rubin EH. Epothilones: mechanism of action and biologic activity. J Clin Oncol. 2004 May 15;22(10):2015–25.
86. Larkin JM, Kaye SB. Epothilones in the treatment of cancer. Expert Opin Investig Drugs. 2006 Jun;15(6):691–702.
87. Simon R, Sauter G. Tissue microarray (TMA) applications: implications for molecular medicine. Expert Rev Mol Med [electronic resource]. 2003 Oct 21;2003:1–12.
88. Agarwal R, Kaye SB. Expression profiling and individualisation of treatment for ovarian cancer. Curr Opin Pharmacol. 2006 Aug;6(4):345–9.
89. Spentzos D, Levine DA, Kolia S, Otu H, Boyd J, Libermann TA, et al. Unique gene expression profile based on pathologic response in epithelial ovarian cancer. J Clin Oncol. 2005 Nov 1;23(31):7911–8.

90. Jazaeri AA, Awtrey CS, Chandramouli GV, Chuang YE, Khan J, Sotiriou C, et al. Gene expression profiles associated with response to chemotherapy in epithelial ovarian cancers. Clin Cancer Res. 2005 Sep 1;11(17):6300–10.
91. Ahmed AA, Brenton JD. Microarrays and breast cancer clinical studies: forgetting what we have not yet learnt. Breast Cancer Res. 2005;7(3):96–9.
92. Simon R, Radmacher MD, Dobbin K, McShane LM. Pitfalls in the use of DNA microarray data for diagnostic and prognostic classification. J Natl Cancer Inst. 2003 Jan 1;95(1):14–8.

Syngeneic Mouse Model of Epithelial Ovarian Cancer: Effects of Nanoparticulate Paclitaxel, Nanotax®

Katherine F. Roby, Fenghui Niu, Roger A. Rajewski, Charles Decedue, Bala Subramaniam, and Paul F. Terranova

1 Ovarian Cancer

In 2006, it has been estimated that 20,180 women in the United States will be diagnosed with ovarian cancer, and 15,310 women will die from the disease (1). Although advances have occurred in treatment strategies, success remains limited (1). The median overall survival for patients with advanced ovarian cancer and receiving the current standards of treatment (surgery and paclitaxel/platinum chemotherapy) is 36–39 months (2). The realities of the overall statistics are sobering. Future success in the treatment of women with ovarian cancer, including those diagnosed with late stage disease, will be dependent on the development of novel approaches to treatment. Until recently, limited progress has been made in the development of new treatments largely because of a lack of laboratory animal model systems of ovarian cancer.

2 Animal Models

Recent work has resulted in the development of several different laboratory models of ovarian cancer. Current rodent models include those of chemical and/or hormonal induction, genetic knockout and transgenics, xenograph, and syngeneic models (3–13). Each model exhibits inherent advantages and disadvantages; however, together these models have provided a means of investigation into multiple aspects of ovarian cancer including initiation, progression and metastasis, and treatment.

The present series of studies will discuss the syngeneic mouse model of ovarian cancer developed in our laboratory several years ago (10) and the use of this model in exploring the efficacy of a novel formulation of paclitaxel.

3 Paclitaxel

Paclitaxel, a naturally occurring diterpenoid isolated from the bark of Pacific yew trees, has exhibited some success in the treatment of several types of cancers including ovarian, breast, and lung cancer. Although several cellular actions have been reported, the most significant effect appears to be binding to the n-terminal region of β-tubulin and formation of depolymerization-resistant microtubules. Paclitaxel prevents the microtubule–kinetichore attachment necessary for chromosome segregation, causing blockage of cell cycling at the G2/M stage. This cell cycle arrest leads to apoptotic death (14).

Paclitaxel is considered to be one of the most promising drugs developed within the last ten years. The current commercially available formulation of paclitaxel is 6 mg ml^{-1} in a 50:50 v/v mixture of Cremophor EL (polyethoxylated caster oil) and dehydrated ethanol. Significant side effects including severe anaphylactic hypersensitivity reactions, hyperlipidaemia, abnormal lipoprotein patterns, aggregation of erythrocytes, and peripheral neuropathy have been observed with the commercial formulation, and have been attributed to the cremophor (15).

Because paclitaxel exhibits significant antitumor properties, the demonstration of a paclitaxel-based therapy without the cremophor-mediated side effects has been the focus of several research groups (16–20). Our studies have focused on the production of nanoparticles of paclitaxel with improved bioavailability. The pharmaceutical industry is increasingly interested in developing technologies for the production of nano/microparticles for drug delivery applications. Such applications require controlled particle size distribution and consistent product quality (crystallinity, purity, morphology). Precipitation with compressed antisolvents (PCA) has been receiving increased attention as a technique to produce particles with such controlled properties (21). An increasing number of drugs processed using PCA technology can be found in many publications (22, 23). In the present series of studies, paclitaxel nanoparticles, termed as Nanotax®, have been produced using a novel technology based on PCA (24, 25). Paclitaxel nanoparticles, ranging in size from 600–800 nm, form stable suspensions in physiological saline and are injected as a suspension. This preparation of Nanotax® is devoid of cremophor and its undesirable side effects. The efficacy of this novel formulation of paclitaxel in a mouse model of ovarian cancer will be described.

4 Intraperitoneal Chemotherapy

Intraperitoneal chemotherapy is an attractive option as many malignancies, including recurrent ovarian cancer, remain confined to the peritoneal cavity. The basic goal of intraperitoneal antineoplastic therapy is to expose the cancer to higher concentrations of drug for longer periods of time than is possible with systemic therapy. Intraperitoneal therapy may be a rational approach against tumors principally confined to the abdominal cavity for most of their natural history, tumors where intraperitoneal spread is the major route of disease progression, and tumors known to be responsive to effective antineoplastic drugs (26). Intraperitoneal therapy is designed to maximize drug delivery to the tumor while sparing the patients many of the systemic toxicities associated with the drug. Agents that have a high level of intrinsic activity against a broad range of tumor types, which are able to diffuse slowly from the peritoneal space, have minimal toxicity when administered into the pleural space, and whose plasma clearance rates substantially exceed their rates of uptake from the peritoneal cavity are especially suited for intraperitoneal administration (26).

Phase I clinical trials have confirmed the safety of intraperitoneal drug delivery and have demonstrated pharmacokinetic advantage (27). Phase II trials using intraperitoneal paclitaxel for ovarian cancer have shown some success (28). Thus far the dose-limiting toxicity is abdominal pain (29), whether this is due to the Cremophor formulation is not clear. Other studies have demonstrated increased efficacy with intraperitoneal vs. intravenous treatment (30, 31). Findings in a recent report (32) were so dramatic that the National Cancer Institute issued a clinical announcement encouraging the use of intraperitoneal chemotherapy (http://ctep.cancer.gov/highlights/ovarian.html). Effects of Nanotax® were further assessed and compared with Taxol® when delivered either intravenously or intraperitoneally.

5 Generation of a Syngeneic Mouse Model of Ovarian Cancer

Our interest in understanding the aspects of ovarian cancer led us to the development of a laboratory mouse model. This model is based on the information provided by other models utilizing human (33) and rat (7, 34) ovarian surface epithelial cells and tumor cell lines (35–37). The hypothesis that multiple passages of ovarian surface epithelial cells in vitro might induce transformation supports the theory of incessant ovulation and the development of ovarian cancer (7, 34, 38, 39). If increased follicular rupture followed by epithelial proliferation to repair the rupture site was a risk factor for cancer, then inducing "repair," or increased cellular proliferation in vitro might also lead to transformation of the epithelial cells (7, 34). It was our mission to establish a model of ovarian cancer in mice with intact immune systems. The ability of ovarian surface epithelial cells to form tumors in normal immune-intact animals would provide a model in which immune interactions in the establishment, progression, and treatment of ovarian cancer could be investigated. In addition, establishment of a mouse model would ultimately allow for the application of gene manipulation strategies.

5.1 Isolation and Culture of MOSEC

Ovarian surface epithelial cells were obtained by gentle trypsinization of mouse ovaries. Mild trypsinization resulted in removal of the surface epithelium without disturbing the underlying stromal tissue. Following isolation, the mouse ovarian surface epithelial cells (MOSEC) were maintained in culture by repeated passaging. Early passage cells exhibited "cobblestone" morphology, typical of epithelial cells and contact inhibition of growth. After more than 20 passages, in vitro cobblestone morphology was no longer apparent, and contact inhibition of growth was lost as evidenced by the growth of multiple layers of cells (Fig. 1) (10).

The initial tumor forming capacity of late passage MOSEC was assessed by injection of the cells into immunocompromised, athymic mice and into syngeneic, immunocompetent mice (C57BL6). In both athymic and immunocompetent mice, subcutaneous injection resulted in relatively slow growth of solid tumors, requiring nearly 4 months. Injection of late passage MOSEC into the peritoneal cavity of both strains of mice resulted in the formation and accumulation of ascitic fluid and the growth of multiple tumor implants over approximately a 90-day period. Within the peritoneal cavity, multiple tumors were present on the omentum, bowel, diaphragm, peritoneal wall, and on the surface of all abdominal organs including the kidneys, pancreas, stomach, and spleen (Fig. 1). Early passage MOSEC, those exhibiting cobblestone morphology and contact inhibition of growth, did not form tumors or ascites fluid following injection into the peritoneal cavity of either athymic or immunocompetent mice.

Late passage MOSEC were cloned by limiting dilution, and ten clonal cell lines were established on the basis of morphology of the cells. The in vivo tumorigenicity of each clonal line was similar; all clones developed intraperitoneal tumors and ascites. One clonal line, ID8 was used for the studies described herein.

The in vivo model is highly reproducible allowing for the assessment of treatment efficacy. ID8 cells are injected intraperitoneally. Macroscopic tumors are present 45 days after cell injection, and the time treatments are initiated. In the absence of treatment, the cancer progresses and mice reach end-stage disease at approximately 90 days. End-stage disease is defined as the accumulation of ascites fluid and the appearance of a scruffy coat.

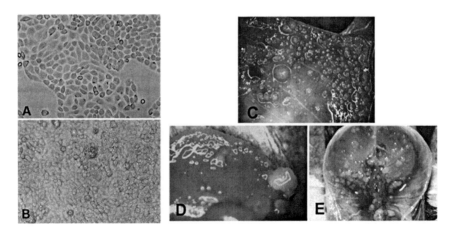

Fig. 1 Growth of MOSEC in vitro and in vivo. Early pass cells prior to (**a**) and after (**b**) spontaneous transformation in vitro. Intraperitoneal injection results in tumor seeds throughout the peritoneal caivity including the body wall (**c**), liver (**d**), and diaphragm (**e**)

5.2 Similarities with Human Ovarian Cancer

The mouse model mimics the progression to late stage ovarian cancer in women. Similar to late stage disease in women, mice produce and accumulate ascites fluid. Tumors are present on multiple tissues within the peritoneal cavity including the diaphragm, intestine, liver, and pancreas. Histologically, the mouse tumors were composed of highly anaplastic malignant appearing cells with a biphasic growth pattern: a carcinomatous component with attempts of glandular formation intermixed with a sarcomatous component composed of fascicles of spindle-shaped anaplastic cells. The cells had large hyperchromatic vesicular nuclei with prominent nucleoli, and occasional tumor giant cells were also noted. At the ultrastructural level, the epithelial nature of the cells was confirmed by the presence of multiple poorly formed desmosomes along the cell surface and by the presence of intracytoplasmic lumina and microvilli. In addition, the cells contained abundant rough endoplasmic reticulum and polyribosomes.

Expression of several genes by the ID8 cells know to be disregulated in ovarian cancer were examined. A few of these are illustrated in Fig. 2 and include the plasminogen activators (tPA and uPA), estrogen receptor alpha (ERα), colony stimulating factor-1 (CSF-1), CSF-1 receptor c-fms, and vascular endothelial cell growth factor (VEGF). Expression of each of these genes has been shown to be altered in ovarian cancer and is thought to play a role in cancer progression (40–43). Therefore, this mouse model of ovarian cancer mimics several aspects of human ovarian cancer including expression of several genes though to be important in cancer metastasis and progression.

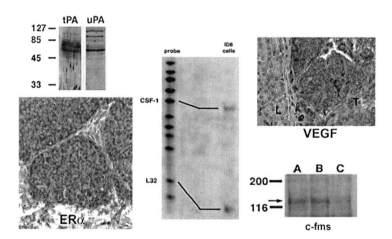

Fig. 2 Expression of several genes by ID8 cells cultured in vitro or in tumors following ID8 cells injection into mice. Expression was assessed by ribonuclease protection, Western blot, and immunohistochemistry

6 Generation of Paclitaxel Nanoparticles, Nanotax®

Clinical problems related with the current formulation of paclitaxel are largely due to the vehicle, i.e., cremophor EL. We have produced a formulation of nanoparticulate paclitaxel consisting of only the paclitaxel and physiological saline. Exclusion of additional compounds from this formulation is anticipated to reduce all potential toxicity related to formulation.

CritiTech Inc. (Lawrence, Kansas, USA) has produced paclitaxel particles by a technique known as precipitation with compressed antisolvent (PCA) (44). Drug dissolved in a suitable organic solvent is sprayed into a flowing stream of supercritical carbon dioxide. The solvent, which is chosen to be one that is fully miscible with supercritical carbon dioxide, is selectively extracted into the CO_2. This causes the drug to precipitate or crystallize out of solution. Figure 3 shows the experimental setup for particle recrystallization from supercritical fluid (SCF).

Drug solution and supercritical CO_2 are mixed in the pressure vessel (crystallizer), where precipitation or crystallization occurs within fractions of a second. The suspension of drug particles and solvent-SCF is swept into the harvesting device (membrane), where the particles are separated from the solvent-SCF mixture in one of two membranes. After the first membrane is filled with particles, the flow is switched to the second while pure CO_2 is pumped through the first to remove all residual solvent. After passing through the membrane, the solvent-CO_2 mixture is depressurized into a condenser where the CO_2 separates as a gas from the solvent. The solvent is recovered for clean up and reuse or for safe disposal. Solvent is not vented to the atmosphere. The CO_2 exiting the condenser may be recompressed to the supercritical state before being fed back into the system. The CritiTech process (*U.S. patent no. 5,874,029; 5,833,891; 6,113,795*) disrupts droplets emerging from the capillary by use of sonic energy resulting in smaller droplets and correspondingly smaller drug particles.

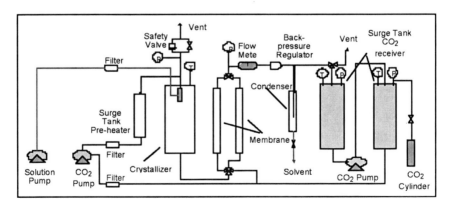

Fig. 3 Experimental apparatus for drug recrystallization with CO_2 recycling

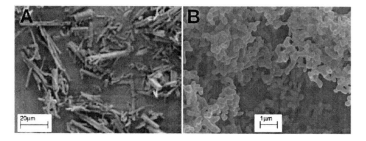

Fig. 4 Unprocessed paclitaxel and Nanotax®. Unprocessed paclitaxel (**a**) is a rectangular crystalline with a broad particle size distribution. Nanotax® (**b**) is a spherical crystalline with a narrow particle size distribution

The particle size of Nanotax® was estimated using an Aerosizer (TSI) particle size analyzer. By number distribution, the mean particle size was 600–700 nm with 95% of all particles measuring smaller than 1 µm. By volume distribution, the mean particle size was 700–1,200 nm with 95% of all particles measuring smaller than 3 µm. Electron micrographic measurement of particle size was consistent with the Aerosizer measurements (Fig. 4).

7 Efficacy of Nanotax® In Vivo

The effects of Nanotax® on survival of mice bearing ovarian cancer were assessed and compared with effects of Taxol®. In addition, the effects of intravenous and intraperioneal delivery were compared. Although the current standard for administration of paclitaxel to patients is intravenous, recent studies reiterate the potential for intraperitoneal therapy and continued exploration of this potential has been encouraged by NCI.

7.1 Effects of Intravenous Delivery

Female C57BL6 mice were injected intraperitoneally with 6×10^6 ID8 mouse ovarian epithelial cancer cells. Forty-five days after tumor cell injection, when macroscopic tumor implants are visible in the peritoneal cavity, treatment was initiated. Mice were injected via the tail vein once every 2 days for a total of three doses with Nanotax® (12 mg kg^{-1}); Taxol® (12 mg kg^{-1}, Bristol-Myers Squibb), cremophor, or saline. Mice were observed daily for signs of toxicity and were sacrificed at "end-stage" disease, when ascites accumulation caused peritoneal swelling and the coat became rough.

Survival of mice treated with Nanotax® or Taxol® intravenously tended to be longer when compared with control treated mice (Fig. 5). Although the effects were

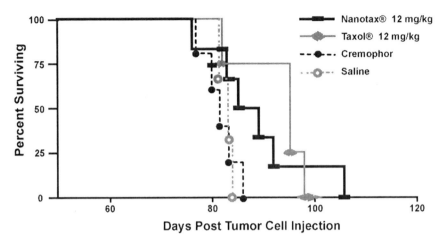

Fig. 5 Effects of intravenous administration of Nanotax® or Taxol® on survival of mice bearing ovarian cancer. Mice were injected intraperitoneally with ID8 cells on day 0 and administered a single intravenous injection of Nanotax® (12 mg kg^{-1}), Taxol® (12 mg kg^{-1}), cremophor, or saline on day 45, 47, and 49. Mice were sacrificed when the cancer progressed to end-stage. Reproduced with permission from (44)

not significant (Kaplan-Meier and Rank Tests), this experiment was carried out three times with a trend toward increased survival in each experiment. It is possible that changing the dose and/or treatment schedule would increase the efficacy of treatment.

7.2 Effects of Intraperitoneal Delivery

Female C57BL6 mice were injected intraperitoneally with 6×10^6 ID8 mouse ovarian epithelial cancer cells. Forty-five days after tumor cell injection, when macroscopic tumor implants are visible in the peritoneal cavity, treatment was initiated. Mice were administered Nanotax® (18, 36, 48 mg kg^{-1}); Taxol® (12, 18, 36 mg kg^{-1}, Bristol-Myers Squibb); cremophor (at the final percent/volume equal to the 36 mg kg^{-1} dose of Taxol®); or saline (at the volume equal to the 48 mg kg^{-1} dose of Nanotax®) intraperitoneally once every 2 days for three doses. Mice were observed daily for signs of toxicity and were sacrificed at "end-stage" disease, when ascites accumulation caused peritoneal swelling and the coat became rough.

Mice administered with Nanotax® survived significantly longer than control and Taxol®-treated mice (Fig. 6); duration of survival was directly related to dose of Nanotax® being administered. In addition, Nanotax® demonstrated reduced toxicity compared with Taxol®. The 36 mg kg^{-1} dose of Taxol® was the ED$_{50}$ (this dose is excluded from Fig. 6). Using the same dosing schedule, mice treated with the 48 mg

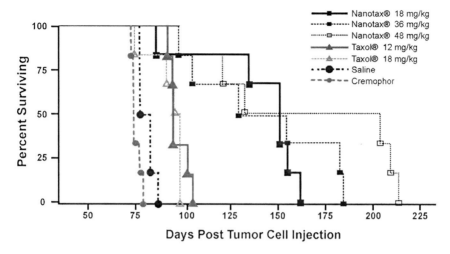

Fig. 6 Effects of intraperitoneal administration of Nanotax® or Taxol® on survival of mice bearing ovarian cancer. Mice were injected intraperitoneally with ID8 cells on day 0 and administered a single intraperitoneal injection of Nanotax®, Taxol®, cremophor, or saline at the indicated dose on day 45, 47, and 49. Mice were killed when the cancer progressed to end-stage. Analysis by Kaplan-Meier and subsequent Rank Tests indicated significant effects of both Nanotax® and taxol®. Reproduced with permission from (44)

kg^{-1} dose of Nanotax® exhibited no signs of toxicity. In addition, mice administered with Taxol® exhibited transient ataxia, while ataxia was not observed with Nanotax® at any dose tested. Together these results indicate Nanotax® is more effective than Taxol® in inhibiting the progression of ovarian cancer and increasing the duration of survival. In addition, Nanotax® exhibited reduced toxicity compared with Taxol®.

7.3 Cancer Progression Following Intraperitoneal Nanotax®

Tumor bearing mice were prepared as already described and treated with Nanotax® (36 mg kg^{-1}) or saline intraperitoneally once every 2 days for three doses, beginning on day 45. A group of mice were killed at the time of treatment to assess tumor progression and all mice were sacrificed when the control-treated mice reached end-stage disease. Tumor progression in each group was compared.

Figures 7 and 8 illustrate the state of tumor progression in control and Nanotax®-treated mice. Nanotax® treatment resulted in significantly reduced tumor burden as evidenced by comparison of panals a and e in Fig. 8. Tumor burden in Nanotax®-treated mice at the time of sacrifice was less compared with the tumor burden at the time of treatment indicating that the treatment resulted in tumor cell killing. In addition, the accumulation of ascites fluids was significantly reduced in Nanotax®-treated mice (Fig. 7). This series of experiments indicate Nanotax® has a significant effect in reducing the progression of ovarian cancer in this model system.

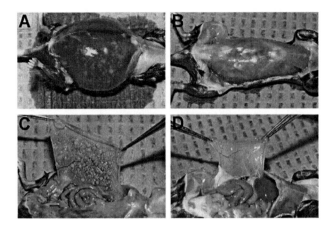

Fig. 7 The state of tumor progression in control and Nanotax®-treated mice at the time controls were killed due to progression to end-stage. Control mice (**a** and **c**) Nanotax®-treated mice (**b** and **d**). At end-stage control mice exhibited accumulation of blood ascites fluids in the peritoneal cavity. The presence of tumors on the inside of the peritoneal wall could be seen from the exterior as white spots (**a**). Upon reflection of the peritoneal wall multiple tumor implants could be visualized (**c**). At the same time, mice treated with Nanotax® had no accumulation of ascites fluid (**b**). Overall tumor load was significantly reduced. Upon reflection of the peritoneal wall only small tumor seeds could be visualized (**d**)

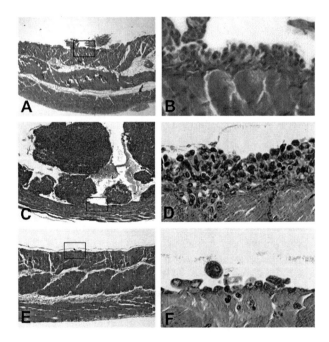

Fig. 8 Tumor load on the peritoneal wall of mice after ID8 cell injection. Representative tumor present: 45 days after cell injection at the time of treatment (**a** and **b**), in control, treated mice at the time of sacrifice due to progression to end-stage disease (**c** and **d**), and Nanotax®-treated mice sacrificed at the time the control mice were sacrificed due to progression to end-stage. The *box* in panels **a, c**, and **e** indicate the region shown at higher magnification in panels **b, d**, and **f**

It is of interest to note that following ID8 cell injection tumors progress to large masses throughout the peritoneal cavity as illustrated in Fig. 8c. However, between these large tumors, on the mesothelial surface, a lawn on cancer cells several layers deep exists, as illustrated in Fig. 8d. This model appears to illustrate one of the difficulties faced in treating women with ovarian cancer. Even with extensive cytoreductive surgery, it is quite likely that cancer cells layering on the mesothelium will remain and ultimately repopulate the peritoneum. The use of intraperitoneal therapies will likely enhance the ability to kill these remaining cancer cells and ultimately improve therapeutic outcomes.

8 Conclusion

Nanotax®, a nanoparticulate formulation of paclitaxel free of toxic diluents, exhibits improved efficacy compared with the formulation currently available. Higher doses were administered without the side effects observed with the cremophor-based formulation. In addition, Nanotax® exhibited greater effects in reducing tumor burden. The present studies further demonstrated improved outcomes with intraperitoneal administration of drug compared with intravenous delivery. These studies demonstrate the utility of the syngeneic mouse model for testing novel treatments for ovarian cancer.

References

1. Jemal, A., Siegel, R., Ward, E., Murray, T., Xu, J., Smigal, C. and Thun, M. (2006). "Cancer statistics, 2006." CA Cancer J Clin 56(2): 106–130.
2. Hogberg, T., Glimelius, B. and Nygren, P. (2001). "A systematic overview of chemotherapy effects in ovarian cancer." Acta Oncol 40(2–3): 340–360.
3. Connolly, D. C., Bao, R., Nikitin, A. Y., Stephens, K. C., Poole, T. W., Hua, X., Harris, S. S., Vanderhyden, B. C. and Hamilton, T. C. (2003). "Female mice chimeric for expression of the simian virus 40 TAg under control of the MISIIR promoter develop epithelial ovarian cancer." Cancer Res 63(6): 1389–1397.
4. Yuspa, S. H. and Poirier, M. C. (1988). "Chemical carcinogenesis: from animal models to molecular models in one decade." Adv Cancer Res 50: 25–70.
5. Tunca, J. C., Erturk, E., Erturk, E. and Bryan, G. T. (1985). "Chemical induction of ovarian tumors in rats." Gynecol Oncol 21(1): 54–64.
6. Massazza, G., Tomasoni, A., Lucchini, V., Allavena, P., Erba, E., Colombo, N., Mantovani, A., D'Incalci, M., Mangioni, C. and Giavazzi, R. (1989). "Intraperitoneal and subcutaneous xenografts of human ovarian carcinoma in nude mice and their potential in experimental therapy." Int J Cancer 44(3): 494–500.
7. Testa, J., Getts, L., Salazar, H., Liu, Z., Handel, L., Godwin, A. and Hamilton, T. (1994). "Spontaneous transformation of rat ovarian surface epithelial cells results in well to poorly differentiated tumors with a parallel range of cytogenetic complexity." Cancer Res 54: 2778–2784.
8. Molpus, K. L., Kato, D., Hamblin, M. R., Lilge, L., Bamberg, M. and Hasan, T. (1996). "Intraperitoneal photodynamic therapy of human epithelial ovarian carcinomatosis in a xenograft murine model." Cancer Res 56(6): 1075–1082.

9. Silva, E. G., Tornos, C., Deavers, M., Kaisman, K., Gray, K. and Gershenson, D. (1998). "Induction of epithelial neoplasms in the ovaries of guinea pigs by estrogenic stimulation." Gynecol Oncol 71(2): 240–246.
10. Roby, K. F., Taylor, C. C., Sweetwood, J. P., Cheng, Y., Pace, J. L., Tawfik, O., Persons, D. L., Smith, P. G. and Terranova, P. F. (2000). "Development of a syngeniec mouse model for events related to ovarian cancer." Carcinogenesis 21(4): 585–591.
11. Selvakumaran, M., Bao, R., Crijns, A. P., Connolly, D. C., Weinstein, J. K. and Hamilton, T. C. (2001). "Ovarian epithelial cell lineage-specific gene expression using the promoter of a retrovirus-like element." Cancer Res 61(4): 1291–1295.
12. Orsulic, S., Li, Y., Soslow, R. A., Vitale-Cross, L. A., Gutkind, J. S. and Varmus, H. E. (2002). "Induction of ovarian cancer by defined multiple genetic changes in a mouse model system." Cancer Cell 1(1): 53–62.
13. Pieretti-Vanmarcke, R., Donahoe, P. K., Szotek, P., Manganaro, T., Lorenzen, M. K., Lorenzen, J., Connolly, D. C., Halpern, E. F. and MacLaughlin, D. T. (2006). "Recombinant human Mullerian inhibiting substance inhibits long-term growth of MIS type II receptor-directed transgenic mouse ovarian cancers in vivo." Clin Cancer Res 12(5): 1593–1598.
14. Spencer, C. M. and Faulds, D. (1994). "Paclitaxel. A review of its pharmacodynamic and pharmacokinetic properties and therapeutic potential in the treatment of cancer." Drugs 48(5): 794–784.
15. Gelderblom, H., Verweij, J., Nooter, K. and Sparreboom, A. (2001). "Cremophor EL: the drawbacks and advantages of vehicle selection for drug formulation." Eur J Cancer 37(13): 1590–1598.
16. Walter, K. A., Cahan, M. A., Gur, A., Tyler, B., Hilton, J., Colvin, O. M., Burger, P. C., Domb, A. and Brem, H. (1994). "Interstitial taxol delivered from a biodegradable polymer implant against experimental malignant glioma." Cancer Res 54(8): 2207–2212.
17. Sharma, A., Sharma, U. S. and Straubinger, R. M. (1996). "Paclitaxel-liposomes for intracavitary therapy of intraperitoneal P388 leukemia." Cancer Lett 107(2): 265–272.
18. Harper, E., Dang, W., Lapidus, R. G. and Garver, R. I. J. (1999). "Enhanced efficacy of a novel controlled release paclitaxel formulation (PACLIMER delivery system) for local-regional therapy of lung cancer tumor nodules in mice." Clin Cancer Res 5(12): 4242–4248.
19. Leung, S. Y., Jackson, J., Miyake, H., Burt, H. and Gleave, M. E. (2000). "Polymeric micellar paclitaxel phosphorylates Bcl-2 and induces apoptotic regression of androgen-independent LNCaP prostate tumors." Prostate 44(2): 156–163.
20. Meerum Terwogt, J. M., ten Bokkel Huinink, W. W., Schellens, J. H., Schot, M., Mandjes, I. A., Zurlo, M. G., Rocchetti, M., Rosing, H., Koopman, F. J. and Beijnen, J. H. (2001). "Phase I clinical and pharmacokinetic study of PNU166945, a novel water-soluble polymer-conjugated prodrug of paclitaxel." Anticancer Drugs 12(4): 315–323.
21. Subramaniam, B., Rajewski, R. and Snavely, K. (1997). "Pharmaceutical processing with supercritical carbon dioxide." J Pharm Sci 86: 885–890.
22. Muhrer, G. and Mazzotti, M. (2003). "Precipitation of lysozyme nanoparticles from dimethyl sulfoxide using carbon dioxide as antisolvent." Biotechnol Prog 19(2): 549–556.
23. Perrut, M. (2003). "Supercritical fluids applications in the pharmaceutical industry." STP Pharma Sci 13(2): 83–91.
24. Subramaniam, B., Saim, S., Rajewski, R. and Stella, V. J. (Issued Nov. 10, 1998). Methods and apparatus for particle precipitation and coating using near-critical and supercritical antisolvents. US Patent 5,833,891.
25. Subramaniam, B., Bochniak, D. J. and Rajewski, R. (Issued Sept. 5, 2000). Methods for continuous particle precipitation and harvesting. U. S. Patent 6,113,795.
26. Markman, M., Bundy, B. N., Alberts, D. S., Fowler, J. M., Clark-Pearson, D. L., Carson, L. F., Wadler, S. and Sickel, J. (2001). "Intraperitoneal drug delivery of antineoplastics." Drugs 61(8): 1057–1065.
27. Francis, P., Rowinsky, E., Schneider, J., Hakes, T., Hoskins, W. and Markman, M. (1995). "Phase I feasibility and pharmacologic study of weekly intraperitoneal paclitaxel: a Gynecologic Oncology Group pilot Study." J Clin Oncol 13(12): 2961–2967.

28. Markman, M., Brady, M. F., Spirtos, N. M., Hanjani, P. and Rubin, S. C. (1998). "Phase II trial of intraperitoneal paclitaxel in carcinoma of the ovary, tube, and peritoneum: a Gynecologic Oncology Group Study." J Clin Oncol 16(8): 2620–2624.
29. Markman, M., Rowinsky, E., Hakes, T., Reichman, B., Jones, W., Lewis, J. L. J., Rubin, S., Curtin, J., Barakat, R. and Phillips, M. e. a. (1992). "Phase I trial of intraperitoneal taxol: a Gynecoloic Oncology Group study." J Clin Oncol 190(9): 1485–1491.
30. Alberts, D. S., Liu, P. Y., Hannigan, E. V., O'Toole, R., Williams, S. D., Young, J. A., Franklin, E. W., Clarke-Pearson, D. L., Malviya, V. K. and DuBeshter, B. (1996). "Intraperitoneal cisplatin plus intravenous cyclophosphamide versus intravenous cisplatin plus for stage III ovarian cancer." N Engl J Med 335(26): 1950–1955.
31. Vermorken, J. B. (2000). "The role of intraperitoneal chemotherapy in epithelial ovariancancer." Int J Gynecol Cancer 10(S1): 26–32.
32. Armstrong, D. K., Bundy, B., Wenzel, L., Huang, H. Q., Baergen, R., Lele, S., Copeland, L. J., Walker, J. L. and Burger, R. A. (2006). "Intraperitoneal cisplatin and paclitaxel in ovarian cancer." N Engl J Med 354(1): 34–43.
33. Dyck, H., Hamilton, T., Godwin, A., Lynch, H., Maines-Bandiera, S. and Auersperg, N. (1996). "Autonomy of the epithelium phenotype in human ovarian surface epithelium: changes with neoplastic progression and with a family history of ovarian cancer." Int J Cancer 69: 429–436.
34. Godwin, A., Testa, J., Handel, L., Liu, Z., Vanderveer, L., Tracey, P. and Hamilton, T. (1992). "Spontaneous transformation of rat ovarian surface epithelial cells: association with cytogenetic changes and implications of repeated ovulation in the etiology of ovarian cancer." J Natl Cancer Inst 84: 592–601.
35. Chen, X., Pine, P., Knapp, A., Tuse, D. and Laderoute, K. (1998). "Oncocidin A1: a novel tubulin-binding drug with antitumor activity against human breast and ovarian carcinoma xenographs in nude mice." Biochem Pharmacol 56(5): 623–633.
36. Mesiano, S., Ferrara, N. and Jaffe, R. (1998). "Role of vascular endothelial growth factor in ovarian cancer: inhibition of ascites formation by immunoneutralization." Am J Pathol 153(4): 1249–1256.
37. Stackhouse, M., Buchsbaum, D., Grizzle, W., Bright, S., Olsen, C., Kancharla, S., Mayo, M. and Curiel, D. (1998). "Radiosensitization mediated by a transfected anti-erbB-2 single-chain antibody in vitro and in vivo." Int J Radiat Oncol Biol Phys 42(4): 817–822.
38. Fathalla, M. (1972). "Factors in the causation and incidence of ovarian cancer." Obstet Gynecol Surv 27: 751–768.
39. Parker, S., Tong, T., Bolden, S. and Wingo, P. (1996). "Cancer statistics 1996." CA Cancer J Clin 46: 5–27.
40. Kacinski, B. M. (1995). "CSF-1 and its receptor on ovarian, endometrial and breast cancer." Ann Med 27: 79–85.
41. Stack, M. S., Ellerbroek, S. M. and Fishman, D. A. (1998). "The role of proteolytic enzymes in the pathology of epithelial ovarian carcinoma." Int J Oncol 12(3): 569–576.
42. Bamberger, E. S. and Perrett, C. W. (2002). "Angiogenesis in epithelial ovarian cancer." Mol Pathol 2002 55(6): 348–359.
43. Cunat, S., Hoffmann, P. and Pujol, P. (2004). "Estrogens and epithelial ovarian cancer." Gynecol Oncol 94(1): 25–32.
44. Niu, F., Roby, K. F., Rajewski, R. A., Decedue, C. and Subramaniam, B. (2006). Paclitaxel Nanoparticles: Productrion using Compressed CO_2 as Antisolvent, Characterization and Animal Model Studies. Polymeric Drug Delivery Volume II - Polymeric Matrices and Drug Particle Engineering. Svenson, S. Washington, DC, American Chemical Society. 924: 262–277.

Individualized Molecular Medicine: Linking Functional Proteomics to Select Therapeutics Targeting the PI3K Pathway for Specific Patients

Mandi M. Murph, Debra L. Smith, Bryan Hennessy, Yiling Lu, Corwin Joy, Kevin R. Coombes, and Gordon B. Mills

1 Introduction

In 1970s, President Nixon declared a war on cancer and signed the National Cancer Act with an expectation of finding a cure. For at least a subset of cancers such as childhood leukemias, we have made remarkable progress to the point where survival is an expectation rather than a rarity. Although we do not have one panacea for all human cancers, 30 years later, we understand that each type of cancer and potentially each person's cancer is different, and it will take the development of rationale combination therapies to cure all cancers. To reach Nixon's idealistic goal, medical oncology care will need to become individualized. Specifically targeted drugs continue to receive FDA approval and guide treatment selection on the basis of the underlying dysfunctional genetic, transcriptional, or protein regulation driving their tumor progression.

For individual treatment selection to become a reality, it is critical to have methods in place to determine where, exactly, the genetic defect in each tumor is and to monitor the response to treatment of the patient's tumor and of the protein product of the genetic defect. Thus, oncology will rely heavily on laboratory analysis of the molecular abnormalities from an individual's biopsy. Repeated biopsies will guide every treatment decision along the way to recovery. Following validation of targets and targeting through biopsies, it may become feasible to move to less invasive approaches such as molecular imaging. Indeed, a major goal is to develop approaches that link the concurrent development and validation of targeted therapeutics, molecular markers, and molecular imaging.

Although the idea of an individually tailored drug regimen has been around for some time and other diseases, such as HIV, are treated in this manner, cancer is such a complex disease that the latest technological developments have only now made it possible to begin unraveling this complexity. This complexity also makes the development of effective personalized therapies both challenging and expensive.

However, personalized therapies are already being used in particular cancers with targeted therapeutics such as gefitinib and imatinib mesylate for tumors that have mutations in the drug targets.

Breast cancer serves as the "poster child" for the implementation and validation of individualized treatment on the basis of biological and genetic abnormalities or molecular markers. After the diagnosis of breast cancer, the expression of estrogen, progesterone, and ErbB2 receptors is measured in all patients to determine the most appropriate course of treatment, either based on hormonal therapy or trastuzumab, combined with other approaches on the basis of the molecular diagnostics results. Here, molecular diagnostics allows the presence of the target to drive the selection of patients who are likely to benefit from a specific treatment. This example demonstrates how molecular medicine can be extended to all cancer types. However, despite the utility of these approaches in breast cancer, only about 40% of patients with the underlying aberration respond to the targeted therapeutic. The presence of the target is not sufficient to faithfully predict response. Thus, it is necessary to develop additional predictive markers to identify likely responders as well as more effective combination therapies. In contrast, the negative predictive value of molecular diagnostics for hormone receptors and ErbB2 is striking with no or extremely few patients without the marker responding.

Although the availability of targeted treatment options like trastuzumab to selectively inhibit ErbB2 receptors is advantageous, the majority of breast cancer patients do not have an amplification of the ErbB2 receptor. Novel inhibitors that target more common protein aberrations without toxic side effects are desperately needed. However, it is essential to note that even within the small population of ErbB2-overexpressing breast cancers, individualized molecular medicine yields a very high patient benefit.

Breast cancer with its frequent early diagnosis presents an additional opportunity for molecular diagnostics and personalized therapy. For patients with small localized disease, the chance of recurrence after local therapy such as surgery and/or radiation is low. Indeed, there is a consensus that patients with low risk disease do not require additional chemotherapy to their management, particularly because of the short and long-term toxicity of treatment. However, the patient and physician are faced with the conundrum of not knowing in which patient the disease is likely to recur. This results in an overtreatment of patients who could have been cured by surgery alone. Recently molecular marker sets with the potential to identify patients who do not require additional chemotherapy or patients who potentially will not respond to chemotherapy have been identified and undergone initial validation. The Oncotype Dx and Agendia approaches are undergoing large scale evaluation to establish their utility in patient management. These studies point the way but markers with high sensitivity and specificity of predicting outcomes and response to therapy are sorely needed.

For effective implementation of new cancer therapies, the development of pharmaceuticals requires two important informational components: who would benefit the most from treatment and what biomarker can be used to measure the response. A biomarker determines whether the appropriate dose is given or can identify early responders allowing triage of nonresponders to alternative therapies. The use of biomarkers can be critically important in clinical trials by increasing the likelihood of success as well as in decreasing the size cost and duration of clinical trials. In the case of tratuzumab, the appropriate selection of patients for "registration" trials proved crucial. Without screening for ErbB2-positive patients only, the low number of responders (about 9% of unselected patients) would not have been detected based on the sample size likely eliminating tratuzumab from FDA approval. In contrast, using a biomarker to select patients, about 30% responded providing an adequate signal to result in approval. Using this model for success, laboratories in industry, clinical, and academic settings are striving to answer both the who and what questions simultaneously using new high through put technologies like genomics, transcriptional profiling, functional proteomics as well as more conventional candidate gene approaches.

1.1 Functional Proteomics

Cancer is a disease of genetic change either at the level of DNA sequence, copy number, or epigenetic change. However, cellular phenotypes and outcomes to stimulii are regulated by protein levels and protein function. A number of robust technologies have been developed to assess the genomic and transcriptional changes in cells, but these do not necessarily accurately reflect protein function. For example, DNA copy number changes resulting in concordant protein level changes in lesser than 30% of cases. In addition, proteins are extensively posttranslationally regulated. To understand how genomic and transcriptional changes affect cell function, we must assess both protein levels and modifications. Thus, there is a need to develop high-throughput functional proteomics approaches that can be readily applied to patient material to select patients likely to respond, to determine whether a biologically relevant dose is being delivered, and to identify individuals responding at an early stage. Mass spectroscopy and related technologies, including nanotechnologies, hold an incredible promise to be able to comprehensively profile the patient material. However, this field is still in its infancy, requiring further development prior to implementation into patient care. Indeed, the NCI recently funded the Clinical Proteomics Technology Assessment Consortium to determine the requirements for the implementation of proteomics technology into patient management.

On the basis of genomic and transcriptional profiling studies, it appears quite likely that a limited number of molecular markers, potentially 10–100 will contain sufficient informational content to be able to predict clinical outcomes. This suggests that the development of robust, quantitative, moderate throughput proteomics technologies able to deal with a limited spectrum of candidate proteins will likely contribute to patient management. To this end, a number of groups have focused on the development of antibody-based protein arrays as both an interim technology and potentially as an implementable technology for the integration of molecular markers into patient care.

This review will focus on the emerging reverse phase protein lysate array (RPPA) approach, which allows the rapid identification of aberrant signaling pathways from patient samples by measuring both the total amounts of proteins and the extent of modification through processes such as phosphorylation. RPPA is the first method that is truly capable of defining functional proteomics of tumors, using a systems biology approach.

2 RPPA

In essence, RPPA is a quantitative, moderately high-throughput, multiplexed ELISA that can assist in the development of molecular signature databases. Two important informatics tools that could be derived from RPPA analysis are intermediary biomarker assessment and protein circuitry maps of the cell. Obtaining as much information about a patient's disease status prior to treatment using, for example, a database categorizing thousands of molecular signatures will be an enormous step toward individualized molecular medicine. RPPA can help reach this goal.

To perform RPPA, lysate from cell lines or patient samples from microdissection or tumor biopsy are spotted onto a glass slide coated with nitrocellulose. Each sample is represented on the slide in a serial microdilution curve; for best results, dilution series are replicated on spatially distant portions of the array. Multiple controls are built into each assay to ensure quality and detection of the linear range of the increasing protein slopes created from the protein dilutions. In addition to the samples, each slide contains positive controls, quantitative peptide, and phosphopeptide controls. Early studies used a robotics arrayer to spot samples onto slides, allowing approximately 192 spots which includes six serial dilutions and controls to be analyzed simultaneously. Using a new arrayer, we have been able to assess up to 1,000 spots in dilution series per slide.

Each slide is then probed for a protein of interest using an antibody against either total or phosphorylated protein. Next, a secondary antibody is used to amplify the signal intensity, followed by further levels of enzymatic amplification leading to the deposition of tyramide on in each spot. The resulting spots are quantified using imaging analysis programs that were initially developed for mRNA expression microarrays, followed by a custom software program designed to detect changes in protein activation levels. Using a newly developed "SuperCurve" method, a common logistic curve is generated by pooling data from all of the samples on the slide. Then, for each sample, the individual dilution series numbers are mapped onto the SuperCurve. In this way, RPPA yields relative quantification of samples. If a control peptide or cell lysate with known amounts of the target is present on the slide, then the absolute levels can be calculated.

RPPA has a number of benefits over immunohistochemistry, ELISA, or genotyping. Perhaps the four most important features are the quantification, cost, sensitivity, and amount of sample material required. However, importantly, these characteristics allow the simultaneous quantitative assessment of multiple different proteins in a single sample providing important information on the status of a protein giving a more comprehensive picture of the activation status of the tumor. The quantitative nature of the approach converts the normal dichotomous or binned protein levels (1+, 2+, 3+) from other assays into a continuous variable. Although RPPA is highly quantitative, it is complementary with immunohistochemistry, as it does not provide spatial organization. The two techniques together have the potential to provide more information. To make the technique widely available for laboratory use, patient benefit, and health insurers, it is adamant that the cost be reasonable. RPPA meets this requirement because the cost averages out to less than $1 per antibody per sample in terms of reagents costs. Labor probably increases the costs by about twofold. Furthermore, a single antibody is applicable to serum or plasma at 100 pg ml^{-1}. Another major advantage of RPPA is that it is sufficiently sensitive to detect around 5 fg of target protein; thus, the amount of total cellular protein required for a sample is as low as 5 ng. The miniscule amounts of protein needed for RPPA are also applicable for needle biopsies, and this is a major advantage over traditional techniques requiring abundant sample (1).

There are additional reasons to use RPPA for individualized molecular medicine. Recent advances in gene chip array technology are limited to DNA and RNA, and there is no equivalent yet for protein analysis with an emphasis on signaling molecules. RPPA is able to fill in this gap in technical ability to quantify protein levels. It appears that the correlation of DNA copy number to RNA is at best 60%, and the correlation of RNA to protein level is 50% or less. Thus RPPA provides a more accurate estimate of protein levels and function. It may also integrate the functional effects of multiple genomic aberrations into a comprehensive and interpretable phenotype.

Even though RPPA is currently an optimal technique to use for protein and phosphorylation status detection, some limitations with this method need to be considered. To successfully perform RPPA, you need robotics equipment and high-quality, validated antibodies that do not produce background or nonspecific binding. Although manageable, issues can arise with the sample loading from the robotics and correcting for protein loading and tumor stroma ratios, particularly from microdissected samples, can introduce a large amount of variability. We use an average of all antibodies analyzed to correct for loading amounts. This approach works well within certain definable limits. The introduction of a high degree of loading correction can result in over or undercorrection. Another limitation of RPPA, in contrast to tissue microarrays, is that the one-dimensional arrayed samples lose spatial organization from the primary tumor (1). Finally, the stability of proteins and phosphoproteins can be problematic, and the half-lives of individual proteins should be considered. The handling of samples prior to being prepared for arrays can be critical for accurate analysis. We are performing systematic analysis to attempt to develop approaches to manage these challenges.

2.1 Utility of RPPA

After using SuperCurve to quantify a set of RPPA slides from different antibodies, the result is a "protein-by-sample" data matrix of concentration estimates, similar to the data structures generated from mRNA expression array experiments. These data can be used to create a heatmap of cellular protein expression and can be used to generate an image of network signaling. Multiple samples from cancer patients can be compared by this analysis through the generation of hierarchical clustering or by other bioinformatics approaches. In theory, hierarchical clustering can facilitate the classification of each sample into specific cancer subtypes, complementing the current tissue and histologic characterization of samples. Ideally, the prediction of patient outcome can also be interpreted, and this is important to guide the selection of appropriate treatments. Signatures that demonstrate metastatic potential, therapeutic response, or prognosis are ultimately of great benefit for treatment. The approach can also be used with serial biopsies to determine whether therapeutics and in particular targeted therapeutics are being given at an optimal dose resulting in the required level of target knockdown.

From the data generated by RPPA, it is possible to create a network map of cellular protein phosphorylation status. Ideally, having information about the tumor's abnormal protein circuitry before treatment could provide well-chosen drug combinations for each patient. This medical protocol would undoubtedly achieve a better long-term outcome than an one-panacea-for-all rationale. Theoretically, targeted therapeutics affecting multiple points within the dysfunctional cellular signaling network will be the most effective means to treat a complex disease like cancer. Drug resistance, the key challenge in cancer therapeutics, may be alleviated by utilizing appropriate therapeutic combinations, based on how the entire system is communicating and responding to treatment.

Currently, few cancer biomarkers are in use, especially when compared against the overwhelming numbers of different cancer types that must be distinguished, categorized, treated, and monitored for recurrence. Although RPPA is not designed to discover new biomarkers, it can monitor intermediary biomarkers to help determine biologically relevant drug doses. In addition, RPPA could potentially provide information about the therapeutic index, which is the ratio of drug efficacy to drug toxicity, by identifying combinations of signaling events that might contribute to cell death. The approach is ideal for identifying on and off-target activity. Structure function studies could be evaluated by an additional criterion of on and off-target activity, greatly facilitating drug development. Furthermore, drug development strategies would benefit from the creation of a cellular signaling map of the network circuitry, as signaling nodes could be identified. This information could rank proteins and nodes in terms of high or low usefulness as inhibition targets within an aberrant signaling pathway. Thus, there is a tremendous opportunity for RPPA technology and the fulfillment of its capabilities. Because of the overwhelming amount of information generated from RPPA and other biomarker-identification technologies, there is an urgent need to integrate proteomics into clinical trials. This integration could rapidly identify patients that are most likely to respond to specific drugs. It could also give information about the biologically relevant and biologically effective doses of a drug within that tumor subtype, highlight early responders, and identify protein phosphorylation changes that evolve as a result of therapy or drug resistance. Factors involved for treatment in this manner are regulatory approval (CAP/CLIA/FDA) and the collection of small amounts of tissue (i.e., breast cancer biopsy) that retains spatial organization and tissue heterogeneity.

2.2 The PI3K Pathway as Proof of Concept

One of the most commonly mutated pathways involved in the etiology of cancer is the phosphatidylinositol 3-kinase (PI3K) pathway. Although there are hundreds of different types of cancer, each protein in this signaling cascade has documented dysfunctional regulation among the various types. Recent success has been achieved using FDA-approved monoclonal antibodies to inhibit cell-surface receptors that initiate the PI3K cascade. Ligand or mutationally activated tyrosine kinase receptors autophosphorylate intracellular tyrosine residues to mediate PI3K signaling. One of the several mechanisms then activates PI3K via its regulatory subunit, p85, or its catalytic subunit, p110, and depending on the context, may involve GRB2, SOS, and Ras complexed together at the phosphorylated receptor (2). At this point activated PI3K phosphorylates phosphatidylinositol (4,5) bisphosphate (PIP_2) converting it to phosphatidylinositol (3,4,5) triphosphate (PIP_3) and actually initiates further intracellular signaling cascades. PIP_3 recruits and activates phosphatidylinositol-dependent kinase 1 (PDK1), which then phosphorylates Akt, setting off multiple downstream protein signaling cascades. The downstream transcription factors that are activated as a result of Akt affect apoptosis, cell-cycle arrest, insulin signaling and metabolism. Any and all of the aforementioned proteins is a potential therapeutic target.

In previous years, the research for a therapeutic target focused on understanding the underlying changes leading to disease. Research concentrated on genetic mutations on genetic mutations or gene alterations that occurred within an individual, whether the change was through inheritance alone, environmental pressure, or random. Although this may still be a worthy goal to pursue, ultimately for the ability of genes to affect the disease process, protein status is altered. Therefore, it is reasonable to favor a focus on the identification of protein alterations to better understand the outcome of genetic changes and how proteins influence disease. For this reason, RPPA is a powerful technique that will enhance understanding of protein deviations in cancer.

3 PI3K Pathways and Inhibitors

The PI3K and p53 pathways are the most frequent aberrant pathways in cancers. The p53 pathway has thus far been recalcitrant to drug intervention, and it has been proven to be difficult to restore function, using gene delivery methods; therefore, development has shifted toward a more suitable target – the PI3K pathway. In addition to PI3K, this pathway includes three isoforms of Akt, PTEN, TSC1/2, mTOR, p70S6k, and other multiple kinases and phosphatases, making it a target-rich microenvironment for drug development. In addition, mutations within the gene that codes for PI3K subunit p110 often correlate with an upregulation in ErbB2 (3), another major cancer target.

The PI3K pathway is responsible for regulating cell proliferation, cell-cycle progression, survival, motility, metabolism, insulin signaling, and morphology. The downstream forkhead box (FOXO) transcription factors controlling many of these functions are activated by PI3K signaling. During oncogenic transformation dominated by aberrant PI3K-Akt signaling, FOXO undergoes marked proteasomal degradation, eliminating this regulatory protein from the cell leading to tumor development (4).

Bernie Weinstein proposed the theory of oncogenic addiction, which refers to the tumor cell's need for the continued presence of a specific oncogenic action to survive (5, 6). We have assessed whether this process applies to the PI3K pathway. The ovarian cancer cell line DOV13 has no obvious aberration in the PI3K pathway, and has a low level of sensitivity to chemical genomic or genomic manipulation of the PI3K pathway. In contrast, following stable introduction of mutationally acativated PI3K or AKT, DOV13 cells become remarkably more sensitive to manipulation of the PI3K pathway. This provides a strong example of adaptive oncogene addiction. However, this increased sensitivity to maniupulation of the PI3K pathway comes at the expense of the cells acquiring resistance to chemotherapeutic agents such as taxanes. This can be reversed by inhibition of the PI3K pathway.

Collectively, altered levels or functions of the proteins in the PI3K pathway materialize in nearly every type of cancer. Gain of function by missense mutation or amplification in the PI3K p110 subunit appears in certain types of breast, ovarian, hepatic, gastric, head and neck, colon, lung, cervical, and brain cancers. Likewise, mutational activation of the p85 regulatory subunit is present in colon and ovarian cancer, Hodgkin's lymphoma, and lymphatic disorders (7). Functional loss of PTEN can be associated with thyroid, endometrial, melanoma (8), breast, prostate, kidney, and brain cancer (9). Overexpression of Akt may occur in gastric, ovarian, pancreatic, lung cancer, and leukemia (7). Phase II clinical trials for drugs targeting mTOR, a downstream component of the pathway, are underway for renal, metastatic breast, endometrial, and lung cancer along with mantel cell lymphoma and melanoma (10). The prevalence of abnormalities among PI3K pathway members in cancer is the reason this pathway will continue to be a major target for cancer treatment and prevention.

A number of new drugs are currently under development to address this need for PI3K pathway-targeting pharmaceuticals. The current repertoire of FDA approved drugs that directly or indirectly target the pathway include imatinib/Gleevec/STI571, which inhibits BCR/ABL; rapamycin inhibits mTOR; and gefitinib, cetuximab, erlotinib and trastuzumab/Herceptin, all of which target receptor tyrosine kinases. Other potential drug candidates targeting PI3K pathway members are at various stages of development. Notable compounds are the nonpeptide Src inhibitor.

AP22408 (11), farnesyltransferase inhibitors that target Ras (12), IC87114, which selectively inhibits PI3K delta (13), AMN107, which is significantly active against some STI571-resistant BCR-ABL cancers (14), CCI-779, RAD001, and AP23573, which each inhibit mTOR (15), QLT-0267, which inhibits ILK (16, 17), KP372–1, a novel multiple Akt/PDK1/FLT3 kinase inhibitor that leads to apoptosis of acute myelogenous leukemia cells (18), and BM-354825 a dual Src and BCR/ABL inhibitor (19).

It was recently shown that PDK1 might be a superior target for inhibition amongst the PI3K pathway proteins (20). Breast cancer cells that overexpress either Akt1 or PDK1 are resistant to treatment with taxol and doxorubicin, while the PDK1 overexpressing cells were more resistant to gemcitabine than Akt1 overexpressors. Furthermore, loss of functional PDK1 appeared to sensitize breast cancer cells to gemcitabine-induced apoptosis, suggesting an enhanced therapeutic benefit through reduction of PDK1-mediated Akt1 activation (20).

Previous studies by Hu et al. demonstrated that simultaneous targeting of multiple parameters in the PI3K pathway was more effective than inhibiting PI3K alone. Gain-of-function mutations of PI3K induce the resistance of ovarian cancer cells to taxol while simultaneously enhancing the sensitivity to the PI3K inhibitor, LY294002. In vivo treatment with both taxol and LY294002 reduced tumor burden to 80% compared with the control group and significantly beyond either drug alone. The combination of these two drugs together resulted in the only group in this study that did not develop ascites from the ovarian cancer cells (21). A number of different PI3K inhibitors are in development and are expected to enter trials in the near future. One of the critical unresolved questions is whether pan-inhibitors or PI3K isoform specific inhibitors will have adequate therapeutic indices for clinical utility. Further it will be critical to develop methods to identify patients likely to respond to targeted therapeutics against the pathway.

Targeting Akt has proven problematic. Pan Akt catalytic domain inhibitors may have a narrow therapeutic index. The Akt isoforms appear to mediate differential functions with Akt2 promoting motility, invasion, and metastasis with Akt1 not mediating these effects and potentially limiting tumor aggressiveness. However, whether effective isoform inhibitors can be developed remains unclear.

Novel alkyl-lysophospholipid drugs that resemble natural phospholipids like Edelfosine, Miltefosine, Perifosine are inhibitors of the PI3K pathway, and have previously been examined for their anticancer properties (22). Although it may still be too early to conclude their usefulness in cancer treatment, their potential should be noted under the combinatorial cancer therapeutic arsenal.

Edelfosine is a proapoptotic mediator in cancer cells and a synthetic analogue for one of the most abundant lipids in human blood circulation, lysophosphatidylcholine. The molecular effects of edelfosine include lipid raft reorganization, which causes the Fas death receptors to translocate to the cell surface membrane, aggregate, and cap into the rafts (23). Miltefosine has antiprotozoal activity and was approved in India in 2002 as an oral drug for use against cutaneous leishmaniasis to reduce parasite burden (24, 25). In spite of its effect on insulin (26), miltefosine showed efficacy when applied to metastatic lesions topically, and has been approved in Europe for the treatment of cutaneous breast cancer (27).

Perifosine inhibits PI3K pathway activity by preventing Akt plasma membrane localization and has already entered phase II clinical trials in combination with trastuzumab (28). It causes cell-cycle arrest through the induction of p21, regardless of p53 status, which makes it therapeutically viable in tumors that lack functional p53 (29). Perifosine is orally bioavailable, but is not without side effects like nausea, vomiting, and diarrhea in patients. It is uncertain whether the in vivo effects of perifosine are due to the PI3K pathway. If perifosine becomes available for treatment in cancer, it will become important to have pharmacodynamic markers available to monitor response to perifosine therapy.

When studying cancer and determining the best approach to rationally design drugs, it is important to also consider the tumor microenvironment. Tumors reside within a very complex space and their interactions within this space are potential targets for therapy because extracellular growth factors enhance metastasis, invasion, growth, and cell survival. For example, LY294002 blocks the signal transduction pathway of VEGF, which inhibits ascites formation associated with ovarian cancer, and LY294002 also inhibits VEGF-mediated angiogenesis (30). Thus, not only it is the cellular proteins within the cell that are tumorigenic but also it is beneficial to consider extracellular influences driving tumor progression.

One of the historical challenges associated with inhibiting the PI3K pathway is the diversity of possible side effects. For example, insulin metabolism in the liver is regulated by the PI3K p85 subunit and PTEN (31). Consequently, resistance to insulin, harmful increase in blood sugar level, and severe diabetes have been demonstrated in mice lacking Akt2 (32, 33). In addition, the PI3K pathway also regulates brain function, and a reduction of Akt1 protein levels was identified in the brains of those with schizophrenia (34). It is clear from these and other examples that cellular proteins perform normal regulatory functions, irrespective of aberrant functions in cancer. Inhibiting the normal function can sometimes contribute to unpleasant or life-threatening side-effects, but these must be weighed within the context of a life-threatening disease.

Because increased PI3K pathway signaling is oftentimes seen during tumor progression, there must exist a normal level of PI3K in an individual's cell that is maintained for normal homeostasis, which does not contribute to diabetes or cancer. This homeostasis could result in unexpected effects of targeted therapeutics. For example, both mTOR and AKT catalytic domain inhibitors result in marked increases in AKT phosphorylation likely because of the activation of potent feedback loops. This in turn could lead to unexpected consequences of inhibitors of the pathway. The RPPA technology described earlier could provide the network and pathway information needed to identify the components of these homeostatic regulatory loops.

4 Conclusion

In this chapter, we have emphasized the importance of individualized medicine for molecular oncology. RPPA is a powerful tool for detecting protein abnormalities within biopsy samples and creating protein circuitry maps. This technology can also create and categorize large database samples of cancer patients, which when combined with information on treatments and outcomes will guide future treatment decisions for similar patients. Response to current therapeutic regimens can additionally be monitored using RPPA. Our overall prediction is that this type of individualized treatment will improve patient outcomes and carries an extremely high benefit to small populations of uncommon cancers. Although we have not yet achieved President Nixon's goal of a cure to cancer, we are certainly on track for discovering novel pharmaceuticals, biomarkers, and detection methods that will bring us closer to accomplishing this objective.

Acknowledgments This work was supported by a training fellowship from the Keck Center Pharmacoinformatics Training Program of the Gulf Coast Consortia, National Institutes of Health Grant No.1 T90 070109-01

References

1. Tibes R, Qiu,Y, Lu Y, Hennessy B, Andreeff M, Mills GB, Kornblau SM (2006) Reverse phase protein array: validation of a novel proteomic technology and utility for analysis of primary leukemia specimens and hematopoietic stem cells. Mol Cancer Ther 5: 2512–21.
2. Cully M, You H, Levine AJ, Mak TW (2006) Beyond PTEN mutations: the PI3K pathway as an integrator of multiple inputs during tumorigenesis. Nat Rev Cancer 6: 184–92.
3. Saal LH, Holm K, Maurer M, Memeo L, Su T, Wang X, Yu JS, Malmstrom PO, Mansukhani M, Enoksson J, Hibshoosh H, Borg A, Parsons R (2005) PIK3CA mutations correlate with hormone receptors, node metastasis, and ERBB2, and are mutually exclusive with PTEN loss in human breast carcinoma. Cancer Res 65: 2554–9.
4. Aoki M, Jiang H, Vogt PK (2004) Proteasomal degradation of the FoxO1 transcriptional regulator in cells transformed by the P3k and Akt oncoproteins. Proc Natl Acad Sci USA 101: 13613–7.
5. Weinstein IB (2000) Disorders in cell circuitry during multistage carcinogenesis: the role of homeostasis. Carcinogenesis 21: 857–64.
6. Weinstein IB, Begemann M, Zhou P, Han EK, Sgambato A, Doki Y, Arber N, Ciaparrone M, Yamamoto H (1997) Disorders in cell circuitry associated with multistage carcinogenesis: exploitable targets for cancer prevention and therapy. Clin Cancer Res 3: 2696–702.
7. Bader AG, Kang S, Zhao L, Vogt PK (2005) Oncogenic PI3K deregulates transcription and translation. Nat Rev Cancer 5: 921–9.
8. Wu H, Goel V, Haluska FG (2003) PTEN signaling pathways in melanoma. Oncogene 22: 3113–22.
9. Steck PA, Pershouse MA, Jasser SA, Yung WK, Lin H, Ligon AH, Langford LA, Baumgard ML, Hattier T, Davis T, Frye C, Hu R, Swedlund B, Teng DH, Tavtigian SV (1997) Identification of a candidate tumour suppressor gene, *MMAC1*, at chromosome 10q23.3 that is mutated in multiple advanced cancers. Nat Genet 15: 356–62.
10. Dancey JE (2006) Therapeutic targets: MTOR and related pathways. Cancer Biol Ther 5: 1065–73.
11. Shakespeare W, Yang M, Bohacek R, Cerasoli F, Stebbins K, Sundaramoorthi R, Azimioara M, Vu C, Pradeepan S, Metcalf C, 3rd, Haraldson C, Merry T, Dalgarno D, Narula S, Hatada M, Lu X, van Schravendijk MR, Adams S, Violette S, Smith J, Guan W, Bartlett C, Herson J, Iuliucci J, Weigele M, Sawyer T (2000) Structure-based design of an osteoclast-selective, nonpeptide src homology 2 inhibitor with in vivo antiresorptive activity. Proc Natl Acad Sci USA 97: 9373–8.
12. Gotlib J (2005) Farnesyltransferase inhibitor therapy in acute myelogenous leukemia. Curr Hematol Rep 4: 77–84.

13. Puri KD, Doggett TA, Douangpanya J, Hou Y, Tino WT, Wilson T, Graf T, Clayton E, Turner M, Hayflick JS, Diacovo TG (2004) Mechanisms and implications of phosphoinositide 3-kinase delta in promoting neutrophil trafficking into inflamed tissue. Blood 103: 3448–56.
14. Weisberg E, Manley P, Mestan J, Cowan-Jacob S, Ray A, Griffin JD (2006) AMN107 (nilotinib): a novel and selective inhibitor of BCR-ABL. Br J Cancer 94: 1765–9.
15. Dancey JE (2005) Inhibitors of the mammalian target of rapamycin. Expert Opin Investig Drugs 14: 313–28.
16. Oloumi A, Syam S, Dedhar S (2006) Modulation of Wnt3a-mediated nuclear beta-catenin accumulation and activation by integrin-linked kinase in mammalian cells. Oncogene 25: 7747–57.
17. Troussard AA, McDonald PC, Wederell ED, Mawji NM, Filipenko NR, Gelmon KA, Kucab JE, Dunn SE, Emerman JT, Bally MB, Dedhar S (2006) Preferential dependence of breast cancer cells versus normal cells on integrin-linked kinase for protein kinase B/Akt activation and cell survival. Cancer Res 66: 393–403.
18. Zeng Z, Samudio IJ, Zhang W, Estrov Z, Pelicano H, Harris D, Frolova O, Hail N, Chen W, Kornblau SM, Huang P, Lu Y, Mills GB, Andreeff M, Konopleva M (2006) Simultaneous inhibition of PDK1/AKT and Fms-like tyrosine kinase 3 signaling by a small molecule KP372-1 induces mitochondrial dysfunction and apoptosis in acute myelogenous leukemia. Cancer Res 66: 3737–46.
19. Lombardo LJ, Lee FY, Chen P, Norris D, Barrish JC, Behnia K, Castaneda S, Cornelius LA, Das J, Doweyko AM, Fairchild C, Hunt JT, Inigo I, Johnston K, Kamath A, Kan D, Klei H, Marathe P, Pang S, Peterson R, Pitt S, Schieven GL, Schmidt RJ, Tokarski J, Wen ML, Wityak J, Borzilleri RM (2004) Discovery of N-(2-chloro-6-methyl-phenyl)-2-(6-(4-(2-hydroxyethyl)-piperazin-1-yl)-2-methylpyrimidin-4-ylamino)thiazole-5-carboxamide (BMS-354825), a dual Src/Abl kinase inhibitor with potent antitumor activity in preclinical assays. J Med Chem 47: 6658–61.
20. Liang K, Lu Y, Li X, Zeng X, Glazer RI, Mills GB, Fan Z (2006) Differential roles of phosphoinositide-dependent protein kinase-1 and akt1 expression and phosphorylation in breast cancer cell resistance to Paclitaxel, Doxorubicin, and gemcitabine. Mol Pharmacol 70: 1045–52.
21. Hu L, Hofmann J, Lu Y, Mills GB, Jaffe RB (2002) Inhibition of phosphatidylinositol 3′-kinase increases efficacy of paclitaxel in in vitro and in vivo ovarian cancer models. Cancer Res 62: 1087–92.
22. Ruiter GA, Zerp SF, Bartelink H, van Blitterswijk WJ, Verheij M (2003) Anti-cancer alkyl-lysophospholipids inhibit the phosphatidylinositol 3-kinase-Akt/PKB survival pathway. Anticancer Drugs 14: 167–73.
23. Gajate C, Mollinedo F (2001) The antitumor ether lipid ET-18-OCH(3) induces apoptosis through translocation and capping of Fas/CD95 into membrane rafts in human leukemic cells. Blood 98: 3860–3.
24. Croft SL, Seifert K, Duchene M (2003) Antiprotozoal activities of phospholipid analogues. Mol Biochem Parasitol 126: 165–72.
25. Schmidt-Ott R, Klenner T, Overath P, Aebischer T (1999) Topical treatment with hexadecylphosphocholine (Miltex) efficiently reduces parasite burden in experimental cutaneous leishmaniasis. Trans R Soc Trop Med Hyg 93: 85–90.
26. Verma NK, Dey CS (2006) The anti-leishmanial drug miltefosine causes insulin resistance in skeletal muscle cells in vitro. Diabetologia 49: 1656–60.
27. Terwogt JM, Mandjes IA, Sindermann H, Beijnen JH, ten Bokkel Huinink WW (1999) Phase II trial of topically applied miltefosine solution in patients with skin-metastasized breast cancer. Br J Cancer 79: 1158–61.
28. Kondapaka SB, Singh SS, Dasmahapatra GP, Sausville EA, Roy KK (2003) Perifosine, a novel alkylphospholipid, inhibits protein kinase B activation. Mol Cancer Ther 2: 1093–103.
29. Patel V, Lahusen T, Sy T, Sausville EA, Gutkind JS, Senderowicz AM (2002) Perifosine, a novel alkylphospholipid, induces p21(WAF1) expression in squamous carcinoma cells

through a p53-independent pathway, leading to loss in cyclin-dependent kinase activity and cell cycle arrest. Cancer Res 62: 1401–9
30. Hu L, Zaloudek C, Mills GB, Gray J, Jaffe RB (2000) In vivo and in vitro ovarian carcinoma growth inhibition by a phosphatidylinositol 3-kinase inhibitor (LY294002). Clin Cancer Res 6: 880–6.
31. Taniguchi CM, Tran TT, Kondo T, Luo J, Ueki K, Cantley LC, Kahn CR (2006) Phosphoinositide 3-kinase regulatory subunit p85alpha suppresses insulin action via positive regulation of PTEN. Proc Natl Acad Sci USA 103: 12093–7.
32. Cho H, Mu J, Kim JK, Thorvaldsen JL, Chu Q, Crenshaw EB, 3rd, Kaestner KH, Bartolomei MS, Shulman GI, Birnbaum MJ (2001) Insulin resistance and a diabetes mellitus-like syndrome in mice lacking the protein kinase Akt2 (PKB beta). Science 292: 1728–31.
33. Garofalo RS, Orena SJ, Rafidi K, Torchia AJ, Stock JL, Hildebrandt AL, Coskran T, Black SC, Brees DJ, Wicks JR, McNeish JD, Coleman KG (2003) Severe diabetes, age-dependent loss of adipose tissue, and mild growth deficiency in mice lacking Akt2/PKB beta. J Clin Invest 112: 197–208.
34. Emamian ES, Hall D, Birnbaum MJ, Karayiorgou M, Gogos JA (2004) Convergent evidence for impaired AKT1-GSK3beta signaling in schizophrenia. Nat Genet 36: 131–7.

Defective Apoptosis Underlies Chemoresistance in Ovarian Cancer

Karen M. Hajra, Lijun Tan, and J. Rebecca Liu

1 Introduction

1.1 Ovarian Cancer and the Development of Chemoresistance

Ovarian cancer is the second most common and the most lethal of malignancies arising in the female reproductive system. In 2006, over 20,000 new cases of ovarian cancer will be diagnosed, with the majority of these being advanced disease (stage III or stage IV). Survival varies by age, with overall 1-year and 5-year survival rates for new ovarian cancer patients of 76% and 45%, respectively (1). Survival rates drop dramatically with increasing stage at the time of diagnosis. The overall poor prognosis of ovarian cancer is largely attributable to both the late stage of diagnosis and the development of chemoresistance that limits treatment for recurrent disease. Although initial responses to chemotherapy are quite good, the majority of patients develop recurrent disease, and over time their tumors become resistant to current treatment modalities (23). Thus, identification and modulation of the mechanisms that underlie chemoresistance is central to improving patient outcomes for ovarian cancer.

The standard adjuvant therapy for ovarian cancer at this time is combination chemotherapy with a platinum-based drug and paclitaxel. The underlying principle of cancer therapy is the selective killing of malignant cells while limiting toxicity to normal cells. Multiple studies have demonstrated that platinum-containing agents, such as cisplatin and carboplatin, kill cancer cells by triggering a pathway of programmed cell death known as the intrinsic pathway of apoptosis (17, 28). Research has identified alterations in the molecules that carry out this cell death cascade in cells that are chemoresistant (17, 28). The specific cellular defects identified in ovarian cancer cells and potential therapeutic approaches to overcome these defects and restore chemosensitivity will be discussed, with specific emphasis on apoptosome defects in ovarian cancers.

1.2 The Apoptotic Cascade

Two major pathways leading to apoptosis have been delineated: the extrinsic or receptor-mediated pathway and the intrinsic or mitochondrial pathway (6). Both pathways involve the activation of a cascade of enzymes called caspases, a family of cysteine proteases that cleave after aspartic acid residues. The extrinsic and intrinsic pathways each have an independent group of "initiator" caspases, and the pathways converge on the same group of "effector" caspases to execute the cell death program.

The extrinsic or receptor-mediated pathway is characterized by the activation of cell surface death receptors following binding of their specific ligand. These death receptors belong to the tumor necrosis factor/nerve growth factor receptor superfamily, and include members such as Fas, TNFR1, TRAMP, and TRAIL receptors (2). Ligand binding to the extracellular domain of the death receptor results in receptor trimerization, with the subsequent recruitment of the adaptor molecule FADD to the death domain on the cytoplasmic face of the receptor. This adaptor molecule recruits initiator procaspase-8 and/or procaspase-10, which then undergo autocatalysis to their active forms. These activated initiator caspases carry out the downstream proteolytic processing of the effector caspases-3, -6, and -7, which execute the cell death program.

The intrinsic or mitochondrial pathway can be executed independent of death receptor signaling, and also results in the activation of effector caspases. Mitochondrial damage results in the leakage of cytochrome c into the cytoplasm. Subsequently, cytochrome c complexes with the cytoplasmic protein Apaf-1, which then oligomerizes and binds to procaspase-9, resulting in the formation of a multimeric complex called the apoptosome. This brings procaspase-9 molecules into proximity with each other, allowing enzymatic self-activation. Caspase-9 is then is able to cleave and activate the downstream effector caspases-3, -6, and -7 (31, 34, 55).

Once activated, the effector caspases degrade vital cellular proteins, leading to cell death. Specific substrates of these caspases include structural proteins such as actin and nuclear lamin, regulatory proteins such as DNA-dependent protein kinase, and inhibitors of deoxyribonuclease (42). The cellular proteolysis carried out by effector caspases results in the biochemical and morphological cellular changes characteristic of apoptosis, including nuclear membrane breakdown, DNA fragmentation, chromatin condensation, and the formation of apoptotic bodies (49).

2 Apoptosis Defects in the Intrinsic Pathway

2.1 Ovarian Cancers Demonstrate Defects Throughout the Intrinsic Pathway

Since the intrinsic, mitochondrial pathway of apoptosis was first described, a number of factors that regulate apoptosome formation and execution of downstream signals have been identified. The regulation of this cascade can be divided into three major steps: inhibition of cytochrome c release, inhibition of apoptosome formation, and modulation of caspase activation (21). Deregulation of these processes has been reported in many tumor types (21), including ovarian cancer (Fig. 1).

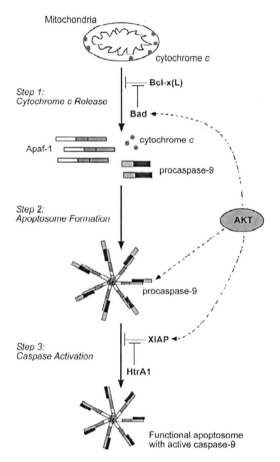

Fig. 1 Apoptosome formation and its deregulation in ovarian cancer. In response to apoptotic stimuli, cytochrome c is released from the mitochondria, allowing Apaf-1 to oligomerize into a heptameric apoptosome. Caspase-9 is then recruited and activated in the apoptosome. This results in a functional apoptosome that cleaves and activates downstream effector caspases to carry out the cell death program. Multiple regulatory steps of this process have been shown to be altered in ovarian cancer, and are shown in *bold* in this figure. Bcl-x(L) overexpression, XIAP overexpression, and HtrA1 downregulation have all been demonstrated in ovarian cancers, and inhibit the formation of an active apoptosome. AKT is also overexpressed in many ovarian cancers, where it phosphorylates and inhibits multiple proteins, including Bad, XIAP, and procaspase-9, resulting in inhibition of this pathway

Mitochondrial cytochrome c release in the initiation of the intrinsic pathway of apoptosis is regulated by the Bcl-2 family of proteins. Bcl-2-related proteins are divided into two groups: the anti-apoptotic Bcl-2 family and the pro-apoptotic Bax and BH3-only proteins (12). In ovarian cancer, the anti-apoptotic Bcl-2 family member Bcl-x(L) is overexpressed, confers resistance to chemotherapy, and is associated with a significantly shorter disease-free interval in patients (51).

Following cytochrome c release, the next step in the intrinsic pathway is the formation of the apoptosome. The primary defects in apoptosome function in ovarian cancer involve alterations in Apaf-1 function, as discussed in detail later. In addition, there are other multiple factors that can inhibit apoptosome formation, including heat shock proteins (4), alterations in physiologic intracellular potassium ion levels (7), and the oncoprotein prothymosin-α (25).

There are a number of proteins that function to regulate the intrinsic pathway of apoptosis through the modulation of caspase activation downstream of apoptosome formation. The inhibitor of apoptosis (IAP) proteins suppress apoptosis by directly inhibiting specific caspases, and include family members XIAP, cIAP1, cIAP2, NIAP, BRUCE, ML-IAP, and Survivin. Overexpression of human IAPs has been shown to suppress apoptosis in response to a number of stimuli, including those which activate the intrinsic pathway of apoptosis (15). There are also proteins that block IAP function, including Smac/DIABLO and the HtrA proteases (21). HtrA1 is downregulated in primary ovarian cancer tumors (11), and in in-vitro studies, the downregulation of HtrA1 causes chemoresistance, and overexpression can promote cisplatin toxicity (23). Finally, caspases can be inhibited by proteins containing caspase-assocaited recruitment domains (CARDs), such as TUCAN, a protein that has been shown to inhibit caspase 9 and is overexpressed in colon cancer and non-small cell lung cancer (39, 9).

The protein kinase AKT plays a central role in tumorigenesis, and exerts multiple effects on the intrinsic pathway of apoptosis. The three human AKT homologs function as serine/threonine kinases with multiple targets that function in cell growth and survival (48). Increased activation of AKT signaling cascades results in the survival of cancer cells that normally undergo apoptosis (48). The importance of AKT in human cancers was first realized with the finding that AKT2 is amplified and overexpressed in ovarian tumors and cell lines (10). Subsequent studies have demonstrated elevated AKT2 kinase activity in approximately 40% of ovarian cancers (48).

Mechanistically, there are multiple effects of AKT activation that lead to inhibition of the intrinsic pathway of apoptosis (Fig. 1). AKT phosphorylates and inactivates the pro-apoptotic Bcl-2 family member Bad to inhibit cytochrome c release from the mitochondria (14). Additionally, AKT phosphorylates procaspase-9 to prevent cleavage and activation (8). Another AKT target is XIAP, with XIAP phosphorylation resulting in inhibition of apoptosis and increased resistance to cisplatin chemotherapy in ovarian cancer cell lines (13). More recent studies have demonstrated that AKT also functions to modulate p53 effects on mitochondrial release of intrinsic pathway factors, and overcoming AKT-mediated cisplatin resistance is dependent on the presence of wild-type p53 (54, 18).

2.2 Apaf-1 Dysfunction in Ovarian Cancer

Defects in apoptosome proteins themselves have been implicated in multiple human malignancies including ovarian cancer. The first report of Apaf-1 inactivation in human cancer was the demonstration that malignant melanomas fail to express this protein (43). The finding that both Apaf-1 alleles are inactivated supports the hypothesis that Apaf-1 functions as a tumor suppressor in malignant melanoma. Apaf-1 loss of function has subsequently been reported in multiple other tumor types including ovarian cancer.

Decreased Apaf-1 function was first reported in ovarian cancer cell lines with the use of a cell-free system in which cell line lysates had decreased cytochrome *c*-dependent caspase activation (52). In these ovarian cancer cell lines, introduction of exogenous Apaf-1 or overexpression of Apaf-1 resulted in restoration of the intrinsic pathway of apoptosis (52). Mutations in Apaf-1 were not identified in the cell lines analyzed. Other studies confirmed dysfunctional apoptosome activation in ovarian cancer cell lines, and demonstrated that primary tumors also have decreased activation of procaspase-9 and downstream effector caspases (32). Further mechanistic insights were gained by the demonstration that in chemoresistant cells, Apaf-1 and procaspase-9 are expressed and the apoptosome is formed, but apoptosome function is impaired (32). Also, expression levels of heat shock proteins or XIAP did not correlate with decreased apoptosome function in the ovarian cancer cell lines or tumor specimens (32). Collectively, these two studies suggest that in ovarian cancer, apoptosome dysfunction leads to decreased activation of the intrinsic pathway of apoptosis.

To further elucidate the specific apoptosome defects in ovarian cancer, studies were carried out in additional cell lines and tumor specimens. A striking correlation was noted in cell lines between apoptosome dysfunction, based on in vitro caspase-9 cleavage and resistance to cisplatin-mediated apoptosis (46). Apaf-1 and caspase-9 were expressed in ovarian cancer cells; however, in response to stimulation with cytochrome *c* and dATP, interaction between these two key apoptosome components was diminished, and caspase-9 cleavage did not occur (46). Through immunodepletion and mixing studies, experiments demonstrated that chemoresistant cell lines have functional caspase-9 and defective Apaf-1. Cell lines with defective apoptosis undergo restoration of the pathway with the introduction of either recombinant Apaf-1 in in vitro assays or overexpression of Apaf-1 in cell culture (46). Additionally, the reintroduction of Apaf-1 results in chemosensitization in ovarian cancer cell lines (46).

In addition to findings in ovarian cancer that Apaf-1 defects contribute to the cancer phenotype by conferring chemoresistance, Apaf-1 has been implicated in tumorigenesis in other human tumors. Apaf-1 deficiency accounts for defective stress-induced apoptosis in human leukemia cell lines and a cervical cancer cell line, with the reintroduction of Apaf-1 into these cells resulting in restoration of apoptosis (24, 27). In a human prostate cancer cell line, an alternatively spliced form of Apaf-1 is expressed and the cells demonstrate impaired apoptosis, but reintroduction of full-length Apaf-1 restores the apoptotic pathway (38). These studies are important in demonstrating a causative role for Apaf-1 defects in therapy resistance, based on the ability of the cell death pathway and drug sensitivity to be restored following reintroduction of Apaf-1.

2.3 Mechanisms Underlying Apaf-1 Dysfunction

Multiple defects in the intrinsic pathway of apoptosis have been identified in ovarian cancer, including apoptosome dysfunction due to Apaf-1 alterations. This prompts the questions of how Apaf-1 is altered and whether novel therapeutic strategies could be undertaken to restore the intrinsic pathway of apoptosis and therefore chemosensitize cells. Conceptually, there are multiple potential mechanisms that could result in loss of Apaf-1 function. At the chromosomal level, potential alterations include gene mutation and allelic loss of heterozygosity. Transcriptional alterations such as decreased transcription from altered transcription factor expression or gene methylation are possible, as well as splicing alterations leading to restricted isoform expression. Once the Apaf-1 protein is synthesized, multiple posttranslational modifications may take place, including phosphorylation and acetylation. Finally, protein–protein interactions with Apaf-1 may impede Apaf-1 function in the apoptosome.

Specific mechanisms underlying alterations in Apaf-1 have been reported in some tumor types. *Apaf-1* gene mutations have been identified in colon and gastric cancers with the microsatellite mutator phenotype (53). In melanoma, Apaf-1 alterations are due to promoter hypermethylation and allelic loss of heterozygosity (19, 43). Loss of heterozygosity has also been reported in colorectal carcinomas (50). Methylation of Apaf-1 has been reported in acute leukemias (20). Additionally, plasma membrane sequestration of the Apaf-1 protein so that it is not available to function in the formation of the apotosome has been reported in human Burkett lymphoma cell lines (44). In contrast, *Apaf-1* gene mutation and hypermethylation have not been observed in ovarian cancers (32, 47, 52). Apaf-1 has multiple isoforms generated by alternative splicing, with some isoforms demonstrating decreased apoptosome function (5), yet restricted isoform expression resulting in altered apoptosis has not been reported in ovarian cancer (32, 52). At the protein level, it does not appear that Apaf-1 is phosphorylated in ovarian or colon cancer cells (8, 52).

Important new insights into the mechanism of Apaf-1 dysfunction in ovarian cancer came with the finding that the histone deacetylase inhibitor trichostatin A (TSA) can restore apoptosome function and induce apoptosis in ovarian cancer cell lines (46). Of note, both chemosensitive and chemoresistant ovarian cancer cell lines undergo apoptosis following TSA treatment, while primary cultures of nontransformed ovarian surface epithelial cells are not sensitive to TSA-induced cell death. Studies using Apaf-1 knock-out mouse embryo fibroblasts demonstrate that TSA-induced apoptotic cell death is dependent on the presence of Apaf-1 (46). In summary, multiple potential mechanisms underlie Apaf-1 defects, and in ovarian cancer the primary mechanism may be one of epigenetic alterations.

3 Histone Deacetylation in Tumorigenesis

The finding that treatment with a histone deacetylase inhibitor such as TSA can induce apoptotic cell death in ovarian cancer cell lines prompts further inquiry into the role of acetylation in regulation of cellular processes. Epigenetic alterations including histone acetylation and DNA methylation have been shown to alter the transcription of oncogenes and tumor suppressor genes, thus playing an important role in tumorigenesis (22). Nuclear DNA is wrapped around an octamer of histones into structures called nucleosomes. Acetylation of lysine residues on the histones alters the chromatin structure to affect the level of gene transcription. Histone acetyltransferases (HATs) and histone deacetylases (HDACs) are the enzymes that regulate the delicate balance of chromatin structure. Histone acetylation leads to an "opening" of the chromatin structure and increased gene transcription. In contrast, deacetylation results in a "closed" conformation of chromatin. Tumorigenesis is characterized by a state of hypo-acetylation (33). Additionally, alterations in HATs and HDACs have been reported in both sporadic and hereditary cancers, further highlighting their role in tumorigenesis (35).

The finding of decreased histone acetylation in tumor cells has led to the hypothesis that histone deacetylase inhibitors may have application in the treatment of human malignancies (16). There are multiple classes of natural and synthetic histone deacetylase inhibitors, all of which function by binding to the catalytic site of the HDAC to inhibit substrate access. Initial studies on the effects of histone deacetylases in transformed cells, both in tissue culture and animal models, have demonstrated that inhibition of HDACs leads to cellular differentiation, growth arrest, and apoptosis (33). Mechanistically, HDAC inhibitors have been shown to induce both apoptotic and autophagic cell death (41). The induction of apoptosis is via the intrinsic pathway, with decreased cell death in mouse embryo fibroblasts null for Apaf-1 (41). This suggests that intact apoptosome function is important for HDAC effects. In addition to the induction of apoptosis, HDACs have been found to inhibit tumor angiogenesis (30).

The growth-inhibitory and cell death-inducing effects of HDAC inhibitors have been observed in both hematologic and solid tumors (35). An important finding is that although increased histone acetylation is observed in both normal and tumor cells, tumor cells are tenfold more sensitive to the HDAC inhibitors (29, 3536). The basis for selective toxicity of HCAC inhibitors in malignant cells when compared with nontransformed cells is unclear. HDAC inhibitors can activate the intrinsic apoptotic pathway in malignant cells through transcriptional activation of proapoptotic proteins including Apaf-1 (40). Indeed, TSA treatment results in transcriptional activation of Apaf-1 in ovarian cancer cells as well; however, TSA treatment has no effect on expression of Apaf-1 in nontransformed ovarian epithelial cells (46). Differences in the acetylome in normal vs. tumor cells may account for differential response to HDAC inhibitors, and Apaf-1 may play an important role in modulating response to treatment.

On the basis of initial studies into the effects of HDAC inhibitors on tumorigenesis, a number of inhibitors are now in clinical trial (37). These are phase I and phase II trials, focused on determining the maximal safe dose of the drug and the efficacy at that dose. Overall, the studies demonstrate low toxicity of HDAC inhibitors in patients, while having a wide therapeutic window (37). Additionally, clinical response has been observed in both in hematologic and solid tumor malignancies (37). Further studies using HDAC inhibitors as both single-agent therapy and in combination with established cancer treatments are underway.

Ovarian cancers are among the solid tumors that demonstrate alterations in histone acetylation patterns (26). Thus, histone deacetylase inhibitors may alter the cellular phenotype in these cancers. In ovarian cancer cell lines, histone deacetylase inhibitors induce activation of caspase-9 and caspase-3, resulting in apoptotic cell death (45,46). Additionally, ovarian cancer tumor growth in nude mice was abrogated, while in mice treated with histone deacetylase inhibitors, no toxic side effects were noted (45). These findings suggest that histone deacetylase inhibitors may be of utility in the treatment of ovarian cancer patients. There are a number of treatment approaches possible, including chemosensitization, combination therapy with standard chemotherapeutic drugs for synergistic effects, and consolidation therapy (3). HDAC inhibitors have been shown in vitro to chemosensitize a number of cancer types, including ovarian cancer cell lines (3).

4 Conclusions

Ovarian cancer treatment is characterized by good initial tumor response to chemotherapeutic agents, but subsequent development of chemoresistance that ultimately limits treatment options. Standard chemotherapeutic regimens trigger the intrinsic pathway of apoptotic cell death. As discussed earlier, ovarian cancers demonstrate defects throughout the intrinsic pathway, from inhibition of cytochrome c release to inhibition of apoptosome formation to modulation of caspase activation. These alterations in the intrinsic pathway lead to defective apoptosis that likely underlies the chemoresistance that is observed in ovarian cancers. Improved treatment options must either restore or circumvent the apoptotic cascade.

Work in our laboratory and that of others suggests that defective Apaf-1 is central to chemoresistance in ovarian cancers. Important insights into how to restore apoptosis in cell lines and primary tumors with alterations in Apaf-1 comes with the finding that treatment with the HDAC inhibitor trichostatin A can restore apoptosome function in ovarian cancer cell lines with resultant apoptotic cell death (46). Apoptotic cell death following treatment with trichostain A is specific to transformed cells, while nontransformed ovarian surface epithelial cells do not undergo apoptosis. The finding that Apaf-1 is necessary for HDAC inhibitor-induced cell death in other studies using mouse embryo fibroblasts null for Apaf-1 (41), and suggests that this protein may be critical for cellular apoptosis triggered by histone deacetylase inhibitors. Continued studies are underway to determine the precise molecular mechanism by which apoptosis is restored in ovarian cancers following treatment with HDAC inhibitors.

Likely HDAC inhibitor treatment results in nonspecific increased acetylation of pro-apoptotic genes and increased gene transcription. However, HDAC inhibitors have also been shown to demonstrate specific effects, including the inhibition of deacetylation of non-histone proteins. For example, HDACs have been shown to remove acetyl groups from p53, GATA-1, E2F, Hsp90, MyoD, and tubulin (16, 35). This suggests that non-histone proteins also undergo acetylation and deacetylation in vivo, as a mechanism of posttranslational modification and regulation of cellular function. Apaf-1 may be a target of regulated non-histone acetylation, and its protein sequence includes potential lysine residues as target sequences for acetylation.

In summary, defective apoptosis underlies the clinically important phenomenon of chemoresistance in ovarian cancers. Apoptosome defects are central to the inability of ovarian cancers to undergo drug-induced apoptotic cell death, and defective Apaf-1 has been observed in many ovarian cancer cell lines and primary tumors. As reviewed earlier, studies suggest that Apaf-1 alterations do not occur at the genetic level in ovarian cancer, suggesting epigenetic regulation of protein function. The understanding of epigenetic alterations underlying tumorigenesis and cancer treatment is making exciting new headway, and it may be the pathways of defective apoptosis and altered acetylation and methylation states in cancer collide at proteins such as Apaf-1. Continued research will illuminate these connections, with clinical trials ultimately allowing for the identification of treatment strategies for ovarian cancer that will lead to longer disease-free intervals and survival rates in patients.

References

1. American Cancer Society Statistics for 2006. (2006), www.cancer.org.
2. Ashkenazi, A. and Dixit, V. M. (1998) Death receptors: signaling and modulation. Science 281, 1305–1308.
3. Balch, C., Huang, T. H., Brown, R. and Nephew, K. P. (2004) The epigenetics of ovarian cancer drug resistance and resensitization. Am. J. Obstet. Gynecol. 191, 1552–1572.
4. Beere, H. M. and Green, D. R. (2001) Stress management – heat shock protein-70 and the regulation of apoptosis. Trends Cell. Biol. 11, 6–10.
5. Benedict, M. A., Hu, Y., Inohara, N. and Nunez, G. (2000) Expression and functional analysis of Apaf-1 isoforms. Extra Wd-40 repeat is required for cytochrome c binding and regulated activation of procaspase-9. J. Biol. Chem. 275, 8461–8468.
6. Budihardjo, I., Oliver, H., Lutter, M., Luo, X. and Wang, X. (1999) Biochemical pathways of caspase activation during apoptosis. Annu. Rev. Cell. Dev. Biol. 15, 269–290.
7. Cain, K., Langlais, C., Sun, X. M., Brown, D. G. and Cohen, G. M. (2001) Physiological concentrations of K + inhibit cytochrome c-dependent formation of the apoptosome. J. Biol. Chem. 276, 41985–41990.
8. Cardone, M. H., Roy, N., Stennicke, H. R., Salvesen, G. S., Franke, T. F., Stanbridge, E., Frisch, S. and Reed, J. C. (1998) Regulation of cell death protease caspase-9 by phosphorylation. Science 282, 1318–1321.
9. Checinska, A., Oudejans, J. J., Span, S. W., Rodriguez, J. A., Kruyt, F. A. and Giaccone, G. (2006) The expression of TUCAN, an inhibitor of apoptosis protein, in patients with advanced non-small cell lung cancer treated with chemotherapy. Anticancer Res. 26, 3819–3824.
10. Cheng, J. Q., Godwin, A. K., Bellacosa, A., Taguchi, T., Franke, T. F., Hamilton, T. C., Tsichlis, P. N. and Testa, J. R. (1992) AKT2, a putative oncogene encoding a member of a

subfamily of protein-serine/threonine kinases, is amplified in human ovarian carcinomas. Proc. Natl. Acad. Sci. USA 89, 9267–9271.

11. Chien, J., Staub, J., Hu, S. I., Erickson-Johnson, M. R., Couch, F. J., Smith, D. I., Crowl, R. M., Kaufmann, S. H. and Shridhar, V. (2004) A candidate tumor suppressor HtrA1 is downregulated in ovarian cancer. Oncogene 23, 1636–1644.
12. Cory, S. and Adams, J. M. (2002) The Bcl2 family: regulators of the cellular life-or-death switch. Nat. Rev. Cancer 2, 647–656.
13. Dan, H. C., Sun, M., Kaneko, S., Feldman, R. I., Nicosia, S. V., Wang, H. G., Tsang, B. K. and Cheng, J. Q. (2004) Akt phosphorylation and stabilization of X-linked inhibitor of apoptosis protein (XIAP). J. Biol. Chem. 279, 5405–5412.
14. Datta, S. R., Dudek, H., Tao, X., Masters, S., Fu, H., Gotoh, Y. and Greenberg, M. E. (1997) Akt phosphorylation of BAD couples survival signals to the cell-intrinsic death machinery. Cell 91, 231–241.
15. Deveraux, Q. L. and Reed, J. C. (1999) IAP family proteins – suppressors of apoptosis. Genes Dev. 13, 239–252.
16. Espino, P. S., Drobic, B., Dunn, K. L. and Davie, J. R. (2005) Histone modifications as a platform for cancer therapy. J. Cell. Biochem. 94, 1088–1102.
17. Fraser, M., Leung, B., Jahani-Asl, A., Yan, X., Thompson, W. E. and Tsang, B. K. (2003) Chemoresistance in human ovarian cancer: the role of apoptotic regulators. Reprod. Biol. Endocrinol. 1, 66.
18. Fraser, M., Leung, B. M., Yan, X., Dan, H. C., Cheng, J. Q. and Tsang, B. K. (2003) p53 is a determinant of X-linked inhibitor of apoptosis protein/Akt-mediated chemoresistance in human ovarian cancer cells. Cancer Res. 63, 7081–7088.
19. Fujimoto, A., Takeuchi, H., Taback, B., Hsueh, E. C., Elashoff, D., Morton, D. L. and Hoon, D. S. (2004) Allelic imbalance of 12q22–23 associated with APAF-1 locus correlates with poor disease outcome in cutaneous melanoma. Cancer Res. 64, 2245–2250.
20. Furukawa, Y., Sutheesophon, K., Wada, T., Nishimura, M., Saito, Y. and Ishii, H. (2005) Methylation silencing of the *Apaf-1* gene in acute leukemia. Mol. Cancer Res. 3, 325–334.
21. Hajra, K. M. and Liu, J. R. (2004) Apoptosome dysfunction in human cancer. Apoptosis 9, 691–704.
22. Herman, J. G. and Baylin, S. B. (2003) Gene silencing in cancer in association with promoter hypermethylation. N. Engl. J. Med. 349, 2042–2054.
23. Herzog, T. J. (2006) The current treatment of recurrent ovarian cancer. Curr. Oncol. Rep. 8, 448–454.
24. Jia, L., Srinivasula, S. M., Liu, F. T., Newland, A. C., Fernandes-Alnemri, T., Alnemri, E. S. and Kelsey, S. M. (2001) Apaf-1 protein deficiency confers resistance to cytochrome *c*-dependent apoptosis in human leukemic cells. Blood 98, 414–421.
25. Jiang, X., Kim, H. E., Shu, H., Zhao, Y., Zhang, H., Kofron, J., Donnelly, J., Burns, D., Ng, S. C., Rosenberg, S. and Wang, X. (2003) Distinctive roles of PHAP proteins and prothymosin-alpha in a death regulatory pathway. Science 299, 223–226.
26. Kalebic, T. (2003) Epigenetic changes: potential therapeutic targets. Ann. N. Y. Acad. Sci. 983, 278–285.
27. Kamarajan, P., Sun, N. K., Sun, C. L. and Chao, C. C. (2001) Apaf-1 overexpression partially overcomes apoptotic resistance in a cisplatin-selected HeLa cell line. FEBS Lett. 505, 206–212.
28. Kaufmann, S. H. and Vaux, D. L. (2003) Alterations in the apoptotic machinery and their potential role in anticancer drug resistance. Oncogene 22, 7414–7430.
29. Kelly, W. K., Richon, V. M., O'connor, O., Curley, T., Macgregor-Curtelli, B., Tong, W., Klang, M., Schwartz, L., Richardson, S., Rosa, E., Drobnjak, M., Cordon-Cordo, C., Chiao, J. H., Rifkind, R., Marks, P. A. and Scher, H. (2003) Phase I clinical trial of histone deacetylase inhibitor: suberoylanilide hydroxamic acid administered intravenously. Clin. Cancer Res. 9, 3578–3588.
30. Kim, M. S., Kwon, H. J., Lee, Y. M., Baek, J. H., Jang, J. E., Lee, S. W., Moon, E. J., Kim, H. S., Lee, S. K., Chung, H. Y., Kim, C. W. and Kim, K. W. (2001) Histone deacetylases induce angiogenesis by negative regulation of tumor suppressor genes. Nat. Med. 7, 437–443.

31. Li, P., Nijhawan, D., Budihardjo, I., Srinivasula, S. M., Ahmad, M., Alnemri, E. S. and Wang, X. (1997) Cytochrome c and dATP-dependent formation of Apaf-1/caspase-9 complex initiates an apoptotic protease cascade. Cell 91, 479–489.
32. Liu, J. R., Opipari, A. W., Tan, L., Jiang, Y., Zhang, Y., Tang, H. and Nunez, G. (2002) Dysfunctional apoptosome activation in ovarian cancer: implications for chemoresistance. Cancer Res. 62, 924–931.
33. Liu, T., Kuljaca, S., Tee, A. and Marshall, G. M. (2006) Histone deacetylase inhibitors: multifunctional anticancer agents. Cancer Treat. Rev. 32, 157–165.
34. Liu, X., Kim, C. N., Yang, J., Jemmerson, R. and Wang, X. (1996) Induction of apoptotic program in cell-free extracts: requirement for dATP and cytochrome c. Cell 86, 147–157.
35. Marks, P., Rifkind, R. A., Richon, V. M., Breslow, R., Miller, T. and Kelly, W. K. (2001) Histone deacetylases and cancer: causes and therapies. Nat. Rev. Cancer 1, 194–202.
36. Marks, P. A., Miller, T. and Richon, V. M. (2003) Histone deacetylases. Curr. Opin. Pharmacol. 3, 344–351.
37. Minucci, S. and Pelicci, P. G. (2006) Histone deacetylase inhibitors and the promise of epigenetic (and more) treatments for cancer. Nat. Rev. Cancer 6, 38–51.
38. Ogawa, T., Shiga, K., Hashimoto, S., Kobayashi, T., Horii, A. and Furukawa, T. (2003) APAF-1-ALT, a novel alternative splicing form of APAF-1, potentially causes impeded ability of undergoing DNA damage-induced apoptosis in the LNCaP human prostate cancer cell line. Biochem. Biophys. Res. Commun. 306, 537–543.
39. Pathan, N., Marusawa, H., Krajewska, M., Matsuzawa, S., Kim, H., Okada, K., Torii, S., Kitada, S., Krajewski, S., Welsh, K., Pio, F., Godzik, A. and Reed, J. C. (2001) TUCAN, an antiapoptotic caspase-associated recruitment domain family protein overexpressed in cancer. J. Biol. Chem. 276, 32220–32229.
40. Peart, M. J., Smyth, G. K., van Laar, R. K., Bowtell, D. D., Richon, V. M., Marks, P. A., Holloway, A. J., and Johnstone, R. W. (2005) Identification and functional significance of genes regulated by structurally different histone deacetylase inhibitors. Proc. Natl. Acad. Sci. USA 102, 3697–3702.
41. Shao, Y., Gao, Z., Marks, P. A. and Jiang, X. (2004) Apoptotic and autophagic cell death induced by histone deacetylase inhibitors. Proc. Natl. Acad. Sci. USA 101, 18030–18035.
42. Shi, Y. (2002) Mechanisms of caspase activation and inhibition during apoptosis. Mol. Cell 9, 459–470.
43. Soengas, M. S., Capodieci, P., Polsky, D., Mora, J., Esteller, M., Opitz-Araya, X., Mccombie, R., Herman, J. G., Gerald, W. L., Lazebnik, Y. A., Cordon-Cardo, C. and Lowe, S. W. (2001) Inactivation of the apoptosis effector Apaf-1 in malignant melanoma. Nature 409, 207–211.
44. Sun, Y., Orrenius, S., Pervaiz, S. and Fadeel, B. (2005) Plasma membrane sequestration of apoptotic protease-activating factor-1 in human B-lymphoma cells: a novel mechanism of chemo resistance. Blood 105, 4070–4077.
45. Takai, N., Kawamata, N., Gui, D., Said, J. W., Miyakawa, I. and Koeffler, H. P. (2004) Human ovarian carcinoma cells: histone deacetylase inhibitors exhibit antiproliferative activity and potently induce apoptosis. Cancer 101, 2760–2770.
46. Tan, L. and Liu, J. R. Unpublished findings.
47. Teodoridis, J. M., Hall, J., Marsh, S., Kannall, H. D., Smyth, C., Curto, J., Siddiqui, N., Gabra, H., Mcleod, H. L., Strathdee, G. and Brown, R. (2005) CpG island methylation of DNA damage response genes in advanced ovarian cancer. Cancer Res. 65, 8961–8967.
48. Testa, J. R. and Bellacosa, A. (2001) AKT plays a central role in tumorigenesis. Proc. Natl. Acad. Sci. USA 98, 10983–10985.
49. Thornberry, N. A. and Lazebnik, Y. (1998) Caspases: enemies within. Science 281, 1312–1316.
50. Umetani, N., Fujimoto, A., Takeuchi, H., Shinozaki, M., Bilchik, A. J. and Hoon, D. S. (2004) Allelic imbalance of APAF-1 locus at 12q23 is related to progression of colorectal carcinoma. Oncogene 23, 8292–8300.
51. Williams, J., Lucas, P. C., Griffith, K. A., Choi, M., Fogoros, S., Hu, Y. Y. and Liu, J. R. (2005) Expression of Bcl-xL in ovarian carcinoma is associated with chemoresistance and recurrent disease. Gynecol. Oncol. 96, 287–295.

52. Wolf, B. B., Schuler, M., Li, W., Eggers-Sedlet, B., Lee, W., Tailor, P., Fitzgerald, P., Mills, G. B. and Green, D. R. (2001) Defective cytochrome c-dependent caspase activation in ovarian cancer cell lines due to diminished or absent apoptotic protease activating factor-1 activity. J. Biol. Chem. 276, 34244–34251.
53. Yamamoto, H., Gil, J., Schwartz, S., Jr. and Perucho, M. (2000) Frameshift mutations in Fas, Apaf-1, and Bcl-10 in gastro-intestinal cancer of the microsatellite mutator phenotype. Cell. Death Differ. 7, 238–239.
54. Yang, X., Fraser, M., Moll, U. M., Basak, A. and Tsang, B. K. (2006) Akt-mediated cisplatin resistance in ovarian cancer: modulation of p53 action on caspase-dependent mitochondrial death pathway. Cancer Res. 66, 3126–3136.
55. Zou, H., Henzel, W. J., Liu, X., Lutschg, A. and Wang, X. (1997) Apaf-1, a human protein homologous to C. elegans CED-4, participates in cytochrome c-dependent activation of caspase-3. Cell 90, 405–413.

Nanoparticle Delivery of Suicide DNA for Epithelial Ovarian Cancer Therapy

Janet A. Sawicki, Daniel G. Anderson, and Robert Langer

1 Introduction

The standard treatment for patients with advanced stage epithelial ovarian cancer is optimal surgical debulking followed by chemotherapy with paclitaxel plus a platinum-based therapy (cisplatin or carboplatin). Although ~80% of patients receiving this therapeutic regimen have an initial favorable response, recurrent disease will occur in a majority of cases. Regrettably, there are currently no effective therapies for those patients with advanced-stage ovarian cancer, who either do not respond to initial therapy or for those who develop recurrent disease. There is an immediate need for a more effective treatment for this deadly disease.

Gene therapy holds great promise as an alternative treatment for metastatic ovarian cancer. Metastatic tumors in this disease are nearly always confined to the peritoneal cavity, so intraperitoneal delivery of therapeutic DNA allows for direct treatment of the tumors. In addition, this delivery route protects healthy organs outside the cavity from harmful side effects. In theory, the ability to target the delivery of DNA to tumor cells, as well as the ability to control its expression once inside the cell, provides an added level of therapeutic efficiency and specificity that is difficult to achieve using chemotherapy. In practice, however, the full potential of these advantages of DNA therapies has yet to be achieved and remains a goal of preclinical and clinical studies.

An important consideration in any gene therapy protocol is the choice of vector used to deliver the DNA to cells. With a few exceptions, viral vectors (either adenoviral or retroviral) have been used in clinical trials for the treatment of ovarian cancer (see http://clinicaltrials.gov). Recently, the use of nonviral vectors for the delivery of therapeutic genes is receiving wide attention by the research community, particularly in light of the serious consequences that have occurred in association with the use of viral vectors in patients.

Another consideration in designing a gene therapy protocol is the nature of the DNA that will be delivered. Gene therapy treatment strategies used in ovarian cancer clinical trials include molecular chemotherapy (prodrugs), mutation compensation, immunotherapy, altered drug sensitivity, antiangiogenic therapy, and virotherapy (37, 39).

In this report, we describe suicide gene therapies for epithelial ovarian cancer that we are developing. These therapies are based on the use of a class of cationic polymers called poly(β-amino ester)s to deliver the so-called suicide genes to tumor cells, resulting in their death. Two mouse models for ovarian cancer that are being used in preclinical testing of therapeutic efficacy are also discussed.

2 Suicide Gene Therapy

2.1 Diphtheria Toxin

Targeting the death of ovarian cancer cells, both in the primary tumor and in metastatic lesions, is an attractive therapeutic option. The naturally occurring toxin made by the bacterium *Corynebacterium diphtheriae* is an especially good choice for use as a therapeutic agent because its mechanism of action is known (13), and the gene encoding the toxin has been cloned, sequenced, and adapted for expression in mammalian cells. This toxin is also extremely potent; a single molecule is sufficient to kill a cell (47). Normally, the toxin is secreted as a precursor peptide that is then enzymatically cleaved into two fragments, A and B chains. The B chain binds to the surface of most eukaryotic cells and is required for delivery of the A chain (DT-A), encoding the toxin, into the cytoplasm. Once inside the cell, DT-A inhibits protein synthesis by catalyzing the ADP ribosylation of EF-2 elongation factor. Diphtheria toxin is especially well-suited to treat many kinds of cancer because, unlike virtually all therapeutic agents in use, it kills cells in a cell-cycle independent manner, so that both dividing and nondividing cells are vulnerable to its deadly effects.

A *DT* gene, engineered for use in mammalian cells, *DT-A*, contains the coding sequence for the DT-A subunit, but not for the DT-B subunit (32). DNA constructs can be engineered so that DT-A expression is controlled by cell-specific, *cis*-acting transcriptional regulatory elements (promoters and enhancers). Use of such constructs restricts expression of the DT-A subunit to target cells. In the absence of the B subunit, even DT-A released from dead cells is not able to enter other neighboring cells. This feature thus allows for death of the tumor cells, but not neighboring healthy cells. It also requires efficient uptake of the therapeutic DNA by the tumor cells in order for the therapy to be effective.

A concern in using a toxin as potent as DT-A is the ability to tightly control gene expression. To address this concern, we have developed a novel genetic strategy that makes use of a site-directed recombinase, Flp recombinase (27, 35, 36). Using this strategy, gene expression is regulated both transcriptionally and by DNA recombination mediated by Flp recombinase (Fig. 1). In this system, a cell-specific promoter controls the expression of the recombinase, and so recombinase-mediated recombination event only takes place in selected cells. In those cells where Flp recombinase is expressed, a gene sequence containing the target sequences for the recombinase undergoes excisional recombination, which results

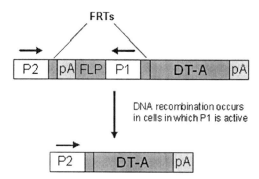

Fig. 1 DNA construct to control DT-A expression. Flp recombinase is produced in tumor cells where P1, a tumor-specific promoter is active. When Flp recombinase is expressed, recombination occurs, allowing a second tumor-specific promoter, P2, to drive expression of DT-A. FRTs are Flp recombinase target sequences. *pA*, polyadenylation sequence. *Horizontal arrows* indicate the direction of transcription

in DT-A expression due to juxtaposition of a second cell-specific promoter to the DT-A coding sequence. Thus, two cell-specific promoters tightly regulate toxin gene expression. By selecting tumor-specific (or tissue-specific) promoters, this system can be applied to target DT-A suicide gene therapy to different kinds of tumors. Delivery of a PSA promoter-regulated DT-A/FLP recombinase DNA construct to human prostate tumor cell xenografts in nude mice resulted in suppression of tumor growth and even tumor regression (6).

In current studies, we are using a promoter sequence of the human mesothelin (*MSLN*) gene to regulate the expression of Flp recombinase. Activity of this promoter is significantly enhanced in ovarian cancer cells relative to normal ovarian cells and cells in other tissues, suggesting its promising use as a cancer-specific promoter in gene therapy strategies (12). The promoter of the gene encoding human epididymis protein 4 (HE4), another promoter having high activity in epithelial ovarian cancer cells, drives expression of DT-A upon DNA recombination (7).

2.2 CD/5-FC + HSV-TK/GCV Gene-Directed Enzyme Prodrug Therapy

The efficacy of cytosine deaminase (CD)/5-flurocytosine (5-FC) and herpes simplex virus type 1 thymidine kinase (HSV-TK)/ganciclovir (GCV) suicide gene-prodrug strategies have been studied extensively in animal models and clinical trials (16, 17). In each of these strategies, a normally innocuous prodrug (5-FC or GCV) is converted by a viral or bacterial enzyme to a toxic compound (5-FU or GCVTP), which causes tumor cells to die. In contrast to diphtheria toxin-based therapy, these prodrug therapies are effective primarily against dividing cells. Production of the toxic drug by cells can also cause neighboring cells to die, a so-called "bystander effect." This may result in the death of more tumor cells, but the trade-off is that it may also result in the death of healthy cells.

Although studies have shown minimal nonspecific toxicity associated with delivery of the genes and administration of the prodrugs, each of these therapies has shown only modest effects on reducing tumor burden and improving clinical outcomes of patients (20). However, when the CD/5-FC and HSV-TK/GCV strategies are administered in combination, therapeutic efficacy is significantly improved when compared with either strategy that is being used alone (1, 8, 19, 40, 46). In a recent report, Boucher et al. demonstrate that sequential administration of the prodrugs (5-FC, then GCV) to cells infected with an adenovirus containing a *CD/HSV-TK* fusion gene in vitro enhanced cytotoxicity above an additive effect by 24- to 35-fold when compared with one to fivefold increase with simultaneous treatment (10). This study suggests that sequential administration of the prodrugs to cancer patients may significantly improve therapeutic outcome.

In preclinical studies in mouse models for ovarian cancer, we are studying the efficacy of sequential prodrug administration following delivery of a *CD/HSV-TK* fusion gene regulated by the MSLN promoter.

3 Poly(β-amino ester)s

The safe and effective delivery of DNA remains a central challenge to the application of gene delivery in the clinic. Currently, the majority of gene therapy protocols employ viral delivery systems, which are associated with serious toxicity and production concerns (43). Nonviral delivery systems offer a number of potential advantages, including ease of production, stability, low immunogenicity and toxicity, and capacity to deliver larger DNA payloads (23). Their use in gene therapy protocols for the treatment of cancer is especially relevant given the low amounts of the requisite receptor for adenoviral infection, CAR, on primary tumors. However, existing nonviral delivery systems are far less efficient than viral vectors (28).

One promising group of nonviral delivery compounds is cationic polymers, which spontaneously bind and condense DNA into nanoparticles (11, 18, 21, 25, 26, 29, 30, 42). A wide variety of cationic polymers that transfect cells in vitro have been characterized; some are natural polymers such as protein (18) and peptide systems (42), while others are synthetic polymers such as poly(ethylene imine) (PEI) (11, 42) and dendrimers (21). Recent advances in polymeric gene delivery have focused in part on the incorporation of biodegradability to decrease toxicity. Typically, these polymers contain both chargeable amino groups, to allow for ionic interaction with the negatively charged DNA phosphate, and a degradable region, such as a hydrolyzable ester linkage. Examples of these include poly[alpha-(4-aminobutyl)-L-glycolic acid] (25), network poly(amino ester) (26), and poly(β-amino ester)s (2, 3, 5, 30). The Langer laboratory has been particularly interested in poly(β-amino ester)s as delivery agents, as they are easily synthesized via the conjugate addition of a primary amine or bis(secondary amine) to a diacrylate, transfect cells with high efficiency in vitro, and generally possess low toxicity.

The efficacy of various cationic polymers in vivo has been demonstrated both in general and therapeutic models (24). To date, most in vivo work has used PEI. Modification of PEI to include both targeting ligands and serum resistance has been demonstrated and shown to be moderately effective at delivering DNA in a targeted fashion in some systems (9, 22, 44). However, PEI is both nondegradable and relatively toxic, and still not as effective at delivery as viral systems. Recently, attention has turned to the development of cationic polymers that are more compatible with in vivo usage. Using high throughput methods, the synthesis and screening of over 2,350 poly(β-amino ester)s was recently completed (5). This initial screening identified 46 new, biodegradable polymers, which transfect cells as well, and in some cases significantly better than, conventional nonviral delivery system such as PEI in vitro. Subsequent scaled-up resynthesis and analysis of over 500 of these initial polymers has identified the critical importance that polymer molecular weight and end-group termination have on transfection potential (2–4). The polymer that transfects cells most efficiently, C32 (Fig. 2), consistently well-outperforms any commercially available compound tested, and is much less toxic than PEI (45). When C32 complexes with DNA at an optimal polymer/DNA ratio, nanoparticles are formed, which have a molecular weight of 18,100 Da and a diameter of 70 nm (4). Recently, we have shown that intraperitoneal gene delivery using C32 polymers containing end-modified amines results in improved expression levels in several abdominal organs and in ovarian tumors (48). On the basis of these studies, C32 and other polymers similar to it merit further investigation as new vehicles for gene delivery for the treatment of ovarian cancer.

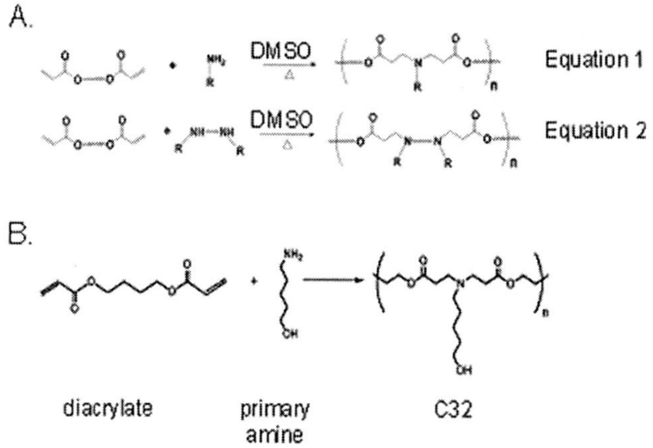

Fig. 2 Synthesis of poly(β-amino ester)s. (**a**) Poly(β-amino ester)s are synthesized by the conjugate addition of primary (*equation 1*) or bis(secondary amines) (*equation 2*) to diacrylates. (**b**) Synthesis of polymer C32

4 The Challenge: Targeted Therapy

Effective treatment of patients with advanced late-stage disease is the major challenge facing clinical oncologists. Although improved detection methods continue to increase the number of patients diagnosed at early stages of disease, the fact remains that, at the time of diagnosis, as many as half of all cancer patients present with metastatic disease. The percentage of ovarian cancer patients presenting with advanced disease is even higher than for other cancers (75%) given the absence of an effective screening for early detection and the relatively asymptomatic nature of the early stages. In addition, many ovarian cancer patients "cured" of their initial malignancy relapse with more aggressive drug-resistant metastatic cancer.

Finding ways to target systemically-delivered therapies to tumor cells and causing minimal toxicity to healthy, nontumorous cells are the key to the development of effective therapies for metastatic disease. Targeting therapy to specific cells can be accomplished by transcriptional targeting, transductional targeting, and ideally, by a combination of both of these approaches.

Transcriptional targeting refers to the use of gene regulatory elements (promoters and enhancers) to restrict gene *expression* to specific cells. The regulatory elements of several ovarian-specific genes have been cloned and characterized. As discussed earlier, we are using two promoter sequences that have enhanced activity in ovarian tumor cells – the promoter of the mesothelin (*MSLN*) gene (12), and the promoter of the gene encoding whey-acidic protein human epididymus protein 4 (*HE4*) (7). Compared with other "ovarian tumor-specific" promoters, these two promoters were recently shown to have the lowest activity in normal tissues (41).

Transductional targeting refers to the *delivery* of DNA to specific cells. Conjugation of vectors (either viral or nonviral) to proteins that have high affinity for specific cells is an effective way to target DNA delivery (15). Proteins that are used for this purpose include ligands, receptors, antibodies, or peptide antagonists. High affinity targeting of the vector to targeted cells will result in reduced sequestration of the particles in nontargeted tissues and more efficient DNA delivery to the targeted population. Successful targeting should reduce the effective dose, thereby reducing any toxicity associated with the therapy.

In our studies, in collaboration with Gregory Adams (Fox Chase Cancer Center), we are conjugating poly(β-amino ester)s with a single chain variable antibody fragment (scFv) of human origin having reactivity to Mullerian Inhibiting Substance II Receptor (MISIIR), a transmembrane serine threonine kinase that is specifically expressed in ovary surface epithelial cells, ovarian tumor cells, in the uterus and Fallopian tubes, and at lower levels in the breast, but not in other tissues in women (31). Its high expression by tumors suggests that it will provide a useful targeting signal for directed therapies.

5 Ovarian Cancer Mouse Models for Preclinical Studies

Mouse models for cancer are very useful for evaluating the efficacy of new gene therapy strategies. Given the importance of identifying possible immunological responses to treatment, and the importance of the microenvironment on tumor development, transgenic mouse models that have an intact immune system and develop organ-specific cancer are preferred over subcutaneous xenograft models or orthotopic tumors in immunocompromized mice. Importantly, refinement of new imaging modalities for small animals including optical imaging, microCT, PET, and MRI makes it possible to monitor accurately the effect of therapies on tumor development.

Transgenic mouse models for epithelial ovarian cancer that recapitulate human disease have only recently been developed [see review of genetically modified mouse models for ovarian cancer (33)]. We are using the MISIIR/Tag transgenic mouse model, developed by Denise Connolly and Thomas Hamilton, to test the efficacy of nanoparticle-delivered suicide gene therapy (14). As a consequence of expression of the transforming region of SV40 under control of the Mullerian inhibitory substance type II receptor gene promoter, 100% of female MISIIR/Tag mice develop bilateral epithelial ovarian tumors. To evaluate the effect of intraperitoneal administration of nanoparticle-delivered DT-A DNA on tumor growth, mice are CT scanned before treatment and then multiple times after treatment. Amira software is used to generate 3D-reconstructions from tumor images, and tumor volumes are then determined using Image J software (Fig. 3). We are also evaluating the effect of this therapy on the lifespan of MISIIR/Tag mice.

Preliminary studies suggest that the lifespan of MISIIR/Tag mice that receive multiple intraperitoneal injections of poly(β-amino ester) nanoparticles to deliver a MSLN promoter/DT-A DNA is significantly increased when compared with mice treated with control DNA.

A second mouse model we are using employs a cell line, MOSEC, derived from mouse ovarian surface epithelial cells that spontaneously transformed in culture

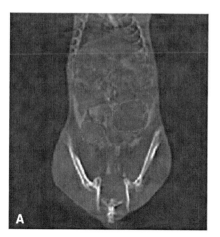 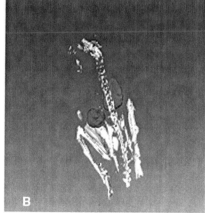

Fig. 3 (a) MicroCT scan of a MISIIR/Tag mouse. *Red dashed lines* delineate bilateral ovarian tumors. (b) 3-D reconstruction of the tumors in the same mouse

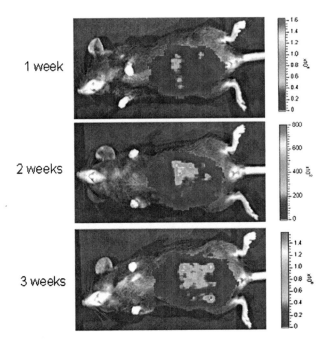

Fig. 4 Optical images of a mouse injected with MOSEC-luc cells taken 1-, 2-, and 3-weeks after injection of the cells into the peritoneum of a C57BL/6 female mouse. Pseudocolor images representing emitted light are superimposed over grayscale reference images of the mouse. RLUs/pixel are indicated in the color scale bar

(38). The properties of this cell line, established by Katherine Roby and Paul Terranova, are described in Chapter 15. When MOSEC cells are injected into the peritoneum of a syngeneic C57BL/6 female mouse, tumors develop throughout the peritoneal cavity. We stably transfected MOSEC cells with DNA encoding firefly luciferase under the control of a strong, ubiquitously expressed promoter/enhancer, CAG (34), and established a clonal cell line, MOSEC-luc. The tumor load of mice injected intraperitoneally with these cells can be quantified by optically imaging these mice to detect bioluminescence following administration of D-luciferin, the substrate of firefly luciferase (Fig. 4). Thus, quantitation of relative light units (RLU) before and after treatment is a convenient and accurate way to assess therapeutic efficacy.

6 Summary

Intraperitoneal administration of polymeric nanoparticles to deliver DNA encoding suicide genes holds much promise as an effective therapy for advanced epithelial ovarian cancer. Poly(β-amino ester)s, a class of cationic, biodegradable polymers complex to DNA to form nanoparticles that deliver DNA to cells in ovarian tumors. Modifications to poly(β-amino ester)s can improve both the efficiency and specificity with which DNA is delivered to tumor cells. Preclinical studies to test therapeutic efficacy of gene therapy strategies that are under development make use of mouse models for epithelial ovarian cancer and new imaging technologies.

Acknowledgments This work is supported by a grant from the Department of Defense (JAS) and NIH grant EB00244 (DGA and RL).

References

1. Aghi, M., Kramm, C.M., Chou, T., Breakefield, X.O., and Chiocca, E.A. (1998) Synergistic anticancer effects of ganciclovir/thymidine kinase and 5-fluorocytosine/cytosine deaminase gene therapies. J. Natl. Cancer Inst. 90, 370–380.
2. Akinc, A.B., Anderson, D.G., Lynn, D.M., and Langer, R. (2003a) Synthesis of poly(β-amino esters) optimized for highly effective gene delivery. Bioconjugate Chem. 14, 979–988.
3. Akinc, A.B., Lynn, D.M., Anderson, D.G., and Langer, R. (2003b) Parallel synthesis and biophysical characterization of a degradable polymer library for gene delivery. J. Am. Chem. Soc. 125, 5316–5323.
4. Anderson, D.G., Akinc, A., Houssain, N., and Langer, R. (2005) Structure/property studies of polymeric gene delivery using a library of poly(β-amino esters). Mol. Ther. 11, 426–434.
5. Anderson, D.G., Lynn, D.M., and Langer, R. (2003) Semi-automated synthesis and screening of a large library of degradable cationic polymers for gene delivery. Angew. Chem. Int. Ed. Engl. 42, 3153–3158.
6. Anderson, D.G., Peng, W., Akinc, A.B., Houssain, N., Kohn, A., Padera, R., Langer, R., and Sawicki, J.A. (2004) A polymer library approach to suicide gene therapy for cancer. Proc. Natl. Acad. Sci. USA 101, 16028–16033.
7. Berry, N.B., Cho, Y.M., Harrington, M.A., Williams, S.D., Foley, J., and Nephew, K.P. (2004) Transcriptional targeting in ovarian cancer cells using the human epididymis protein 4 promoter. Gynecol. Oncol. 92, 896–904.
8. Blackburn, R.V., Galoforo, S.S., Corry, P.M., and Lee, Y.J. (1999) Adenoviral transduction of a cytosine deaminase/thymidine kinase fusion gene into prostate carcinoma cells enhances prodrug and radiation sensitivity. Int. J. Cancer 82, 293–297.
9. Blessing, T., Kursa, M., Holzhauser, R., Kircheis, R., and Wagner, E. (2001) Different strategies for formation of pegylated EGF-conjugated PEI/DNA complexes for targeted gene delivery. Bioconjugate Chem. 12, 529–537.
10. Boucher, P.D., Im, M.M., Freytag, S.O., and Shewach, D.S. (2006) A novel mechanism of synergistic cytotoxicity with 5-fluorocytosine and ganciclovir in double suicide gene therapy. Cancer Res. 66, 3230–3237.
11. Boussif, O., Lezoualc'h, F., Zanta, M.A., Mergny, M.D., Scherman, D., Demeneix, B., and Behr, J.P. (1995) A versatile vector for gene and oligonucleotide transfer into cells in culture and in vivo – polyethylenimine. Proc. Natl. Acad. Sci. USA 92, 7297–7301.
12. Breidenbach, M., Rein, D.T., Evert, M., Glasgow, J.N., Wang, M., Passineau, M.J., Alvarez, R.D., Korokhov, N., and Curiel, D.T. (2005) Mesothelin-mediated targeting of adenoviral vectors for ovarian cancer gene therapy. Gene Therapy 12, 187–193.
13. Collier, R.J. (1975) Diphtheria toxin: mode of action and structure. Bacteriol. Rev. 39, 54–85.
14. Connolly, D.C., Bao, R., Nikitin, A.Y., Stephens, K.C., Poole, T.W., Hua, H., Harris, S.S., Vanderhyden, B.C., and Hamilton, T.C. (2003) Female mice chimeric for expression of the simian virus 40 TAg under control of the MISIIR promoter develop epithelial ovarian cancer. Cancer Res. 63, 1389–1397.

15. Cristiano, R.J. (2002) Protein/DNA polyplexes for gene therapy. Surg. Oncol. Clin. N. Am. 11, 697–716.
16. Dachs, G.U., Tupper, J., and Tozer, G.M. (2005) From bench to bedside for gene-directed enzyme prodrug therapy of cancer. Anticancer Drugs 16, 349–359.
17. Fillat, C., Carrio, M., Cascante, A., and Sangro, B. (2003) Suicide gene therapy mediated by the herpes simplex virus thymidine kinase gene/ganciclovir system: fifteen years of application. Curr. Gene Ther. 3, 13–26.
18. Fominaya, J., and Wels, W. (1996) Target cell-specific DNA transfer mediated by a chimeric multidomain protein – novel non-viral gene delivery system. J. Biol. Chem. 271, 10560–10568.
19. Freytag, S.O., Dang, C.V., and Lee, W.M.F. (1990) Definition of the activities and properties of c-*myc* required to inhibit cell differentiation. Cell Growth Diff. 1, 339–343.
20. Greco, O. and Dachs, G.U. (2001) Gene directed enzyme/prodrug therapy for cancer: historical appraisal and future prospectives. J. Cell Physiol. 187, 22–36.
21. Kabanov, A.V., Felgner, P.L., and Seymour, L.W. (1998) Self-assembling complexes for gene delivery: from laboratory to clinical trial. Wiley, New York.
22. Kursa, M., Walker, G.F., Roessler, V., Ogris, M., Roedl, W., Kircheis, R., and Wagner, E. (2003) Novel shielded transferrin-polyethylene glycol-polyethylenimine/DNA complexes for systemic tumor-targeted gene transfer. Bioconjug. Chem. 14, 222–231.
23. Ledley, F.D. (1995) Nonviral gene therapy: the promise of genes as pharmaceutical products. Human Gene Ther. 6, 1129–1144.
24. Lemkine, G.F., and Demeneix, B.A. (2001) Polyethylenimines for in vivo gene delivery. Curr. Opin. Mol. Ther. 3, 178–182.
25. Lim, Y.B., Kim, C.H., Kim, K., Kim, S.W., and Park, J.S. (2000) Development of a safe gene delivery system using biodegradable polymer, poly[alpha-(4-aminobutyl)-L-glycolic acid]. J. Am. Chem. Soc. 122, 6524–6525.
26. Lim, Y.B., Kim, S.M., Suh, H., and Park, J.S. (2002) Biodegradable, endosome disruptive, and cationic network-type polymer as a highly efficient and nontoxic gene delivery carrier. Bioconjugate Chem. 13, 952–957.
27. Logie, C. and Stewart, F. (1995) Ligand-regulated site-specific recombination. Proc. Natl. Acad. Sci. USA 92, 5940–5944.
28. Luo, D. and Saltzman, W.M. (2000) Synthetic DNA delivery systems. Nat. Biotechnol. 18, 33–37.
29. Lynn, D.M., Anderson, D.G., Putnam, D., and Langer, R. (2001) Accelerated discovery of synthetic transfection vectors: parallel synthesis and screening of a degradable polymer library. J. Am. Chem. Soc. 123, 8155–8156.
30. Lynn, D.M. and Langer, R. (2000) Degradable poly(beta-amino esters): synthesis, characterization, and self-assembly with plasmid DNA. J. Am. Chem. Soc. 122, 10761–10768.
31. Masiakos, P.T., MacLaughlin, D.T., Maheswaran, S., Teixeira, J., Fuller, A.F., Jr., Shah, P.C., Kehas, D.J., Kenneally, M.K., Dombkowski, D.M., Ha, T.U., Preffer, F.I., and Donahoe, P.K. (1999) Human ovarian cancer, cell lines, and primary acites cells express the human Mullerian inhibiting substance (MIS) type II receptor, bind, and are responsive to MIS. Clin. Cancer Res. 5, 3488–3499.
32. Maxwell, I.H., Maxwell, F., and Glode, L.M. (1986) Regulated expression of a diphtheria toxin A-chain gene transfected into human cells: possible strategy for inducing cancer cell suicide. Cancer Res. 46, 4660–4664.
33. Nikitin, A.Y., Connolly, D.C., and Hamilton, T.C. (2004) Pathology of ovarian neoplasms in genetically modified mice. Comp. Med. 54, 26–28.
34. Niwa, H., Yamamura, K., and Miyazaki, J. (1991) Efficient selection for high-expression transfectants with a novel eukaryotic vector. Gene 108, 193–200.
35. O'Gorman, S., Fox, D.T., and Wahl, G.M. (1991) Recombinase-mediated gene activation and site-specific integration in mammalian cells. Science 251, 1351–1355.
36. Peng, W., Verbitsky, A., Bao, Y., and Sawicki, J.A. (2002) Regulated expression of diphtheria toxin in prostate cancer cells. Mol. Ther. 6, 537–545.

37. Raki, M., Rein, D.T., Kanerva, A., and Hemminki, A. (2006) Gene transfer approaches for gynecological diseases. Mol. Ther. 14, 154–163.
38. Roby, K.F., Taylor, C.C., Sweetwood, J.P., Cheng, Y., Pace, J.L., Tawfik, O., Persons, D.L., Smith, P.G., and Terranova, P.F. (2000) Development of a syngenic mouse model for events related to ovarian cancer. Carcinogenesis 21, 585–591.
39. Rocconi, R.P., Numnum, T.M., Stoff-Khalili, M., Makhiga, S., Alvarez, R.D., and Curiel, D.T. (2005) Targeted gene therapy for ovarian cancer. Curr. Gene Ther. 5, 643–653.
40. Rogulski, K.R., Kim, J.H., Kim, S.H., and Freytag, S.O. (1997) Glioma cells transduced with an *Escherichia coli* CD/HSV-1TK fusion gene exhibit enhanced metabolic suicide and radio-sensitivity. Hum. Gene Ther. 8, 73–85.
41. Rosen, D.G., Wang, L., Atkinson, J.N., Yu, Y., Lu, K.H., Diamandis, E.P., Hellstrom, I., Mok, S.C., Liu, J., and Bast, R.C.J. (2005) Potential markers that complement expression of CA125 in epithelial ovarian cnacer. Gynecol. Oncol. 99, 267–277.
42. Schwartz, J.J., and Zhang, S. (2000) Peptide-mediated cellular delivery. Curr. Opin. Mol. Ther. 2, 162–167.
43. Somia, N., and Verma, I.M. (2000) Gene therapy trials and tribulations. Nat. Rev. Genet. 1, 91–99.
44. Suh, W., Han, S.O., Yu, L., and Kim, S.W. (2002) An angiogenic, endothelial-cell-targeted polymeric gene carrier. Mol. Ther. 6, 664–672.
45. Trubetskoy, V.S., Wong, S.C., Subbotin, V., Budker, V.G., Loomis, A., Hagstrom, J.E., and Wolff, J.A. (2003) Recharging cationic DNA complexes with highly charged polyanions for in vitro and in vivo gene delivery. Gene Ther. 10, 261–271.
46. Uckert, W., Kammertons, T., Haack, K., Qin, Z., Gebert, J., Schendel, D.J., and Blankenstein, T. (1998) Double suicide gene (cytosine deaminase and herpes simplex virus thymidine kinase) but not single gene transfer allow reliable elimination of tumor cells *in vivo*. Hum. Gene Ther. 9, 855–865.
47. Yamaizumi, M., Mekada, E., Uchida, T., and Okada, Y. (1978) One molecule of diphtheria toxin fragment A introduced into a cell can kill the cell. Cell 15, 245–250.
48. Zugates, G.T., Peng, W., Zumbuchl, A., Jhunjhunwala, S., Langer, R., Sawicki, J.A., and Anderson, D.G. (2007) Rapid optimization of gene delivery by parallel end modification of poly(β-amino ester)s. Mol. Ther. 15, 1306–1312.

Biological Therapy with Oncolytic Herpesvirus

Fabian Benencia and George Coukos

1 Introduction

Oncolytic virus therapy refers to the biological therapy of tumors, using live viruses with relative tumor selectivity. Replication-restricted virus strains have been genetically engineered, which replicate selectively within tumor cells. Examples include replication-competent mutants of herpes virus, adenovirus, vesicular stomatitis virus, reovirus, and measles virus (2, 17, 37, Martuza, 2000; 56, 63). In particular, replication-competent recombinant HSV strains may offer distinct advantages in oncolytic therapy of epithelial tumors: (a) HSV is highly infectious to tumors of epithelial origin, resulting in high efficacy (70); (b) there is considerable redundancy in HSV receptors, which makes the loss of HSV receptors by tumors due to mutations less likely; (c) antiherpetic drugs are commercially available, which may be used clinically to control undesired side effects, should local or systemic spread of the virus occur; and (d) because of its large genome, HSV offers ample packaging opportunities – up to 30 kb – without affecting viral replication in cancer cells.

2 Oncolytic HSV Strains

HSV-1 is an enveloped virus containing approximately 152 kb of genomic DNA, which codify for about 80 identified viral genes. Detailed investigation of the HSV genome led to the generation of mutants lacking specific genes that are unable to replicate readily in normal diploid cells, but retain the ability to replicate selectively in malignant cells owing to compensatory overexpression of homologue eukaryotic genes in tumor cells. A number of replication-restricted mutants have been generated including the isolated or combined deletion or mutation of *UL23* gene encoding thymidine kinase gene (*TK*), *UL39* gene encoding the large subunit of HSV ribonucleotide reductase (RR), or the *g134.5* gene encoding infected cell protein (ICP) 34.5 protein. TK-deficient strains are insensitive to two of the most potent antiherpetic agents, acyclovir and ganciclovir, and ICP34.5-deficient mutants retain their sensitivity to these drugs. Interestingly, RR negative mutants were shown to have increased sensitivity to acyclovir and ganciclovir (53). Replication-competent HSV strains lacking TK, RR, or ICP34.5 were shown to selectively replicate in neuronal tumor cells and induce their death without exerting cytotoxicity on differentiated neuronal cells (35).

3 Efficacy of Oncolytic HSV

A large amount of preclinical data has been accumulated in immunodeficient as well as immunocompetent rodent models demonstrating the antitumor efficacy and safety of intracranial intratumoral inoculation of replication-selective HSV-1 mutants (3, 4, 12, 13, 15, 42, 43, 49, 53, 54, 58, 73, 14). Safety data have also been collected from HSV-sensitive *Aotus nancymae* monkeys, in which intracerebral inoculation of HSV-G207 led to no acute or long-term toxicity (38). No viral spread could be documented from treated monkeys (67). Results from two dose-escalation phase-I clinical trials of intracerebral stereotactic inoculation of ICP34.5-deleted HSV-1716 or ICP34.5-deleted/RR- mutated HSV-G207 for the treatment of malignant glioblastoma reported no toxicity in the human (48, 59). A growing bulk of in vitro and in vivo evidence in experimental models indicates that oncolytic HSV is efficacious against a variety of nonneuronal tumors (23). Importantly, intravascular systemic or locoregional administration of oncolytic HSV resulted in no toxicity in preclinical models (14, 57).

4 Oncolytic Therapy of Epithelial Ovarian Carcinoma (EOC)

A variety of tumors that metastasize to the peritoneal cavity are also susceptible to HSV oncolytic therapy including ovarian (23, 26, 27), cervical (11), colorectal (57), prostate (69); mesothelioma, and breast carcinoma (65). Epithelial tumor cells express cell surface receptors for HSV-1. For example, we found that A2780 EOC cells express strongly HVE-B and small amounts of HVE-A by flow cytometry and are readily infected by ICP34.5-deficient HSV-1. Furthermore, different ICP34.5-deficient HSV-1 and ICP34.5/RR-deficient HSV exerted a dose-dependent killing effect against established EOC cell lines SKOV3, A2780, OVCAR, Caov3, the ovarian teratocarcinoma PA-1 line, and primary cultures obtained from ascites of patients with stage III EOC (25, 27). To assess the efficacy of oncolytic ICP34.5-deficient HSV-1 against EOC in vivo, we utilized xenograft and syngeneic mouse models. Severe combined immune deficiency (SCID) mice were injected with A2780 EOC cells i.p., resulting in intraperitoneal tumor nodules resembling stage III ovarian carcinoma 2 weeks later. The mice were then treated with HSV-1716 i.p., while control mice received a mock administration. Virus was administered both to mice with small volume disease (1 week following tumor inoculation) and animals bearing bulky disease (4 weeks following tumor inoculation). Histologic examination of tumors from treated animals revealed large areas of tissue necrosis displaying cytopathic effects and a brisk inflammatory infiltrate composed of polymorphonuclear leukocytes that extended deep into tumor nodules. Immunohistochemical detection of HSV whole proteins utilizing a polyclonal antibody revealed several areas of HSV positivity within or adjacent to areas displaying cytopathic effects. In the SCID mouse, viral antigen expression persisted within tumor for up to 8 weeks following a single i.p. administration. A single administration of virus resulted in arrest of tumor growth; treated mice displayed tumors of similar weight 4 or 8 weeks after viral inoculation compared with the tumors immediately before administration of the virus. This was significantly smaller compared with the disease of untreated animals at the end of the experiment. Similar results were obtained with both early and late virus administration. Single i.p. administration of the virus resulted also in significant prolongation of animal survival (25). We further tested the efficacy of HSV-1716 by using a syngeneic mouse model of ovarian carcinoma developed in our laboratory (74). This model allows for the development of flank as well as orthotopic intraperitoneal tumors that generate ascites after inoculation of ID8 mouse ovarian cancer cells over-expressing VEGF (ID8-VEGF) (74). HSV-1716 was able to infect and kill ID8-VEGF cells in vitro. Moreover, HSV-1716 was able to replicate in ID8-VEGF cells given the typical viral growth curve obtained upon infection of tumor cells in vitro. Intraperitoneal administration of HSV-1716 to mice bearing i.p. tumors significantly prolonged animal survival compared with mock treated animals (9). Moreover, intratumoral injection of solid tumors with HSV-1716 induced significant reduction in tumor growth. Interestingly, tumor endothelial cells seemed to be susceptible to oncolytic HSV infection (10). Thus, similar to what we have observed in human ovarian cancer, HSV-1716 exerts oncolytic activities against murine ovarian cancer.

5 Antitumor Immune Response

In a significant number of patients, antitumor immune response is undetectable or compromised because of immunologic ignorance or tolerance of tumor antigens. On the other hand, we have shown that in EOC, the presence of intratumoral T cells independently correlated with delayed recurrence or delayed death and was associated with increased expression of interferon-gamma, interleukin-2, and lymphocyte-attracting chemokines within the tumor (76). This shows that in some patients an antitumor immune response is triggered, setting the ground for immunotherapeutic approaches against EOC. Cancer vaccination has been attempted to circumvent the lack of tumor antigen presentation, and highly encouraging results have been reported in several occasions. During the past few years, a great emphasis has been put on the feasibility of DC-based vaccinations (6). DCs are the most potent antigen-presenting cells (APCs) known to date. Monocyte DC precursors differentiate into immature DCs, which are very efficient in taking up antigen but unable to stimulate T cells. DC maturation (associated with upregulation of costimulatory molecules CD40, CD80, and CD86) is induced by specific cytokines or inflammatory signals, and allows DCs to present antigen owing to the expression of surface major histocompatibility class II (MHC-II) and dendritic cell lysosome-associated membrane protein (DC-LAMP). Following activation by CD40 ligand (CD40L) or inflammatory signals, mature DCs migrate toward T cell areas, where they activate antigen-specific T cells in a MHC-restricted fashion through the secretion of IL-1a, IL-2, and IL-12, and the upregulation of CD83 and costimulatory/adhesion molecules (7). Multiple subsets of DCs are recognized to date (7, 68), including myeloid and plasmacytoid DCs (CD11c + IL-3R− and CD11c-IL-3R +, respectively, in the human). Although DCs may be generated from CD14− monocyte precursors, CD14 + macrophages can also differentiate into DCs under the appropriate cytokine stimulation (5, 77). Macrophages also have the ability to engulf dying cells and process antigenic material (16). Experimental evidence suggests that in the absence of known tumor antigens, vaccination with whole tumor antigen may be a viable alternative (1, 36). Moreover, therapeutic approaches involving new costimulatory molecules may prove effective against EOC (20, 21). EOC is characterized by the presence of high numbers of resident APCs. Thus, in situ tumor vaccination using recombinant viruses may offer an appealing alternative to classical vaccination.

6 Immunosuppressive Properties of EOC

EOC milieu is characterized by elevated levels of immunosuppressive and growth-stimulating factors including IL-4, IL-10, transforming growth factor beta (TGF-b), and vascular endothelial growth factor (VEGF) (55, 61, 75), where they exert multiple negative functions including downregulation of MHC-I expression by tumor cells and inactivation of immune cells, allowing tumor escape from immune attack by cytotoxic T lymphocytes (CTLs) (28). For example, IL-4, IL-10, and TGF-b suppress macrophage function (33), while IL-10 and TGF-b may induce type-2 polarization of DCs. VEGF suppresses the differentiation and maturation of DCs (34), thus impairing antigen presentation. Moreover, immature DCs may present antigen in vivo inducing tolerance or energy of T cells (31, 39, 47). High levels of IL-4, IL-10, and TGF-b are quite likely responsible for suppression of T cell function through suppression of expression of the T cell receptor zeta chain (TCR-ξ) (41). An excessive influx of plasmacytoid DCs due to adverse chemokine environment may be partly responsible for induction of tolerance mechanisms (78). Moreover, we have demonstrated that in EOC the presence of a population of dendritic cell precursors with angiogenic properties may further contribute to tumor growth and immunosuppression (19–21). The origin of immunosuppressive cytokines in ascites may be multiple. Tumor cells have been reported to express high levels of VEGF as well as TGF-b. In addition, HLADR-monocytes producing high amounts of IL-10 have been isolated from the peritoneal cavity of patients with ovarian carcinoma (46). Moreover, tumor-infiltrating macrophages have been associated with T cell suppression and induction of tolerance toward tumor antigens (33). Finally, high amounts of CD4 + CD25 + regulatory T cells were described in ovarian carcinoma and ascites, which may also produce elevated amounts of IL-10 and TGF-b (72). We have shown that the presence of T regulatory cells in EOC positively correlates with poor prognosis (30). Thus, the immune milieu of EOC is adverse to antigen presentation and immune effector cell function. However, evidence suggests that immune suppression in the tumor microenvironment may be reversed by stimulatory cytokines. For example, macrophages procured from malignant ascites were induced to differentiate into potent DCs presenting tumor antigen ex vivo with GM-CSF and IL-4 (18).

7 Immunological Effects of Oncolytic HSV

It has been previously demonstrated that the immune response contributes to the viral-mediated tumor destruction and the increase in survival during HSV oncolytic therapy (51, 52, 64, 66). The mechanisms underlying tumor vaccination triggered by HSV are unknown. Viruses are potent stimulators of the immune system triggering a strong immune response against viral and, occasionally, "self" antigens. In part, viruses may achieve such potent immune stimulation through the generation of strong "danger" signals, including cytokines and chemokines released by infected cells. For example, heat shock proteins (HSPs), cytoplasmic chaperones that facilitate antigen up-take by APCs and correlate with tumor immunogenicity, are upregulated by viruses (71). Important information may be derived from wild type (wt) HSV. Herpetic infections in the eye trigger a potent antiviral response mediated by CD4+ and CD8+ cells and are often followed by a postherpetic T cell response against corneal stromal cells, resulting in autoimmune stromal keratitis (32). These manifestations are evidence of efficient presentation not only of viral but also of host antigens. This evidence is in contrast with reports indicating that HSV suppresses myeloid DC maturation and the ability to stimulate naïve T cells (45, 60). Although human peripheral blood monocytes are resistant to HSV infection (62), immature myeloid DCs express the herpes receptors HVE-A and HVE-C and are susceptible to infection by HSV, which results in functional inactivation. However, HSV may potently stimulate other APCs, namely macrophages and interferon-producing plasmacytoid DCs. In fact, wt HSV induces secretion of interferon-alpha (IFN-a), interferon-beta (IFN-b), and TNF-a in CD4 + CD11c- plasmacytoid DC precursors, promoting their survival, differentiation and maturation (40). Furthermore, HSV infection induces high levels of CC chemokine macrophage inflammatory protein-beta (MIP-b) and HSPs (50). Interestingly, supernatants of HSV-infected cells strongly induce DC maturation, providing a stimulus as strong as bacterial lipopolysaccharide (LPS) (60).

8 In Situ Tumor Vaccination

We have found that similar inflammatory mechanisms as those triggered by wt HSV are activated by oncolytic HSV. Oncolytic HSV-1716 therapy in our mouse model of ovarian carcinoma induced expression of IFN-γ, MIG, and IP-10 in the tumor milieu (9). This was accompanied by a significant increase in the number of tumor-associated NK and CD8 + T cells expressing CXCR3 and CD25, and a significantly higher frequency of tumor-reactive IFN-γ producing T cells. Ascites from HSV-1716-treated animals efficiently induced in vitro migration of NK and CD8+ T cells, which was dependent on the presence of MIG and IP-10. Murine monocytes and dendritic cells produce MIG and IP-10 upon HSV-1716 infection. This effect was partially abrogated by neutralizing antibodies against IFN-α and β, thus indicating a role of type-1 IFNs in the reported effect (9). Upon HSV-1716 infection, mouse ovarian tumor cells showed high levels of expression viral glycoproteins B and D and were highly phagocyted by DCs. Interestingly, increased phagocytosis of tumor-infected cells by DCs was impaired by heparin, and anti-HSV glycoproteins B and D, indicating that viral infection enhances adhesive interactions between DCs and tumor apoptotic bodies. Moreover, HSV-1716 infected cells expressed high levels of heat shock proteins 70 and GRP94 (Fig. 1). We have reported similar effects upon infection of tumor cells with other recombinant HSV strains (8, 29). After phagocytosis of tumor-infected cells, DCs acquired a mature status in vitro and in vivo, upregulated the expression of costimulatory molecules and increased migration toward MIP-3β (Fig. 2). Furthermore, HSV-1716 oncolytic treatment markedly reduced VEGF levels in tumor-bearing animals, thus abrogating a tumor immunosuppressive milieu (Fig. 3). These mechanisms may account for the highly enhanced antitumoral immune responses observed in HSV-1716 treated animals (Fig. 4).

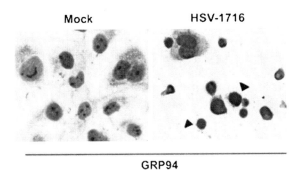

Fig. 1 Immunohistochemistry of GRP94 antigen in ID8-VEGF cells infected with live or UV-inactivated (mock) HSV-1716, 36h p.i., 1 MOI. *Arrowheads* show typical apoptotic cells strongly staining for GRP94

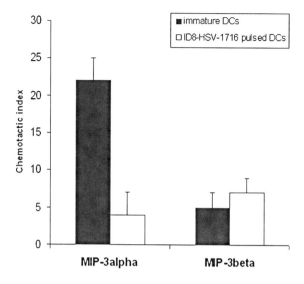

Fig. 2 Maturation of dendritic cells upon phagocytosis of HSV-1716 infected cells. Chemotaxis analysis of DCs 48 h after phagocytosis of ID8-VEG Fcells killed by HSV-1716 compared with immature DCs

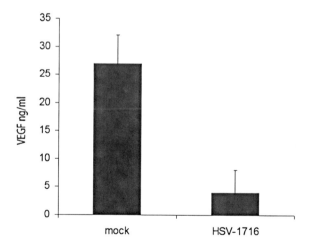

Fig. 3 Oncolytic treatment reduces VEGF expression. ELISA in ascites of i.p. ID8-VEGF tumors showed a decrease in the levels of VEGF protein 5 days after HSV-1716 treatment in the intraperitoneal model of murine ovarian carcinoma

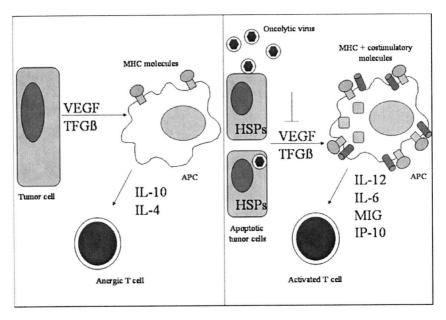

Fig. 4 The in situ tumor vaccination mechanism. *Left panel*: typical immunosuppressive EOC milieu. APCs show an immature status with low expression of MHC molecules and little to no expression of costimulatory molecules. Cytokines produce by both tumor cells and resident APCs contribute to immunosuppressive status, rendering T cells anergic or increasing the amount of regulatory T cells. *Right panel*: Upon oncolytic therapy, death of tumor cells abrogate cytokine production from these cells, decreasing VEGF levels. Also, apoptotic cells expressing HSP are actively phagocyte by APCs which, in the absence of VEGF, turn into mature APCs and are able to activate specific antitumor T cells

9 Conclusion

We hypothesized that one mechanism underlying tumor vaccination triggered by HSV-1716 could be the upregulation of the immunogenic properties of infected cells. We detected overexpression of danger signals such as Hsp-70 and GRP-94 in HSV-infected tumor cells. These proteins play a critical role in promoting antigen cross-presentation by DCs and eliciting tumor-specific protective immunity, in part through cytokine production and increased maturation of DCs. Consistently, phagocytosis of ID8 cells killed by HSV-1716 induced phenotypical and functional maturation of myeloid DCs. This effect on DC maturation was observed both in vitro and in vivo after HSV-1716 treatment of tumors. Tumor-associated VEGF has been reported as playing a suppressive role in DC maturation. We observed that upon HSV-1716 treatment, VEGF levels were downregulated in vivo. Thus, HSV oncolysis induces complex modifications of the tumor microenvironment, which are conducive to enhanced DC function. Collectively, this work shows that HSV oncolytic therapy triggers local and systemic immune response and may be a convenient therapeutic tool for combinatorial biological cancer therapy.

Acknowledgments This work was supported through grants by the National Institutes of Health RO-1 CA098951 and Ovarian SPORE PO1-CA083638

References

1. Alberti, M.L., Sauteri, B. and Bhardwaji, N. (1998) Dendritic cells acquire antigen from apoptotic cells and induce class I-restricted CTLs. Nature 392: 86–89.
2. Alemany, R., Balague, C. and Curiel, D.T. (2000) Replicative adenoviruses for cancer therapy. Nat. Biotechnol. 18: 723–727.
3. Andreansky, S.S., He, B., Gillespie, G.Y., Soroceanu, L., Markert, J., Chou, J., Roizman, B. and Whitley, R.J. (1996) The application of genetically engineered herpes simplex viruses to the treatment of experimental brain tumors. Proc. Natl. Acad. Sci. USA 93: 11313–11318.
4. Andreansky, S., He, B., van Cott, J., McGhee, J., Markert, J.M., Gillespie, G.Y., Roizman, B. and Whitley, R.J. (1998) Treatment of intracranial gliomas in immunocompetent mice using herpes simplex viruses that express murine interleukins. Gene Ther. 5: 121–130.
5. Andreesen, R., Hennemann, B. and Krause, S.W. (1998) Adoptive immunotherapy of cancer using monocyte-derived macrophages: rationale, current status, and perspectives. J. Leukoc. Biol. 64: 419–426.
6. Banchereau, J. and Steinman, R.M. (1998) Dendritic cells and the control of immunity. Nature 392: 245–252.
7. Banchereau, J., Briere, F., Caux, C., Davoust, J., Lebecque, S., Liu, Y.J., Pulendran, B. and Palucka, K. (2000) Immunobiology of dendritic cells. Annu. Rev. Immunol. 18: 767–811.
8. Benencia, F., Courreges, M.C., Conejo Garcia, J.R., Mohammed-Hadley, A. and Coukos, G. (2006) Direct vaccination with tumor cells killed with ICP4-deficient HSVd120 elicits effective antitumor immunity. Cancer Biol. Ther. 5(7): 867–874.
9. Benencia, F., Courreges, M.C., Conejo-Garcia, J.R., Mohamed-Hadley, A., Zhang, L., Buckanovich, R.J., Carroll, R., Fraser, N. and Coukos, G. (2005a) HSV oncolytic therapy upregulates interferon-inducible chemokines and recruits immune effector cells in ovarian cancer. Mol. Ther. 12(5): 789–802.
10. Benencia, F., Courreges, M.C., Conejo-Garcia, J.R., Buckanovich, R.J., Zhang, L., Carroll, R.H., Morgan, M.A. and Coukos, G. (2005b) Oncolytic HSV exerts direct antiangiogenic activity in ovarian carcinoma. Hum. Gene Ther. 16(6): 765–778.
11. Blank, S.V., Rubin, S.C., Coukos, G., Amin, K.M., Albelda, S.M. and Molnar-Kimber, K.L. (2002) Replication-selective herpes simplex virus type 1 mutant therapy of cervical cancer is enhanced by low-dose radiation. Hum. Gene Ther. 13: 627–639.
12. Brandt, C.R., Imesch, P.D., Robinson, N.L., Syed, N.A., Untawale, S., Darjatmoko, S.R., Chappell, R.J., Heinzelman, P. and Albert, D.M. (1997) Treatment of spontaneously arising retinoblastoma tumors in transgenic mice with an attenuated herpes simplex virus mutant. Virology 229: 283–291.
13. Brown, S.M., Harland, J., MacLean, A.R., Podlech, J. and Clements, J.B. (1994) Cell type and cell state determine differential in vitro growth of non-neurovirulent ICP34.5-negative herpes simplex virus types 1 and 2. J. Gen. Virol. 75: 2367–2377.
14. Carroll, N.M., Chiocca, E.A., Takahashi, K. and Tanabe, K.K. (1996) Enhancement of gene therapy specificity for diffused colon carcinoma liver metastases with recombinant herpes simplex virus. Ann. Surg. 224: 323–330.
15. Chambers, R., Gillespie, G.Y., Soroceanu, L., Andreansky, S., Chatterjee, S., Chou, J., Roizman, B. and Whitley, R.J. (1995) Comparison of genetically engineered herpes simplex viruses for the treatment of brain tumors in a scid mouse model of human malignant glioma. Proc. Natl. Acad. Sci. USA. 92: 1411–1415.
16. Chaperot, L., Chokri, M., Jacob, M.C., Drillat, P., Garban, F., Egelhofer, H., Molens, J.P., Sotto, J.J., Bensa, J.C. and Plumas, J. (2000) Differentiation of antigen-presenting cells (den-

dritic cells and macrophages) for therapeutic application in patients with lymphoma. Leukemia 14: 1667–1677.
17. Chiocca, E.A. and Smith, E.R. (2000) Oncolytic viruses as novel anticancer agents: turning one scourge against another. Expert Opin. Investig. Drugs 9: 311–327.
18. Chu, C.S., Woo, E.Y., Toll, A.J., Rubin, S.C., June, C.H., Carroll, R.G. and Schlienger, K. (2002) Tumor associated macrophages as a source of functional dendritic cells in ovarian cancer patients. Clin. Immunol. 102: 281–301.
19. Conejo-Garcia, J.R., Buckanovich, R.J., Benencia, F., Courreges, M.C., Rubin, S.C., Carroll, R.G. and Coukos, G. (2005) Vascular leukocytes contribute to tumor vascularization. Blood 105(2): 679–681.
20. Conejo-Garcia, J.R., Benencia, F., Courreges, M.C., Kang, E., Mohamed-Hadley, A., Buckanovich, R.J., Holtz, D.O., Jenkins, A., Na, H., Zhang, L., Wagner, D.S., Katsaros, D., Caroll, R. and Coukos, G. (2004b) Tumor-infiltrating dendritic cell precursors recruited by a beta-defensin contribute to vasculogenesis under the influence of Vegf-A. Nat. Med. 10(9): 950–958.
21. Conejo-Garcia, J.R., Benencia, F., Courreges, M.C., Gimotty, P.A., Khang, E., Buckanovich, R.J., Frauwirth, K.A., Zhang, L., Katsaros, D., Thompson, C.B., Levine, B. and Coukos, G. (2004a). Ovarian carcinoma expresses the NKG2D ligand Letal and promotes the survival and expansion of CD28- antitumor T cells. Cancer Res. 64(6): 2175–2182.
22. Conejo-Garcia, J.R., Benencia, F., Courreges, M.C., Khang, E., Zhang, L., Mohamed-Hadley, A., Vinocur, J.M., Buckanovich, R.J., Thompson, C.B., Levine, B. and Coukos, G. (2003) Letal, A tumor-associated NKG2D immunoreceptor ligand, induces activation and expansion of effector immune cells. Cancer Biol. Ther. 2(4): 446–451.
23. Coukos, G., Rubin, S.C. and Molnar-Kimber, K.L. (1999a) Application of recombinant herpes simplex virus-1 (HSV-1) for the treatment of malignancies outside the central nervous system. Gene Ther. Mol. Biol. 3: 78–89.
24. Coukos, G., Benencia, F., Buckanovich, R.J. and Conejo-Garcia, J.R. (2005) The role of dendritic cell precursors in tumour vasculogenesis. Br. J. Cancer. 92(7): 1182–1187.
25. Coukos, G., Makrigiannakis, A., Kang, E.H., Caparelli, D., Benjamin, I., Kaiser, L.R., Rubin, S.C., Albelda, S.M. and Molnar-Kimber, K.L. (1999b) Use of carrier cells to deliver a replicationselective Herpes Simplex Virus-1 mutant for the intraperitoneal therapy if epithelial ovarian cancer. Clin. Cancer Res. 5: 1523–1537.
26. Coukos, G., Makrigiannakis, A., Montas, S., Kaiser, L.R., Toyozumi, T., Benjamin, I., Albelda, S.M., Rubin, S.C. and Molnar-Kimber, K.L. (2000a) Multi-attenuated herpes simplex virus-1 mutant G207 exerts cytotoxicity against epithelial ovarian cancer but not normal mesothelium and is suitable for intraperitoneal oncolytic therapy. Cancer Gene Ther., 7: 275–283.
27. Coukos, G., Makrigiannakis, A., Kang, E.H., Rubin, S.C., Albelda, S.M. and Molnar-Kimber, K.L. (2000b) Oncolytic herpes simplex virus-1 lacking ICP34.5 induces p53-independent death and is efficacious against chemotherapy-resistant ovarian cancer. Clin. Cancer Res., 6: 3342–3353.
28. Couldwell, W.T., Dore-Duffy, P., Apuzzo, M.L. and Antel, J.P. (1991) Malignant glioma modulation of immune function: relative contribution of different soluble factors. J. Neuroimmunol. 33: 89–96.
29. Courreges, M.C., Benencia, F., Conejo-Garcia, J.R., Zhang, L. and Coukos, G. (2006) Preparation of apoptotic tumor cells with replication-incompetent HSV augments the efficacy of dendritic cell vaccines. Cancer Gene Ther. 13(2): 182–193.
30. Curiel, T.J., Coukos, G., Zou, L., Alvarez, X., Cheng, P., Mottram, P., Evdemon-Hogan, M., Conejo-Garcia, J.R., Zhang, L., Burow, M., Zhu, Y., Wei, S., Kryczek, I., Daniel, B., Gordon, A., Myers, L., Lackner, A., Disis, M.L., Knutson, K.L., Chen, L. and Zou, W. (2004) Specific recruitment of regulatory T cells in ovarian carcinoma fosters immune privilege and predicts reduced survival. Nat. Med. 10(9): 942–949.
31. Dhodapkar, M.V., Steinman, R.M., Krasovsky, J., Munz, C. and Bhardwaj, N. (2001) Antigen-specific inhibition of effector T cell function in humans after injection of immature dendritic cells. J. Exp. Med. 193: 233–238.

32. Doymaz, M.Z. and Rouse, B.T. (1992) Immunopathology of herpes simplex virus infections. Curr. Top. Microbiol. Immunol. 179: 121–136.
33. Elgert, K.D., Alleva, D.G. and Mullins, D.W. (1998) Tumor-induced immune dysfunction: the macrophage connection. J. Leukoc. Biol. 64: 275–290.
34. Gabrilovich, D.I., Chen, H.L., Girgis, K.R., Cunningham, H.T., Meny, G.M., Nadaf, S., Kavanaugh, D. and Carbone, D.P. (1996) Production of vascular endothelial growth factor by human tumors inhibits the functional maturation of dendritic cells. Nat. Med. 2: 1096–10103.
35. Glorioso, J., Bender, M.A., Fink, D. and DeLuca, N. (1995) Herpes simplex virus vectors. Mol. Cell Biol. Hum. Dis. Ser. 5: 33–63.
36. Hart, I. and Colaco, C. (1997) Immunotherapy. Fusion induces tumour rejection. Nature 388: 626–627.
37. Heise, C., Sampson-Johannes, A., Williams, A., McCormick, F., Von Hoff, D.D. and Kirn, D.H. (1997) ONYX-015, an $E1B$ gene attenuated adenovirus, causes tumor-specific cytolysis and antitumoral efficacy that can be augmented by standard chemotherapeutic agents. Nat. Med. 3: 639–645.
38. Hunter, W.D., Martuza, R.L., Feigenbaum, F., Todo, T., Mineta, T., Yazaki, T., Toda, M., Newsome, J.T., Platenberg, R.C., Manz, H.J. and Rabkin, S.D. (1999) Attenuated, replication-competent herpes simplex virus type 1 mutant G207: safety evaluation of intracerebral injection in nonhuman primates. J. Virol. 73: 6319–6326.
39. Jonuleit, H., Schmitt, E., Schuler, G., Knop, J. and Enk, A.H. (2000) Induction of interleukin 10-producing, nonproliferating CD4(+) T cells with regulatory properties by repetitive stimulation with allogeneic immature human dendritic cells. J. Exp. Med. 192: 1213–1222.
40. Kadowaki, N., Antonenko, S., Lau, J.Y. and Liu, Y.J. (2000) Natural interferon alpha/beta-producing cells link innate and adaptive immunity. J. Exp. Med. 192: 219–226.
41. Kalinski, P., Schuitemaker, J.H., Hilkens, C.M., Wierenga, E.A. and Kapsenberg, M.L. (1999) Final maturation of dendritic cells is associated with impaired responsiveness to IFN-gamma and to bacterial IL-12 inducers: decreased ability of mature dendritic cells to produce IL-12 during the interaction with Th cells. J. Immunol. 162: 3231–3236.
42. Kaplitt, M.G., Tjuvajev, J.G., Leib, D.A., Berk, J., Pettigrew, K.D., Posner, J.B., Pfaff, D.W., Rabkin, S.D. and Blasberg, R.G. (1994) Mutant herpes simplex virus induced regression of tumors growing in immunocompetent rats. J. Neurooncol. 19: 137–147.
43. Kesari, S., Randazzo, B.P., Valyi-Nagy, T., Huang, Q.S., Brown, S.M., MacLean, A.R., Lee, V.M., Trojanowski, J.Q. and Fraser, N.W. (1995) Therapy of experimental human brain tumors using a neuroattenuated herpes simplex virus mutant. Lab. Invest. 73: 636–648.
44. Kucharczuk, J.C., Randazzo, B., Chang, M.Y., Amin, K.M., Elshami, A.A., Sterman, D.H., Rizk, N.P., Molnar-Kimber, K.L., Brown, S.M., MacLean, A.R., Litzky, L.A., Fraser, N.W., Albelda, S.M. and Kaiser, L.R. (1997) Use of a "replication-restricted" herpes virus to treat experimental human malignant mesothelioma. Cancer Res. 57: 466–471.
45. Kruse, M., Rosorius, O., Kratzer, F., Stelz, G., Kuhnt, C., Schuler, G., Hauber, J. and Steinkasserer, A. (2000) Mature dendritic cells infected with herpes simplex virus type 1 exhibit inhibited T-cell stimulatory capacity. J. Virol. 74: 7127–7136.
46. Loercher, A.E., Nash, M.A., Kavanagh, J.J., Platsoucas, C.D. and Freedman, R.S. (1999) Identification of an IL-10-producing HLADR- negative monocyte subset in the malignant ascites of patients with ovarian carcinoma that inhibits cytokine protein expression and proliferation of autologous T cells. J. Immunol. 163: 6251–6260.
47. Mahnke, K., Guo, M., Lee, S., Sepulveda, H., Swain, S.L., Nussenzweig, M. and Steinman, R.M. (2000) The dendritic cell receptor for endocytosis, DEC-205, can recycle and enhance antigen presentation via major histocompatibility complex class II positive lysosomal compartments. J. Cell Biol. 151: 673–684.
48. Markert, J.M., Medlock, M.D., Rabkin, S.D., Gillespie, G.Y., Todo, T., Hunter, W.D., Palmer, C.A., Feigenbaum, F., Tornatore, C., Tufaro, F. and Martuza, R.L. (2000) Conditionally replicating herpes simplex virus mutant, G207 for the treatment of malignant glioma: results of a phase I trial. Gene Ther. 7: 867–874.

49. McMenamin, M.M., Byrnes, A.P., Pike, F.G., Charlton, H.M., Coffin, R.S., Latchman, D.S., and Wood, M.J. (1998) Potential and limitations of a gamma 34.5 mutant of herpes simplex 1 as a gene therapy vector in the CNS. Gene Ther. 5: 594–604.
50. Mikloska, Z., Danis, V.A., Adams, S., Lloyd, A.R., Adrian, D.L. and Cunningham, A.L. (1998) In vivo production of cytokines and beta (C-C) chemokines in human recurrent herpes simplex lesions – do herpes simplex virus-infected keratinocytes contribute to their production? J. Infect. Dis. 177: 827–838.
51. Miller, C.G., Fraser, N.W. (2000) Role of the immune response during neuro-attenuated herpes simplex virus-mediated tumor destruction in a murine intracranial melanoma model. Cancer Res. 60: 5714–5722.
52. Miller, C.G. and Fraser, N.W. (2003) Requirement of an integrated immune response for successful neuroattenuated HSV-1 therapy in an intracranial metastatic melanoma model. Mol. Ther. 7: 741–747.
53. Mineta, T., Rabkin, S.D. and Martuza, R.L. (1994) Treatment of malignant gliomas using ganciclovir-hypersensitive, ribonucleotide reductase-deficient herpes simplex viral mutant. Cancer Res. 54: 3963–3966.
54. Mineta, T., Rabkin, S.D., Yazaki, T., Hunter, W.D. and Martuza, R.L. (1995) Attenuated multi-mutated herpes simplex virus-1 for the treatment of malignant gliomas. Nat. Med. 1: 938–943.
55. Nash, M.A., Ferrandina, G., Gordinier, M., Loercher, A. and Freedman, R.S. (1999) The role of cytokines in both the normal and malignant ovary. Endocr. Relat. Cancer 6: 93–107.
56. Norman, K.L. and Lee, P.W. (2000) Reovirus as a novel oncolytic agent. J. Clin. Invest. 105: 1035–1038.
57. Pawlik, T.M., Nakamura, H., Yoon, S.S., Mullen, J.T., Chandrasekhar, S., Chiocca, E.A. and Tanabe, K.K. (2000) Oncolysis of diffuse hepatocellular carcinoma by intravascular administration of a replication-competent, genetically engineered herpesvirus. Cancer Res. 60: 2790–2795.
58. Pyles, R.B., Warnick, R.E., Chalk, C.L., Szanti, B.E. and Parysek, L.M. (1997) A novel multiply-mutated HSV-1 strain for the treatment of human brain tumors. Hum. Gene Ther. 8: 533–544.
59. Rampling, R., Cruickshank, G., Papanastassiou, V., Nicoll, J., Hadley, D., Brennan, D., Petty, R., MacLean, A., Harland, J., McKie, E., Mabbs, R. and Brown, M. (2000) Toxicity evaluation of replication-competent herpes simplex virus (ICP 34.5 null mutant 1716) in patients with recurrent malignant glioma. Gene Ther. 7: 859–866.
60. Salio, M., Cella, M., Suter, M. and Lanzavecchia, A. (1999) Inhibition of dendritic cell maturation by herpes simplex virus. Eur. J. Immunol. 29: 3245–3253.
61. Santin, A.D., Hermonat, P.L., Ravaggi, A., Cannon, M.J., Pecorelli, S. and Parham, G.P. (1999) Secretion of vascular endothelial growth factor in ovarian cancer. Eur. J. Gynaecol. Oncol. 20: 177–181.
62. Sarmiento, M. and Kleinerman, E.S. (1990) Innate resistance to herpes simplex virus infection. Human lymphocyte and monocyte inhibition of viral replication. J. Immunol. 144: 1942–1953.
63. Stojdl, D.F., Lichty, B., Knowles, S., Marius, R., Atkins, H., Sonenberg, N. and Bell, J.C. (2000) Exploiting tumor-specific defects in the interferon pathway with a previously unknown oncolytic virus. Nat. Med. 6: 821–825.
64. Thomas, D.L. and Fraser, N.W. (2003) HSV-1 therapy of primary tumors reduces the number of metastases in an immune-competent model of metastatic breast cancer. Mol. Ther. 8: 543–551.
65. Toda, M., Rabkin, S.D. and Martuza, R.L. (1998a) Treatment of human breast cancer in a brain metastatic model by G207, a replication-competent multimutated herpes simplex virus 1. Hum. Gene Ther. 9: 2177–2185.
66. Toda, M., Martuza, R.L., Kojima, H. and Rabkin, S.D. (1998b) In situ cancer vaccination: an IL-12 defective vector/replicationcompetent herpes simplex virus combination induces local and systemic antitumor activity. J. Immunol. 160: 4457–4464.

67. Todo, T., Feigenbaum, F., Rabkin, S.D., Lakeman, F., Newsome, J.T., Johnson, P.A., Mitchell, E., Belliveau, D., Ostrove, J.M. and Martuza, R.L. (2000) Viral shedding and biodistribution of G207, a multimutated, conditionally replicating herpes simplex virus type 1, after intracerebral inoculation in aotus. Mol. Ther. 2: 588–595.
68. Vandenabeele, S. and Wu, L. (1999) Dendritic cell origins: puzzles and paradoxes. Immunol. Cell Biol. 77: 411–9.
69. Walker, J.R., McGeagh, K.G., Sundaresan, P., Jorgensen, T.J., Rabkin, S.D. and Martuza, R.L. (1999) Local and systemic therapy of human prostate adenocarcinoma with the conditionally replicating herpes simplex virus vector G207. Hum. Gene Ther. 10: 2237–2243.
70. Wang, M., Rancourt, C., Navarro, J.G., Krisky, D., Marconi, P., Oligino, T., Alvarez, R.D., Siegal, G.P., Glorioso, J.C. and Curiel, D.T. (1998) High-efficacy thymidine kinase gene transfer to ovarian cancer cell lines mediated by herpes simplex virus type 1 vector. Gynecol. Oncol. 71: 278–287.
71. Wells, A.D. and Malkovsky, M. (2000) Heat shock proteins, tumor immunogenicity and antigen presentation: an integrated view. Immunol. Today 21: 129–132.
72. Woo, E.Y., Chu, C.S., Goletz, T.J., Schlienger, K., Yeh, H., Coukos, G., Rubin, S.C., Kaiser, L.R. and June, C.H. (2001) Regulatory CD4 + CD25 + T cells in tumors from patients with early stage non-small cell lung cancer and late-stage ovarian cancer. Cancer Res. 61: 4766–4772.
73. Yazaki, T., Manz, H.J., Rabkin, S.D. and Martuza, R.L. (1995) Treatment of human malignant meningiomas by G207, a replication-competent multimutated herpes simplex virus 1. Cancer Res. 55: 4752–4756.
74. Zhang, L., Yang, N., Garcia, J.R., Mohamed, A., Benencia, F., Rubin, S.C., Allman, D. and Coukos, G. (2002) Generation of a syngeneic mouse model to study the effects of vascular endothelial growth factor in ovarian carcinoma. Am. J. Pathol. 161: 2295–2309.
75. Zhang, L., Yang, N., Conejo-Garcia, J.R., Katsaros, D., Mohamed-Hadley, A., Fracchioli, S., Schlienger, K., Toll, A., Levine, B., Rubin, S.C. and Coukos, G. (2003b) Expression of endocrine gland-derived vascular endothelial growth factor in ovarian carcinoma. Clin. Cancer Res. 9 (1): 264–272.
76. Zhang, L., Conejo-Garcia, J.R., Katsaros, D., Gimotty, P.A., Massobrio, M., Regnani, G., Makrigiannakis, A., Gray, H., Schlienger, K., Liebman, M.N., Rubin, S.C. and Coukos, G. (2003a) Intratumoral T cells, recurrence, and survival in epithelial ovarian cancer. N. Engl. J. Med. 348(3): 203–213.
77. Zou, W., Borvak, J., Marches, F., Wei, S., Galanaud, P., Emilie, D. and Curiel, T.J. (2000) Macrophage-derived dendritic cells have strong Th1-polarizing potential mediated by beta-chemokines rather than IL-12. J. Immunol. 165: 4388–4396.
78. Zou, W., Machelon, V., Coulomb-L'Hermin, A., Borvak, J., Nome, F., Isaeta, T., Wei, S., Krzysiek, R., Durand-Gasselin, I., Gordon, A., Pustilnik, T., Curiel, T., Galanaud, P., Capron, F., Emilie, D. and Curiel, T.J. (2001) Stromal-derived factor-1 in human tumors recruits and alters the function of plasmacytoid precursor dendritic cells. Nat. Med. 7: 1339–1346.

Cancer Immunotherapy: Perspectives and Prospects

Sonia A. Perez and Michael Papamichail

1 Introduction

During the last 15–20 years, immunotherapy has emerged as an alternative or adjuvant approach to cancer treatment. The immunotherapy of cancer holds promise in harnessing the host immune response to specifically target tumor cells. Although the potential of this strategy remains auspicious, the approach requires optimization. In this review, we intend to delineate basic aspects of the different immunotherapeutic modalities, their advantages and drawbacks, as well as ways to improve the efficacy of each, either alone or in combination with others.

A key advance in immunology in the past decade has been the elucidation of the antigenic basis of tumor-cell recognition and destruction. As in normal cells, tumor cells express MHC-peptide antigen complexes, and thus can elicit specific HLA-restricted immune responses. The result of such responses is the generation of specific $CD4^+$ and $CD8^+$ T lymphocytes, as well as antibodies, against peptide epitopes derived from tumor-associated antigens (TAA). TAA fall into four categories: unique antigens (mutated, alternatively processed, idiotypic antibodies), shared antigens (cancer/testis antigens, normally expressed during development but aberrantly expressed in adult somatic cells, i.e., NY-ESO-1), differentiation antigens (cell-lineage-specific, i.e., Melan-A/MART-1, gp100), overexpressed antigens (expressed at higher levels than in normal cells, i.e., HER-2/neu), and viral antigens (HPV, HBV, EBV, etc.) (72).

To develop an effective immune response, the cooperation of different cell types is required: antigen-presenting cell (APC), $CD4^+$ helper T cell, cytotoxic $CD8^+$ T cell, and antibody-producing B cell. Other cells, of the innate immune system, contribute to the regulation of the immune response, among which natural killer (NK) cells play a major role because they can also directly kill tumor cells. Termination of an immune response on the one hand and prevention of autoimmunity on the other hand are processes that can be regulated by suppressor mechanisms, which include soluble factors (such as TGF-β, IL-10, IDO, etc.) and cell populations (such as myeloid suppressor cells, regulatory T cells, etc.). It is only in the last few years after extensive research on the role of these regulatory mechanisms in cancer that they need to be addressed when considering the immunotherapeutic treatment of these diseases.

The knowledge acquired from tumor immunology and from animal models led to the design of clinical protocols for cancer immunotherapy. Current approaches in cancer immunotherapy include the use of monoclonal antibodies, cytokines, tumor vaccines, dendritic cells (DCs), and adoptive transfer of T cells or NK cells.

2 Antibody-Based Immunotherapy of Cancer

The use of monoclonal antibodies (mAbs) in cancer treatment has emerged following the development of hybridoma technology by Köhler and Milstein (55). The design of mAbs is at the moment one of the favorite preferences of pharmaceutical companies. Antibodies with the potential for use in cancer immunotherapy are either directed against the tumor or the tumor microenvironment, or are intended to function as immune modulators by inducing responses against cancer (93, 98). Targeting of intracellular molecules (such as signal transducers, transcription factors, hormone receptors) with mAbs is also feasible by means of cell penetrating peptides (CPP) (74, 77). Antibody-based immunotherapy has the overall advantage of not interfering with suppressor mechanisms that may affect its successful treatment capacity, as is the case with cancer vaccines and adoptive cellular therapy.

The mechanisms of action of monoclonal antibodies targeting tumor cells themselves include (1) the induction of death pathways by the engaging and modulating of receptors on the tumor cell, (2) antibody-dependent cellular cytotoxicity (ADCC), (3) blockade of factors necessary for tumor growth, and (4) the delivery of cytotoxic agents to the tumor cells.

The inhibition of tumor neovascularization or the disruption of the tumor surrounding stroma is among the targets of mAbs directed against the tumor microenvironment.

MAbs targeting immunosuppressive cytokines (such as neutralizing mAbs for IL-10 and TGF-β), immunosuppressive populations (CD25, CTLA-4, and GITR on regulatory T cells), costimulatory molecules on effector cells (CD28, 4-1BB, and CD40), or factors affecting DC, T cell, or NK cell trafficking are intended to directly promote antitumor immune responses. On the other hand, mAbs directed against TAA can enhance the processing and presentation of these antigens, leading to more effective T and B antigen-specific immune responses.

There are currently a number of FDA-approved mAbs employed in cancer treatment and others that are still under investigation in clinical trials (98, 110).

Although no mAb has yet been introduced into clinical practice for ovarian cancer treatment, Trastuzumab has a potential for application in HER-2/neu expressing ovarian tumors, especially under the light of new data indicating that patients with tumors expressing low/intermediate HER-2/neu and high heregulin might also benefit from Trastuzumab therapy (68). Oregovomab (OvaRex), a murine mAb against CA125 belongs to an innovative group of antibodies, targeting TAA in the bloodstream rather than directly binding to antigens on tumor cells themselves. Oregovomab forms complexes with circulating CA125 antigens that are recognized as foreign, as they contain the murine mAb, thereby leading to enhanced processing of the autologous antigen-mAb complex (9, 10). Bevacizumab (Avastin), a mAb preventing VEGF receptor binding and inhibiting angiogenesis and tumor growth, and Cetuximab (Erbitux), targeting EGFR, are also under evaluation for possible clinical application in patients with ovarian cancer (2, 24).

3 Cytokines in Cancer Immunotherapy

Cytokines are substances secreted by immune cells that act as positive or negative regulators of both the innate and the adaptive immune system. There are several positively acting cytokines currently being used in cancer immunotherapy. IFN-α2b (Intron A) and IL-2 (Proleukin) are two FDA-approved cytokines for the treatment of cancer.

IFN-α2b has been approved for the treatment of malignant melanoma, renal and kidney carcinoma, Kaposi's sarcoma, follicular lymphoma, hairy cell leukemia, and CML (94). The efficacy of IFN-α2b in ovarian cancer is still under investigation in clinical trials (10, 34). Furthermore, administration of IFN-γ along with first-line chemotherapy has also been reported to improve progression-free survival in ovarian cancer patients (10, 67).

IL-2 has been approved for the treatment of malignant melanoma, renal and kidney carcinoma, leukemia, and lymphoma (94). However, apart from its role in the initial activation of T and NK cells, IL-2 is also essential for the maintenance of self tolerance, either by activation-induced cell death (AICD) (61) to eliminate self-reactive T cells or by generating and maintaining $CD4^+$ $CD25^+$ regulatory T cells (97). Thus, its replacement by IL-15, a cytokine with antiapoptotic action, very effective in the activation of T, NK, and NKT cells, the generation and maintenance of $CD8^+$ memory T cells, has been proposed as more effective for cancer immunotherapy (109, 110).

GM-CSF (Sargramostim) is indicated for use following induction chemotherapy in older adult patients with acute myelogenous leukemia (AML) to shorten time to neutrophil recovery and in autologous bone marrow transplantation for acceleration of myeloid recovery. On the other hand, GM-CSF is also used in clinical trials as an adjuvant, acting on dendritic cells, to enhance T cell immune responses.

Other cytokines, such as IL-12 (51) and IL-21 (76), are also under evaluation for their potential use in cancer immunotherapy. However, in contrast to mAb development, the field of cytokine application in cancer immunotherapy is not moving very quickly.

4 Cancer Vaccines

Generating an immune response directed against a tumor antigen has several potential clinical advantages. Vaccination, if effective, would stimulate immunologic memory and could result in the prevention of relapse after standard therapy such as surgery and radiation has been administered.

Active immunization can be performed with tumor antigens, synthetic tumor antigen peptides, whole tumor cells (autologous or allogeneic), tumor cell lysates, naked DNA, or viral vectors (88). There are many ongoing clinical trials worldwide using injection of synthetic peptides in combination with GM-CSF and/or adjuvant proteins, such as KLH, to elicit antitumor T cell responses. In another series of clinical protocols, DCs are used as APCs loaded with synthetic peptides, whole tumor proteins, antigen RNA, immune-complexes of the relevant TAA with antibodies, through activating FcγR (95) or they are genetically modified to encode tumor antigen or fused with tumor cells.

TAAs tested in clinical trials or in preclinical models for vaccination in ovarian cancer include MUC-1, STn, TAG-72, HER-2/neu, CA125, CEA, p53, mesothelin (44), and NY-ESO-1. Vaccination with whole tumor cells and gene-modified DCs has also been explored (37, 39, 88).

Immunomonitoring of patients is required following active immunization. This includes specific T cell responses: in vivo by DTH responses and in vitro by specific T cell frequency enumeration (ELISPOT or tetramers/pentamers), cytokine production (ELISA), and cytotoxic potential against autologous tumor cells. Since clinical and preclinical data show that effective vaccination may result in epitope spreading (16, 71, 89) and specific antibody responses (46, 75) their evaluation might also be required.

4.1 Peptide Vaccines

Peptide vaccines for cancer immunotherapy have many practical and theoretical advantages compared with other forms of active immunization: (1) they can be appropriately selected to be immunogenic and to avoid autoimmunity, (2) peptides are easy and cheap to produce, (3) synthetic peptides can be modified to improve their antigenicity compared with their native counterparts, (4) the use of peptides overcomes defects in antigen-processing machinery, (5) immune responses to specific peptides are easy to monitor, and (6) they can be combined to produce multiepitope vaccines, thus mimicking the advantages of whole-cell-based vaccines.

On the other hand, there are substantial limitations while using peptide vaccination (27, 100):

- Peptides are MHC restricted, especially MHC class I restricted peptides, while there are many promiscuous MHC class II restricted peptides. Thus, their application is restricted to a limited patient group expressing the relevant MHC alleles. Furthermore, given the fact that HLA loss is a common phenomenon in cancer (5), a peptide might be irrelevant even for the same individual at a later stage of the disease.
- There is great heterogeneity in TAA expression among patients with the same neoplastic disease.
- TAAs are often downregulated in tumor cell escape-variants, and so the corresponding peptides are no longer expressed.

- Many TAAs represent overexpressed self antigens (such as HER-2/neu, CEA, telomerase, etc.); thus the immune system is tolerized against these antigens. Consequently, the remaining antigen specific T cell clones are usually of low affinity and the generation of an effective immune response to them might lead to autoimmunity.
- We still do not know the best way to vaccinate with peptides (timing, adjuvants, repetitions, etc.)
- Finally, although antigen specific immune responses have been reported in many patients enrolled in a large number of clinical trials using peptide vaccines, only limited objective clinical responses have been documented.

4.2 Dendritic Cell-Based Cancer Vaccines

DCs are powerful antigen-presenting cells able to process and present antigens to $CD4^+$ and $CD8^+$ T cells, while delivering the costimulatory signals necessary for effective T cell activation. On the other hand, DCs also have the potential to actively downregulate an immune response or to induce immune tolerance (30). Since the first clinical trial using a DC vaccine reported in 1996 (41), DCs are considered a valuable tool in cancer immunotherapy and are widely used in many clinical trials worldwide.

Because DCs are found at trace levels in tissues or in circulation, methods to generate them in vitro are of critical value for their application in immunotherapy. DCs can be produced from either peripheral blood monocytes or $CD34^+$ hemopoietic progenitor cells. Culture of monocytes in the presence of GM-CSF and IL-4 results in the generation of immature DCs (iDCs). Similarly, culture of $CD34^+$ cells with an appropriate combination of cytokines leads to the generation of iDCs. Further maturation of DCs (mDCs) can be achieved by different stimuli, such as LPS, TNF-a, Poly I-C, or CD40L.

Functionally, iDCs are specialized to uptake and process antigens, whereas mDCs lose this capability. On the other hand, iDCs induce T cell tolerance, while mDCs can effectively prime T_H1 $CD4^+$ and $CD8^+$ T cells (87). Additionally, mature DCs have the capacity to induce T_H2 $CD4^+$ cells, de novo regulatory T cells, or clonal expansion of natural regulatory T cells, or even to induce deletional tolerance (87). The differentiation, maturation, and effector functions, as well as the trafficking capacity of DCs, depend on the signals delivered by the microenvironment, by either soluble factors or cell-to-cell interactions.

All these issues have to be taken into consideration when applying DC vaccines for cancer therapy. Although there is a growing body of information arising from studies on the biology of DCs and from their application in clinical trials, there is still a necessity for standardization of culture conditions, phenotypic characteristics, antigen loading, maturation procedure, dose, time and route of administration as well as establishment of quality criteria for DC-based cancer immunotherapy (29).

There are only few clinical trials evaluating the efficacy of dendritic cell-based vaccination in ovarian cancer patients. Nevertheless, the experience acquired from clinical trials using DCs for several other cancer types would facilitate the application of this immunotherapeutic modality in ovarian cancer patients.

5 Adoptive T Cell Transfer for Cancer Immunotherapy

Adoptive T cell therapy, using specific CD4 (helper or cytotoxic) and/or CD8 cytotoxic cells against tumor antigens, may be very efficacious for the eradication of existing malignancies. These cells need to be highly active against tumor cells and can be generated by either in vitro stimulation of patient's peripheral blood mononuclear cells (PBMC) or ex vivo expansion of specific tumor infiltrating lymphocytes (TIL) or from PBMC of prevaccinated individuals (25, 88).

In a standard rapid expansion protocol for CTL, either from TIL or from patient's PBMC, cells selected for transfer are expanded with anti-CD3 mAb or PHA, a high dose of IL-2 and allogeneic irradiated feeder cells (8, 26, 107). This procedure generates tumor-specific CTL differentiated to an intermediate to late effector state. Monitoring of cancer patients, either immunocompetent or immunosuppressed, receiving adoptively transferred T cells, reveals that blood levels of tumor-specific T cells decrease rapidly following vaccination, thereby creating the need for repeated vaccinations (25). Consequently, it is of the utmost importance to improve existing protocols in terms of both, in vitro expansion of tumor-specific T cells as well as acquisition of the appropriate phenotype (effector-memory), to ensure their long term in vivo persistence (42).

Adoptive T cell transfer after host preconditioning by lymphodepletion has been shown to induce clear and reproducible responses in about 50% of the treated melanoma patients (26). The mechanisms underlying this improved efficacy of adoptively transferred tumor-reactive T cells include the elimination of regulatory T cells, the depletion of endogenous cells competing for activating cytokines, and the increased function and availability of APCs (32). It is worth mentioning that the vast majority of clinical trials using adoptive transfer of T cells have been conducted in melanoma patients, and the reported beneficial clinical results refer to this particular group of patients. This could be either attributed to the difficulty in obtaining large numbers of TIL-derived antigen-specific T cells from other malignancies or to some particular features of melanoma as a disease.

Genetic modification of T cells to be adoptively transferred is another recently emerging approach. In a recent report, 15 melanoma patients adoptively received autologous PBMC-derived T cells, transduced to express a tumor specific TCR. The transferred cells persisted for at least 2 months in all patients, representing 10% of their PBMC. Two of the patients demonstrated objective clinical responses (73) (currently, one more patient has responded positively, Morgan, personal communication). In another phase I study, autologous T cells were genetically modified to express a chimeric receptor against the α-folate receptor and then adoptively transferred to patients with metastatic ovarian cancer, but these cells did not persist for long in large numbers and no clinical responses were observed (50).

Although adoptive transfer of tumor-specific T cells constitutes a promising immunotherapeutic approach, it is a very expensive and laborious procedure, undertaken only by a few laboratories, and which also needs further improvement so as to yield appropriate numbers of nonexhausted functional T cells.

6 NK Cell-Based Immunotherapy

Adoptive NK-cell therapy has held great promise for the immunotherapy of cancer for over three decades. However, to date, only modest clinical success has been achieved by manipulating the NK cell compartment in patients with malignant disease. Progress in the field of NK cell receptors, inhibiting or activating NK cells [killer cell immunoglobulin-like receptor (KIR), C-type lectins, and natural cytotoxicity receptors (NCR)], will assist the development of novel approaches to manipulate NK receptor–ligand interactions for the potential benefit of patients with cancer.

NK cells are essential elements of the immune system and possess a key role in immune reaction against pathogens and tumors. In humans, NK cells are characterized by expression of the CD56 receptor and lack of CD3. Their rapid activation and cytotoxic activity, without need of prior sensitization, as well as the release of cytokines and chemokines, such as IFN-γ, TNF-α, IL-10, GM-CSF, and MIP-1a, reveals their significance in the progress of an immune response (81).

The ability of NK cells to spontaneously attack and kill transformed cells in an MHC-unrestricted fashion renders them suitable candidates for use in immunotherapeutic protocols. Early studies aiming to improve the antitumor effect of NK cells include endogenous activation of the patients' NK cells, through administration of cytokines (15, 20, 62, 69), or adoptive transfer of ex vivo stimulated autologous lymphokine-activated killer cells (LAK) (59, 90, 102). Initial clinical trials based on adoptive transfer of LAK cells in combination with high doses of IL-2, apart from side effects such as high toxicity, resulted in poor clinical outcome. To overcome drawbacks caused by high IL-2 doses, IL-2-activated autologous LAK cells and daily administration of low doses of IL-2 were applied as a novel immunotherapeutic approach, but again with limited success (35). Regional administration of IL-2 activated LAK cells appeared to be more effective, and less toxic, than systemic IL-2 administration (13, 36, 49).

LAK-based immunotherapy has also been applied in ovarian cancer patients in a few clinical trials (103, 104, 116), although with poor clinical results.

In the last 3–4 years, NK cells have reemerged as a powerful tool in cancer immunotherapy (6, 101), following new studies on their biology and function. Data obtained from haploidentical hematopoietic stem cell (HSC) transplantation, following T cell depletion in patients with acute myeloid leukemia (AML), have revealed that alloreactive KIR-incompatible NK cells protected patients from disease relapse, thus indicating the therapeutic role of allogeneic NK cells in AML (91). This observation led to the design of new cell therapy clinical protocols on the basis of the administration of mismatched (haploidentical) allogeneic NK cells, activated with IL-2, in patients with different types of malignancies, as recently reported (70). In this phase I clinical study, patients with metastatic melanoma, metastatic renal carcinoma, refractory Hodgkin disease, and poor-prognosis AML received up to 2×10^7 NK cells per kg (the higher tolerated dose could not be defined, due to limitations in NK cell numbers), following an immunosuppressive regimen. Haploidentical NK cell administration has been proven safe, with minimal toxicity, mostly attributable to the concomitant administration of a low dose of IL-2. A very interesting finding in this study was the fact that conditioning of the AML patients with high-dose cyclophosphamide and fludarabine resulted in an increased endogenous production of IL-15, which was correlated with the prolonged in vivo expansion and persistence (up to 28 days) of the engrafted allogeneic NK cells. Furthermore, although the purpose of the study was only safety testing, it was found that 5 of 19 poor-prognosis patients with AML achieved complete hematologic remission, with a significantly higher complete remission rate when KIR ligand mismatched donors were used.

The limiting factor in such protocols remains the number of appropriately activated NK that can be obtained (6). Usually the donor undergoes a 4- to 5-h leukapheresis, which is a protracted and cumbersome procedure. We have recently reported (85) novel data on the effect of glucocorticoids (GCs), along with stimulatory signals from IL-15 or IL-2, on NK cell expansion and function. Although GCs have been reported to exert an inhibitory effect on NK cells, we have demonstrated that NK cells activated with IL-15 (or IL-2) in the presence of hydrocortisone (HC) not only retain their cytotoxic activity but are also protected from apoptosis, and their proliferative rate is significantly elevated. HC plus IL-15 is capable of inducing up to a 400-fold expansion of human NK cells within 20 days of culture compared with about a 50-fold with IL-15 alone, thus making feasible their large scale production from a limited volume of peripheral blood (200 ml of peripheral blood could give rise to more than 10^{10} NK cells, using this protocol) and their potential application in clinical protocols. Furthermore, expanding NK cells in the presence of HC, thus rendering them preconditioned to the presence of GCs, is advantageous, as their intravenous injection into a recipient exposes them to a microenvironment where active cortisol is naturally present at doses similar to those used in vitro.

Autologous NK cells, activated and expanded in vitro in the presence of IL-15 and HC, have been found to be effective in vivo in a lung metastasis mouse model. These NK cells retained their ability to efficiently infiltrate tumor-bearing lung tissue and to persist for more than 3 days after NK cell infusion. Furthermore, survival of animals with preestablished lung tumors, treated with NK cells activated and expanded with IL-15 and HC, was more than sixfold extended compared with untreated animals (unpublished data). We plan to initiate a phase I clinical study in patients with advanced nonsmall cell lung cancer, who will be adoptively transferred with allogeneic NK cells expanded ex vivo with IL-2 and HC, to take advantage of both, i.e., the novel NK cell-expansion protocol to obtain large numbers of appropriate effector cells and the KIR mismatch between donor and recipient. The selection of lung cancer is based on the observation that the majority of intravenously administered NK cells accumulate in the lung and are retained at the tumor site (4).

Adoptive transfer of established NK cell lines, with broad antitumor activity, such as NK-92 (6, 14, 19), represents an alternative for use in clinical trials. In an attempt to overcome HLA-mediated inhibitory signals, thus endowing NK cells with cytotoxicity against otherwise NK-resistant cells, genetic modification with chimeric receptors (consisting of a tumor antigen-targeting moiety, such as a single chain Fv against HER-2/neu, and a signaling part, i.e., a CD3 ζ chain), has already been tested in preclinical models and might be proven effective in a clinical setting of NK cell-based immnunotherapy.

HLA class I antigens are often downregulated in ovarian cancer cells (60, 78, 108). This would imply that ovarian cancer cells are more susceptible to NK cell attack. It has been recently shown that freshly isolated ovarian tumor cells are directly recognized by resting allogeneic NK cells (17), further strengthening the notion that NK cell-based immunotherapy for ovarian cancer patients requires reevaluation.

7 Regulatory Cells and Cancer Immunotherapy

Cancer progression and metastasis has been correlated with the presence of immunosuppressive factors. Reports on the implication of suppressor/regulatory cells in cancer development and progression date back more than three decades (33, 105). Populations with immunosuppressive properties include the following:

- Tumor cells: These have been found to produce a broad range of factors conferring immunosuppression, including cytokines (TGF-β, IL-10, etc.), TNF family ligands (FasL, TRAIL, etc.), small molecules (prostaglandin E2, INOS, etc.), and enzymes (indoleamine 2,3-dioxygenase (IDO), arginase I), as well as cytokine receptors competing for cytokines with effector cells (i.e., IL-2R) (82, 112).
- Stromal cells: Cells surrounding the tumor (including endothelial cells, fibroblasts, macrophages, etc.) possess immunosuppressive properties, either through soluble factors or by cell-to-cell contact interactions (47, 115).
- Dendritic cells: As already discussed in a previous section ("dendritic cell-based immunotherapy"), under some conditions, DCs induce and maintain tolerance.
- Myeloid suppressor cells: Myeloid suppressor cells (MSCs) represent a heterogenous cell population comprising immature macrophages, granulocytes, DCs, and other myeloid cells at early stages of differentiation that can be identified in humans as $CD34^+$, $CD33^+$, $CD15^-$, and $CD13^+$. MSCs have been detected in the tumor microenvironment and have been found to inhibit tumor-specific T cell functions through several mechanisms, mainly through arginase and/or nitric oxide synthase (NOS) activation (96).

- T cells: There are several subpopulations of CD4+ T cells, which have immunosuppressive capacity (45, 80), either by producing suppressive cytokines, such as T_H3 and Tr1 cells, exerting their function through TGF-β and IL-10 production respectively, as well as subpopulations of CD8+ cells (117).

The identification of a better characterized regulatory subpopulation among CD4+ cells, constitutively expressing high levels of CD25, CTLA-4, GITR, Foxp3, being anergic when stimulated by TCR cross-linking in vitro and actively inhibiting CD4+ CD25- T cells, CD8+ T cells, DC, NK, NKT, and B cells in a cell-to-cell contact and dose-dependent manner (12) led to a recent intensification of the investigation on the role of these cells on tumor development, growth, and escape from immunosurveillance.

Increased numbers of Tregs, either at the tumor site, the tumor-draining lymph nodes, or in the circulation, have been reported in patients with several types of malignancy, including gastrointestinal malignancies (56, 63, 92), ovarian cancer (7, 22, 113, 114), nonsmall-cell lung carcinoma (79, 114), melanoma (18), renal cell carcinoma (18), glioma (28), Hodgkin lymphoma (66), and chronic lymphocytic leukemia (CLL) (11). Increased numbers of Tregs have been correlated with greater disease burden and poorer overall survival (92, 113). Therapeutic approaches in cancer treatment may affect Tregs cell frequency and function. IL-2 therapy has been found to induce expansion of Tregs in melanoma and renal carcinoma patients treated with high dose IL-2 (3, 18). On the contrary, some chemotherapeutic agents, such as cyclophosphamide and fludarabine, have been observed to reduce the frequency and function of Tregs (12).

Increased Treg frequency has been correlated with advanced stages and bad prognosis in cancer patients (53), although differences among patients with different malignancies or even the same type of cancer have not yet been explained (12). We have recently (84) demonstrated that within a defined group of breast cancer patients with progressed disease and similar clinical outcome, there is, at least, one factor that separates these patients into two different subgroups, according to circulating Treg frequency: the overexpression of HER-2/*neu*, a self antigen expressed by normal epithelial cells and several types of cancer cells, which is released by the cells into the systemic circulation (83). The specificity of Tregs in cancer patients remains unclear (54), although some evidence supports that Tregs associated with tumors are tumor antigen-specific Tregs (111). Whether systemically circulating tumor-associated self-antigens at increased concentrations, such as HER-2/*neu*, CEA, PSA, and others, induce the expansion of circulating antigen-specific Tregs in cancer patients, as described for endogenous systemic antigens (52, 64), remains to be elucidated.

Treg cell depletion or functional inhibition can lead to the induction of efficient antitumor activity. This can be achieved either by pharmacological agents such as cyclophosphamide (65), fludarabine (11), COX2 inhibitors (99), temozolomide (1), or by directly targeting Tregs with antibodies against the IL-2 receptor or against CTLA-4, or with IL-2 conjugated with toxin (denileukin diftitox, ONTAK). Furthermore, it has been recently documented that peptide vaccination (43) and antibody-based immunotherapy with Trastuzumab (84) in breast cancer patients can also result in a substantial decrease in Tregs.

Since elimination of Tregs has been proven promising, either alone or in combination with other immunotherapeutic interventions, in preclinical models (53) or in pilot clinical studies (23), a number of clinical trials are currently in progress to further evaluate this approach in cancer immunotherapy. There are currently ongoing clinical trials using ONTAK to eliminate Tregs in ovarian cancer patients with very encouraging initial results (7, 21).

8 Prospects and Perspectives for the Improvement of Cancer Immunotherapy

The use of the immune system in cancer therapy, with the entirety of its repertoire modalities, i.e., active immunization or adoptive transfer, stimulating innate or adaptive immune mechanisms, targeting the humoral or the cellular immune compartment, has proven or has the potential to be effective. However, from the experience gained so far, it is clear that focusing on a single immunotherapeutic strategy might not confer success in combating cancer, since tumors elicit diverse mechanisms to escape from immunosurveillance.

- The use of multiantigen vaccines, targeting peptides associated with multiple MHC molecules and encompassing both MHC class I and class II peptide sequences, would overcome the heterogeneity of antigen expression and MHC downregulation by tumor cells, as well as the concomitant stimulation of $CD4^+$ cells to provide help for $CD8^+$ T cell priming and antibody production.
- The exploitation of more effective vaccine adjuvants (57) such as CpG and other TLR agonists, allogeneic or xenogeneic cells or molecules, would generate stronger stimulatory signals for T cells, thus overcoming the inhibitory signals delivered from regulatory cells and suppressor soluble factors.
- Targeting of tumor stroma and neovascularization, along with the tumor itself, is a currently exploited approach, which seems very effective (38, 40, 86).
- Elimination or inactivation of regulatory cells (myeloid suppressor cells or regulatory T cells) prior to other immunotherapeutic interventions, including vaccination or adoptive therapy is of great importance.
- Blockade of soluble suppressor factors (i.e., TGF-β, VEGF, IL-10, IDO, etc.) or inhibitory receptors (i.e., PD-1) represents another approach to improve immunotherapy.
- The lack of costimulatory molecules from tumor cells or from DCs represents another obstacle to effective cancer immunotherapy. Genetic modification either by in vivo gene delivery to tumor cells (48) or by genetically engineered DCs (31), or other strategies effectively arming APCs with costimulatory molecules might overcome this obstacle.
- The fact that the immune system consists of an integral network, where the function of each component population, positively or negatively, affects the other

immune cell populations, the joined function of the nonspecifically acting antigen-presenting cells with NK cells, followed by the cells of the adaptive immune compartment, the T and B cells, need to be combined in the immunotherapy of cancer to effectively destroy tumor cells and to generate a persistent memory antitumor response.

- The combination of different therapeutic modalities, including classical cancer treatments, with immunotherapeutic approaches, seems indispensable. Surgery would rid the body of the tumor mass, thus diminishing the immunomodulatory effects from the tumor burden, and chemotherapy and/or radiotherapy might lead to cross-presentation of tumor antigens by inducing tumor cell death (58, 106). If this cross-presentation will result in cross-tolerance or in cross-priming will depend upon the presence or absence of inflammatory signals provided by the appropriate cytokines, as well as costimulatory signals. On the one hand, the immunotherapeutic intervention at this point would drive the immune response toward an effective antitumor cross-priming. On the other hand, as already mentioned, many chemotherapeutic agents lead to regulatory cell depletion/inhibition, further conferring a successful antitumor response.

Thus, the new therapeutic modalities for the treatment of cancer should combine surgery, chemotherapy, and/or radiotherapy with immunotherapeutic intervention. Furthermore, similarly to conventional, established adjuvant chemotherapy, "adjuvant" immunotherapy in patients carrying the minimum tumor load needs to be seriously considered. It is unfair and misleading to evaluate the efficacy of cancer immunotherapy treatments in patients with advanced cancer, as currently happens, where the tumor burden is already large and the immune system highly deregulated.

Finally, the development of prophylactic cancer vaccines, whenever and wherever possible, is also a promising immunotherapeutic intervention, already applied in the case of some virally induced tumors, such as HBV and, more recently, HPV vaccination.

9 Conclusion

Immunotherapy remains a very promising approach for the treatment of cancer. From the many clinical trials conducted worldwide, it has become clear that cancer immunotherapy is safe, with limited side effects, but still with only modest overall clinical efficacy.

The new generation of cancer immunotherapy protocols requires consideration and potential application of multiple new modalities exploited so far, such as multiantigen vaccines, effective adjuvants, costimulatory molecules, lymphodepletion and/or elimination, or blockade of inhibitory populations and factors.

It is critical for investigators in cancer immunotherapy to join forces with clinical oncologists for the development of novel immunotherapeutic regimens, integrating different compartments of the immune system. Emerging strategies should rely on the knowledge acquired from preclinical models and basic immunology and use the latest technological achievements, along with established cancer treatment modalities (surgery, chemotherapy, radiotherapy, as well as the new generation of anticancer drugs), for the new era in cancer therapy.

References

1. Aamdal, S.D., Engebraaten, O., Owre, K., Dyrhaug, M., Trachsel, S. and Gaudernack, G. (2006) A phase I/II study of telomerase peptide vaccination in combination with chemotherapy in patients with stage IV malignant melanoma. *J Clin Oncol*, 24, 8031.
2. Aghajanian, C. (2006) The role of bevacizumab in ovarian cancer – an evolving story. *Gynecol Oncol*, 102, 131–133.
3. Ahmadzadeh, M., Rosenberg, S.A. (2006) IL-2 administration increases CD4+ CD25(hi) Foxp3+ regulatory T cells in cancer patients. *Blood*, 107, 2409–2414.
4. Albertsson, P.A., Basse, P.H., Hokland, M., Goldfarb, R.H., Nagelkerke, J.F., Nannmark, U. and Kuppen, P.J. (2003) NK cells and the tumour microenvironment: implications for NK-cell function and anti-tumour activity. *Trends Immunol*, 24, 603–609.
5. Algarra, I., Garcia-Lora, A., Cabrera, T., Ruiz-Cabello, F. and Garrido, F. (2004) The selection of tumor variants with altered expression of classical and nonclassical MHC class I molecules: implications for tumor immune escape. *Cancer Immunol Immunother*, 53, 904–910.
6. Arai, S. and Klingemann, H.G. (2005) Natural killer cells: can they be useful as adoptive immunotherapy for cancer? *Expert Opin Biol Ther*, 5, 163–172.
7. Barnett, B., Kryczek, I., Cheng, P., Zou, W. and Curiel, T.J. (2005) Regulatory T cells in ovarian cancer: biology and therapeutic potential. *Am J Reprod Immunol*, 54, 369–377.
8. Benlalam, H., Vignard, V., Khammari, A., Bonnin, A., Godet, Y., Pandolfino, M.C., Jotereau, F., Dreno, B. and Labarriere, N. (2006) Infusion of Melan-A/Mart-1 specific tumor-infiltrating lymphocytes enhanced relapse-free survival of melanoma patients. *Cancer Immunol Immunother*, 56, 515–526.
9. Berek, J.S. (2004) Immunotherapy of ovarian cancer with antibodies: a focus on oregovomab. *Expert Opin Biol Ther*, 4, 1159–1165.
10. Berek, J.S., Dorigo, O., Schultes, B. and Nicodemus, C. (2003) Specific keynote: immunological therapy for ovarian cancer. *Gynecol Oncol*, 88, S105–109; Discussion S110–S103.
11. Beyer, M., Kochanek, M., Darabi, K., Popov, A., Jensen, M., Endl, E., Knolle, P.A., Thomas, R.K., von Bergwelt-Baildon, M., Debey, S., Hallek, M. and Schultze, J.L. (2005) Reduced frequencies and suppressive function of CD4 + CD25hi regulatory T cells in patients with chronic lymphocytic leukemia after therapy with fludarabine. *Blood*, 106, 2018–2025.
12. Beyer, M. and Schultze, J.L. (2006) Regulatory T cells in cancer. *Blood*, 108, 804–811.
13. Boiardi, A., Silvani, A., Ruffini, P.A., Rivoltini, L., Parmiani, G., Broggi, G. and Salmaggi, A. (1994) Loco-regional immunotherapy with recombinant interleukin-2 and adherent lymphokine-activated killer cells (A-LAK) in recurrent glioblastoma patients. *Cancer Immunol Immunother*, 39, 193–197.
14. Boyiadzis, M. and Foon, K.A. (2006) Natural killer cells: from bench to cancer therapy. *Expert Opin Biol Ther*, 6, 967–970.
15. Caligiuri, M.A., Murray, C., Soiffer, R.J., Klumpp, T.R., Seiden, M., Cochran, K., Cameron, C., Ish, C., Buchanan, L., Perillo, D., et al. (1991) Extended continuous infusion low-dose recombinant interleukin-2 in advanced cancer: prolonged immunomodulation without significant toxicity. *J Clin Oncol*, 9, 2110–2119.
16. Cannon, M.J., Santin, A.D. and O'Brien, T.J. (2004) Immunological treatment of ovarian cancer. *Curr Opin Obstet Gynecol*, 16, 87–92.

17. Carlsten, M., Björkström, N.K., Norell, H., Bryceson, Y., van Hall, T., Baumann, B.C., Hanson, M., Schedvins, K., Kiessling, R., Ljunggren, H.G. and Malmberg, K.J. (2007) DNAX accessory molecule-1 mediated recognition of freshly isolated ovarian carcinoma by resting natural killer cells. *Cancer Res*, 67, 1317–1325.
18. Cesana, G.C., DeRaffele, G., Cohen, S., Moroziewicz, D., Mitcham, J., Stoutenburg, J., Cheung, K., Hesdorffer, C., Kim-Schulze, S. and Kaufman, H.L. (2006) Characterization of CD4 + CD25+ regulatory T cells in patients treated with high-dose interleukin-2 for metastatic melanoma or renal cell carcinoma. *J Clin Oncol*, 24, 1169–1177.
19. Chen, G., Ling, B., Zhu, H.P., Zhao, W.D., Wang, Q.H., Zhang, H.Y., Wu, A.D., Wei, H.M. and Tian, Z.G. (2005) Effect of natural killer cell line NK-92 against human ovarian carcinoma cells in vitro and in vivo. *Zhonghua Fu Chan Ke Za Zhi*, 40, 476–479.
20. Cortes, J.E., Kantarjian, H.M., O'Brien, S., Giles, F., Keating, M.J., Freireich, E.J. and Estey, E.H. (1999) A pilot study of interleukin-2 for adult patients with acute myelogenous leukemia in first complete remission. *Cancer*, 85, 1506–1513.
21. Curiel, T.J., Barnett, B., Kryczek, I., Cheng, P., Zou, W. (2006) Regulatory T cells in ovarian cancer: biology and therapeutic potential. *Cancer Immunity*, 6, 20–21.
22. Curiel, T.J., Coukos, G., Zou, L., Alvarez, X., Cheng, P., Mottram, P., Evdemon-Hogan, M., Conejo-Garcia, J.R., Zhang, L., Burow, M., Zhu, Y., Wei, S., Kryczek, I., Daniel, B., Gordon, A., Myers, L., Lackner, A., Disis, M.L., Knutson, K.L., Chen, L. and Zou, W. (2004) Specific recruitment of regulatory T cells in ovarian carcinoma fosters immune privilege and predicts reduced survival. *Nat Med*, 10, 942–949.
23. Dannull, J., Su, Z., Rizzieri, D., Yang, B.K., Coleman, D., Yancey, D., Zhang, A., Dahm, P., Chao, N., Gilboa, E. and Vieweg, J. (2005) Enhancement of vaccine-mediated antitumor immunity in cancer patients after depletion of regulatory T cells. *J Clin Invest*, 115, 3623–3633.
24. del Carmen, M.G., Rizvi, I., Chang, Y., Moor, A.C., Oliva, E., Sherwood, M., Pogue, B. and Hasan, T. (2005) Synergism of epidermal growth factor receptor-targeted immunotherapy with photodynamic treatment of ovarian cancer in vivo. *J Natl Cancer Inst*, 97, 1516–1524.
25. Dudley, M.E. and Rosenberg, S.A. (2003) Adoptive-cell-transfer therapy for the treatment of patients with cancer. *Nat Rev Cancer*, 3, 666–675.
26. Dudley, M.E., Wunderlich, J.R., Yang, J.C., Sherry, R.M., Topalian, S.L., Restifo, N.P., Royal, R.E., Kammula, U., White, D.E., Mavroukakis, S.A., Rogers, L.J., Gracia, G.J., Jones, S.A., Mangiameli, D.P., Pelletier, M.M., Gea-Banacloche, J., Robinson, M.R., Berman, D.M., Filie, A.C., Abati, A. and Rosenberg, S.A. (2005) Adoptive cell transfer therapy following non-myeloablative but lymphodepleting chemotherapy for the treatment of patients with refractory metastatic melanoma. *J Clin Oncol*, 23, 2346–2357.
27. Durrant, L.G. and Ramage, J.M. (2005) Development of cancer vaccines to activate cytotoxic T lymphocytes. *Expert Opin Biol Ther*, 5, 555–563.
28. Fecci, P.E., Mitchell, D.A., Whitesides, J.F., Xie, W., Friedman, A.H., Archer, G.E., Herndon, J.E., 2nd, Bigner, D.D., Dranoff, G. and Sampson, J.H. (2006) Increased regulatory T-cell fraction amidst a diminished CD4 compartment explains cellular immune defects in patients with malignant glioma. *Cancer Res*, 66, 3294–3302.
29. Figdor, C.G., de Vries, I.J., Lesterhuis, W.J. and Melief, C.J. (2004) Dendritic cell immunotherapy: mapping the way. *Nat Med*, 10, 475–480.
30. Fricke, I. and Gabrilovich, D.I. (2006) Dendritic cells and tumor microenvironment: a dangerous liaison. *Immunol Invest*, 35, 459–483.
31. Garnett, C.T., Greiner, J.W., Tsang, K.Y., Kudo-Saito, C., Grosenbach, D.W., Chakraborty, M., Gulley, J.L., Arlen, P.M., Schlom, J. and Hodge, J.W. (2006) TRICOM vector based cancer vaccines. *Curr Pharm Des*, 12, 351–361.
32. Gattinoni, L., Powell, D.J., Jr., Rosenberg, S.A. and Restifo, N.P. (2006) Adoptive immunotherapy for cancer: building on success. *Nat Rev Immunol*, 6, 383–393.
33. Gershon, R.K. (1975) A disquisition on suppressor T cells. *Transplant Rev*, 26, 170–185.
34. Hall, G.D., Brown, J.M., Coleman, R.E., Stead, M., Metcalf, K.S., Peel, K.R., Poole, C., Crawford, M., Hancock, B., Selby, P.J. and Perren, T.J. (2004) Maintenance treatment with

interferon for advanced ovarian cancer: results of the Northern and Yorkshire gynaecology group randomised phase III study. *Br J Cancer*, 91, 621–626.
35. Hayakawa, M., Hatano, T., Ogawa, Y., Gakiya, M., Ogura, H. and Osawa, A. (1994) Treatment of advanced renal cell carcinoma using regional arterial administration of lymphokine-activated killer cells in combination with low doses of rIL-2. *Urol Int*, 53, 117–124.
36. Hayes, R.L., Koslow, M., Hiesiger, E.M., Hymes, K.B., Hochster, H.S., Moore, E.J., Pierz, D.M., Chen, D.K., Budzilovich, G.N. and Ransohoff, J. (1995) Improved long term survival after intracavitary interleukin-2 and lymphokine-activated killer cells for adults with recurrent malignant glioma. *Cancer*, 76, 840–852.
37. Hodge, J.W., Tsang, K.Y., Poole, D.J. and Schlom, J. (2003) General keynote: vaccine strategies for the therapy of ovarian cancer. *Gynecol Oncol*, 88, S97–104; Discussion S110–S103.
38. Hofmeister, V., Vetter, C., Schrama, D., Brocker, E.B. and Becker, J.C. (2006) Tumor stroma-associated antigens for anti-cancer immunotherapy. *Cancer Immunol Immunother*, 55, 481–494.
39. Holmberg, L.A. and Sandmaier, B. (2004) Vaccination as a treatment for breast or ovarian cancer. *Expert Rev Vaccines*, 3, 269–277.
40. Holmgren, L., Ambrosino, E., Birot, O., Tullus, C., Veitonmaki, N., Levchenko, T., Carlson, L.M., Musiani, P., Iezzi, M., Curcio, C., Forni, G., Cavallo, F. and Kiessling, R. (2006) A DNA vaccine targeting angiomotin inhibits angiogenesis and suppresses tumor growth. *Proc Natl Acad Sci USA*, 103, 9208–9213.
41. Hsu, F.J., Benike, C., Fagnoni, F., Liles, T.M., Czerwinski, D., Taidi, B., Engleman, E.G. and Levy, R. (1996) Vaccination of patients with B-cell lymphoma using autologous antigen-pulsed dendritic cells. *Nat Med*, 2, 52–58.
42. Huang, J., Khong, H.T., Dudley, M.E., El-Gamil, M., Li, Y.F., Rosenberg, S.A. and Robbins, P.F. (2005) Survival, persistence, and progressive differentiation of adoptively transferred tumor-reactive T cells associated with tumor regression. *J Immunother*, 28, 258–267.
43. Hueman, M.T., Stojadinovic, A., Storrer, C.E., Foley, R.J., Gurney, J.M., Shriver, C.D., Ponniah, S. and Peoples, G.E. (2006) Levels of circulating regulatory CD4 + CD25+ T cells are decreased in breast cancer patients after vaccination with a HER2/neu peptide (E75) and GM-CSF vaccine. *Breast Cancer Res Treat*, 98, 17–29.
44. Hung, C.F., Calizo, R., Tsai, Y.C., He, L. and Wu, T.C. (2006) A DNA vaccine encoding a single-chain trimer of HLA-A2 linked to human mesothelin peptide generates anti-tumor effects against human mesothelin-expressing tumors. *Vaccine*, 25, 127–135.
45. Iliopoulou, E.G., Karamouzis, M.V., Missitzis, I., Ardavanis, A., Sotiriadou, N.N., Baxevanis, C.N., Rigatos, G., Papamichail, M., Perez, S.A. (2006) Increased frequency of CD4+ cells expressing CD161 in cancer patients. *Clin Cancer Res*, 12, 6901–6909.
46. Jager, E., Karbach, J., Gnjatic, S., Neumann, A., Bender, A., Valmori, D., Ayyoub, M., Ritter, E., Ritter, G., Jager, D., Panicali, D., Hoffman, E., Pan, L., Oettgen, H., Old, L.J. and Knuth, A. (2006) Recombinant vaccinia/fowlpox NY-ESO-1 vaccines induce both humoral and cellular NY-ESO-1-specific immune responses in cancer patients. *Proc Natl Acad Sci USA*, 103, 14453–14458.
47. Kammertoens, T., Schuler, T. and Blankenstein, T. (2005) Immunotherapy: target the stroma to hit the tumor. *Trends Mol Med*, 11, 225–231.
48. Kaufman, H.L., Cohen, S., Cheung, K., DeRaffele, G., Mitcham, J., Moroziewicz, D., Schlom, J. and Hesdorffer, C. (2006) Local delivery of vaccinia virus expressing multiple costimulatory molecules for the treatment of established tumors. *Hum Gene Ther*, 17, 239–244.
49. Keilholz, U., Scheibenbogen, C., Brado, M., Georgi, P., Maclachlan, D., Brado, B. and Hunstein, W. (1994) Regional adoptive immunotherapy with interleukin-2 and lymphokine-activated killer (LAK) cells for liver metastases. *Eur J Cancer*, 30A, 103–105.
50. Kershaw, M.H., Westwood, J.A., Parker, L.L., Wang, G., Eshhar, Z., Mavroukakis, S.A., White, D.E., Wunderlich, J.R., Canevari, S., Rogers-Freezer, L., Chen, C.C., Yang, J.C., Rosenberg, S.A. and Hwu, P. (2006) A phase I study on adoptive immunotherapy using gene-modified T cells for ovarian cancer. *Clin Cancer Res*, 12, 6106–6115.

51. Kikuchi, T., Akasaki, Y., Abe, T., Fukuda, T., Saotome, H., Ryan, J.L., Kufe, D.W. and Ohno, T. (2004) Vaccination of glioma patients with fusions of dendritic and glioma cells and recombinant human interleukin 12. *J Immunother*, 27, 452–459.
52. Knoechel, B., Lohr, J., Kahn, E., Bluestone, J.A. and Abbas, A.K. (2005) Sequential development of interleukin 2-dependent effector and regulatory T cells in response to endogenous systemic antigen. *J Exp Med*, 202, 1375–1386.
53. Knutson, K.L., Dang, Y., Lu, H., Lukas, J., Almand, B., Gad, E., Azeke, E. and Disis, M.L. (2006a) IL-2 immunotoxin therapy modulates tumor-associated regulatory T cells and leads to lasting immune-mediated rejection of breast cancers in neu-transgenic mice. *J Immunol*, 177, 84–91.
54. Knutson, K.L., Disis, M.L. and Salazar, L.G. (2006b) CD4 regulatory T cells in human cancer pathogenesis. *Cancer Immunol Immunother*.
55. Kohler, G. and Milstein, C. (1975) Continuous cultures of fused cells secreting antibody of predefined specificity. *Nature*, 256, 495–497.
56. Kono, K., Kawaida, H., Takahashi, A., Sugai, H., Mimura, K., Miyagawa, N., Omata, H. and Fujii, H. (2006) CD4(+)CD25high regulatory T cells increase with tumor stage in patients with gastric and esophageal cancers. *Cancer Immunol Immunother*, 55, 1064–1071.
57. Kornbluth, R.S. and Stone, G.W. (2006) Immunostimulatory combinations: designing the next generation of vaccine adjuvants. *J Leukoc Biol*, 80, 1084–1102.
58. Lake, R.A. and Robinson, B.W. (2005) Immunotherapy and chemotherapy – a practical partnership. *Nat Rev Cancer*, 5, 397–405.
59. Law, T.M., Motzer, R.J., Mazumdar, M., Sell, K.W., Walther, P.J., O'Connell, M., Khan, A., Vlamis, V., Vogelzang, N.J. and Bajorin, D.F. (1995) Phase III randomized trial of interleukin-2 with or without lymphokine-activated killer cells in the treatment of patients with advanced renal cell carcinoma. *Cancer*, 76, 824–832.
60. Le, Y.S., Kim, T.E., Kim, B.K., Park, Y.G., Kim, G.M., Jee, S.B., Ryu, K.S., Kim, I.K. and Kim, J.W. (2002) Alterations of HLA class I and class II antigen expressions in borderline, invasive and metastatic ovarian cancers. *Exp Mol Med*, 34, 18–26.
61. Lenardo, M.J., Boehme, S., Chen, L., Combadiere, B., Fisher, G., Freedman, M., McFarland, H., Pelfrey, C. and Zheng, L. (1995) Autocrine feedback death and the regulation of mature T lymphocyte antigen responses. *Int Rev Immunol*, 13, 115–134.
62. Lim, S.H., Newland, A.C., Kelsey, S., Bell, A., Offerman, E., Rist, C., Gozzard, D., Bareford, D., Smith, M.P. and Goldstone, A.H. (1992) Continuous intravenous infusion of high-dose recombinant interleukin-2 for acute myeloid leukaemia – a phase II study. *Cancer Immunol Immunother*, 34, 337–342.
63. Liyanage, U.K., Moore, T.T., Joo, H.G., Tanaka, Y., Herrmann, V., Doherty, G., Drebin, J.A., Strasberg, S.M., Eberlein, T.J., Goedegebuure, P.S. and Linehan, D.C. (2002) Prevalence of regulatory T cells is increased in peripheral blood and tumor microenvironment of patients with pancreas or breast adenocarcinoma. *J Immunol*, 169, 2756–2761.
64. Lohr, J., Knoechel, B. and Abbas, A.K. (2006) Regulatory T cells in the periphery. *Immunol Rev*, 212, 149–162.
65. Lutsiak, M.E., Semnani, R.T., De Pascalis, R., Kashmiri, S.V., Schlom, J. and Sabzevari, H. (2005) Inhibition of CD4(+)25+ T regulatory cell function implicated in enhanced immune response by low-dose cyclophosphamide. *Blood*, 105, 2862–2868.
66. Marshall, N.A., Christie, L.E., Munro, L.R., Culligan, D.J., Johnston, P.W., Barker, R.N. and Vickers, M.A. (2004) Immunosuppressive regulatory T cells are abundant in the reactive lymphocytes of Hodgkin lymphoma. *Blood*, 103, 1755–1762.
67. Marth, C., Windbichler, G.H., Hausmaninger, H., Petru, E., Estermann, K., Pelzer, A. and Mueller-Holzner, E. (2006) Interferon-gamma in combination with carboplatin and paclitaxel as a safe and effective first-line treatment option for advanced ovarian cancer: results of a phase I/II study. *Int J Gynecol Cancer*, 16, 1522–1528.
68. Menendez, J.A., Mehmi, I. and Lupu, R. (2006) Trastuzumab in combination with heregulin-activated Her-2 (erbB-2) triggers a receptor-enhanced chemosensitivity effect in the absence of Her-2 overexpression. *J Clin Oncol*, 24, 3735–3746.

69. Meropol, N.J., Barresi, G.M., Fehniger, T.A., Hitt, J., Franklin, M. and Caligiuri, M.A. (1998) Evaluation of natural killer cell expansion and activation in vivo with daily subcutaneous low-dose interleukin-2 plus periodic intermediate-dose pulsing. *Cancer Immunol Immunother*, 46, 318–326.
70. Miller, J.S., Soignier, Y., Panoskaltsis-Mortari, A., McNearney, S.A., Yun, G.H., Fautsch, S.K., McKenna, D., Le, C., Defor, T.E., Burns, L.J., Orchard, P.J., Blazar, B.R., Wagner, J.E., Slungaard, A., Weisdorf, D.J., Okazaki, I.J. and McGlave, P.B. (2005) Successful adoptive transfer and in vivo expansion of human haploidentical NK cells in patients with cancer. *Blood*, 105, 3051–3057.
71. Mittendorf, E.A., Gurney, J.M., Storrer, C.E., Shriver, C.D., Ponniah, S. and Peoples, G.E. (2006) Vaccination with a HER2/neu peptide induces intra- and inter-antigenic epitope spreading in patients with early stage breast cancer. *Surgery*, 139, 407–418.
72. Mocellin, S., Mandruzzato, S., Bronte, V., Lise, M. and Nitti, D. (2004) Part I: Vaccines for solid tumours. *Lancet Oncol*, 5, 681–689.
73. Morgan, R.A., Dudley, M.E., Wunderlich, J.R., Hughes, M.S., Yang, J.C., Sherry, R.M., Royal, R.E., Topalian, S.L., Kammula, U.S., Restifo, N.P., Zheng, Z., Nahvi, A., de Vries, C.R., Rogers-Freezer, L.J., Mavroukakis, S.A. and Rosenberg, S.A. (2006) Cancer regression in patients after transfer of genetically engineered lymphocytes. *Science*, 314, 126–129.
74. Muller, S., Zhao, Y., Brown, T.L., Morgan, A.C. and Kohler, H. (2005) TransMabs: cell-penetrating antibodies, the next generation. *Expert Opin Biol Ther*, 5, 237–241.
75. Nagorsen, D. and Thiel, E. (2006) Clinical and immunologic responses to active specific cancer vaccines in human colorectal cancer. *Clin Cancer Res*, 12, 3064–3069.
76. Nakano, H., Kishida, T., Asada, H., Shin-Ya, M., Shinomiya, T., Imanishi, J., Shimada, T., Nakai, S., Takeuchi, M., Hisa, Y. and Mazda, O. (2006) Interleukin-21 triggers both cellular and humoral immune responses leading to therapeutic antitumor effects against head and neck squamous cell carcinoma. *J Gene Med*, 8, 90–99.
77. Niarchos, D.K., Perez, S.A. and Papamichail, M. (2006) Characterization of a novel cell penetrating peptide derived from Bag-1 protein. *Peptides*, 27, 2661–2669.
78. Norell, H., Carlsten, M., Ohlum, T., Malmberg, K.J., Masucci, G., Schedvins, K., Altermann, W., Handke, D., Atkins, D., Seliger, B. and Kiessling, R. (2006) Frequent loss of HLA-A2 expression in metastasizing ovarian carcinomas associated with genomic haplotype loss and HLA-A2-restricted HER-2/neu-specific immunity. *Cancer Res*, 66, 6387–6394.
79. Okita, R., Saeki, T., Takashima, S., Yamaguchi, Y. and Toge, T. (2005) CD4 + CD25+ regulatory T cells in the peripheral blood of patients with breast cancer and non-small cell lung cancer. *Oncol Rep*, 14, 1269–1273.
80. Orentas, R.J., Kohler, M.E. and Johnson, B.D. (2006) Suppression of anti-cancer immunity by regulatory T cells: back to the future. *Semin Cancer Biol*, 16, 137–149.
81. Papamichail, M., Perez, S.A., Gritzapis, A.D. and Baxevanis, C.N. (2004) Natural killer lymphocytes: biology, development, and function. *Cancer Immunol Immunother*, 53, 176–186.
82. Pawelec, G. (2004) Tumour escape: antitumour effectors too much of a good thing? *Cancer Immunol Immunother*, 53, 262–274.
83. Payne, R.C., Allard, J.W., Anderson-Mauser, L., Humphreys, J.D., Tenney, D.Y. and Morris, D.L. (2000) Automated assay for HER-2/neu in serum. *Clin Chem*, 46, 175–182.
84. Perez, S.A., Karamouzis, M.V., Skarlos, D.V., Ardavanis, A., Sotiriadou, N.N., Iliopoulou, E.G., Salagianni, M.L., Sotiropoulou, P.A., Orphanos, G., Baxevanis, C.N., Rigatos, G., Papamichail, M. (2007) CD4 + CD25+ regulatory T cell frequency in HER-2/neu positive and negative advanced stage breast cancer patients. *Clin Cancer Res*, 13, 2714–2721.
85. Perez, S.A., Mahaira, L.G., Demirtzoglou, F.J., Sotiropoulou, P.A., Ioannidis, P., Iliopoulou, E.G., Gritzapis, A.D., Sotiriadou, N.N., Baxevanis, C.N. and Papamichail, M. (2005) A potential role for hydrocortisone in the positive regulation of IL-15-activated NK-cell proliferation and survival. *Blood*, 106, 158–166.
86. Petrulio, C.A., Kim-Schulze, S. and Kaufman, H.L. (2006) The tumour microenvironment and implications for cancer immunotherapy. *Expert Opin Biol Ther*, 6, 671–684.
87. Reis e Sousa, C. (2006) Dendritic cells in a mature age. *Nat Rev Immunol*, 6, 476–483.

88. Ribas, A., Butterfield, L.H., Glaspy, J.A. and Economou, J.S. (2003a) Current developments in cancer vaccines and cellular immunotherapy. *J Clin Oncol*, 21, 2415–2432.
89. Ribas, A., Timmerman, J.M., Butterfield, L.H. and Economou, J.S. (2003b) Determinant spreading and tumor responses after peptide-based cancer immunotherapy. *Trends Immunol*, 24, 58–61.
90. Rosenberg, S.A., Lotze, M.T., Muul, L.M., Leitman, S., Chang, A.E., Ettinghausen, S.E., Matory, Y.L., Skibber, J.M., Shiloni, E., Vetto, J.T., et al. (1985) Observations on the systemic administration of autologous lymphokine-activated killer cells and recombinant interleukin-2 to patients with metastatic cancer. *N Engl J Med*, 313, 1485–1492.
91. Ruggeri, L., Mancusi, A., Perruccio, K., Burchielli, E., Martelli, M.F. and Velardi, A. (2005) Natural killer cell alloreactivity for leukemia therapy. *J Immunother*, 28, 175–182.
92. Sasada, T., Kimura, M., Yoshida, Y., Kanai, M. and Takabayashi, A. (2003) CD4 + CD25+ regulatory T cells in patients with gastrointestinal malignancies: possible involvement of regulatory T cells in disease progression. *Cancer*, 98, 1089–1099.
93. Schultes, B.C. and Nicodemus, C.F. (2004) Using antibodies in tumour immunotherapy. *Expert Opin Biol Ther*, 4, 1265–1284.
94. Schuster, M., Nechansky, A. and Kircheis, R. (2006) Cancer immunotherapy. *Biotechnol J*, 1, 138–147.
95. Schuurhuis, D.H., van Montfoort, N., Ioan-Facsinay, A., Jiawan, R., Camps, M., Nouta, J., Melief, C.J., Verbeek, J.S. and Ossendorp, F. (2006) Immune complex-loaded dendritic cells are superior to soluble immune complexes as antitumor vaccine. *J Immunol*, 176, 4573–4580.
96. Serafini, P., Borrello, I. and Bronte, V. (2006) Myeloid suppressor cells in cancer: recruitment, phenotype, properties, and mechanisms of immune suppression. *Semin Cancer Biol*, 16, 53–65.
97. Setoguchi, R., Hori, S., Takahashi, T. and Sakaguchi, S. (2005) Homeostatic maintenance of natural Foxp3(+) CD25(+) CD4(+) regulatory T cells by interleukin (IL)-2 and induction of autoimmune disease by IL-2 neutralization. *J Exp Med*, 201, 723–735.
98. Sharkey, R.M. and Goldenberg, D.M. (2006) Targeted therapy of cancer: new prospects for antibodies and immunoconjugates. *CA Cancer J Clin*, 56, 226–243.
99. Sharma, S., Yang, S.C., Zhu, L., Reckamp, K., Gardner, B., Baratelli, F., Huang, M., Batra, R.K. and Dubinett, S.M. (2005) Tumor cyclooxygenase-2/prostaglandin E2-dependent promotion of FOXP3 expression and CD4 + CD25+ T regulatory cell activities in lung cancer. *Cancer Res*, 65, 5211–5220.
100. Slingluff, C.L., Jr. and Speiser, D.E. (2005) Progress and controversies in developing cancer vaccines. *J Transl Med*, 3, 18.
101. Smyth, M.J., Hayakawa, Y., Takeda, K. and Yagita, H. (2002) New aspects of natural-killer-cell surveillance and therapy of cancer. *Nat Rev Cancer*, 2, 850–861.
102. Soiffer, R.J., Murray, C., Gonin, R. and Ritz, J. (1994) Effect of low-dose interleukin-2 on disease relapse after T-cell-depleted allogeneic bone marrow transplantation. *Blood*, 84, 964–971.
103. Steis, R.G., Urba, W.J., VanderMolen, L.A., Bookman, M.A., Smith, J.W., 2nd, Clark, J.W., Miller, R.L., Crum, E.D., Beckner, S.K., McKnight, J.E., et al. (1990) Intraperitoneal lymphokine-activated killer-cell and interleukin-2 therapy for malignancies limited to the peritoneal cavity. *J Clin Oncol*, 8, 1618–1629.
104. Stewart, J.A., Belinson, J.L., Moore, A.L., Dorighi, J.A., Grant, B.W., Haugh, L.D., Roberts, J.D., Albertini, R.J. and Branda, R.F. (1990) Phase I trial of intraperitoneal recombinant interleukin-2/lymphokine-activated killer cells in patients with ovarian cancer. *Cancer Res*, 50, 6302–6310.
105. Umiel, T. and Trainin, N. (1974) Immunological enhancement of tumor growth by syngeneic thymus-derived lymphocytes. *Transplantation*, 18, 244–250.
106. van der Most, R.G., Currie, A., Robinson, B.W. and Lake, R.A. (2006) Cranking the immunologic engine with chemotherapy: using context to drive tumor antigen cross-presentation towards useful antitumor immunity. *Cancer Res*, 66, 601–604.

107. Vignard, V., Lemercier, B., Lim, A., Pandolfino, M.C., Guilloux, Y., Khammari, A., Rabu, C., Echasserieau, K., Lang, F., Gougeon, M.L., Dreno, B., Jotereau, F. and Labarriere, N. (2005) Adoptive transfer of tumor-reactive Melan-A-specific CTL clones in melanoma patients is followed by increased frequencies of additional Melan-A-specific T cells. *J Immunol*, 175, 4797–4805.
108. Vitale, M., Pelusi, G., Taroni, B., Gobbi, G., Micheloni, C., Rezzani, R., Donato, F., Wang, X. and Ferrone, S. (2005) HLA class I antigen down-regulation in primary ovary carcinoma lesions: association with disease stage. *Clin Cancer Res*, 11, 67–72.
109. Waldmann, T.A. (2006a) The biology of interleukin-2 and interleukin-15: implications for cancer therapy and vaccine design. *Nat Rev Immunol*, 6, 595–601.
110. Waldmann, T.A. (2006b) Effective cancer therapy through immunomodulation. *Annu Rev Med*, 57, 65–81.
111. Wang, H.Y., Lee, D.A., Peng, G., Guo, Z., Li, Y., Kiniwa, Y., Shevach, E.M. and Wang, R.F. (2004) Tumor-specific human CD4+ regulatory T cells and their ligands: implications for immunotherapy. *Immunity*, 20, 107–118.
112. Whiteside, T.L. (2006) Immune suppression in cancer: effects on immune cells, mechanisms and future therapeutic intervention. *Semin Cancer Biol*, 16, 3–15.
113. Wolf, D., Wolf, A.M., Rumpold, H., Fiegl, H., Zeimet, A.G., Muller-Holzner, E., Deibl, M., Gastl, G., Gunsilius, E. and Marth, C. (2005) The expression of the regulatory T cell-specific forkhead box transcription factor FoxP3 is associated with poor prognosis in ovarian cancer. *Clin Cancer Res*, 11, 8326–8331.
114. Woo, E.Y., Chu, C.S., Goletz, T.J., Schlienger, K., Yeh, H., Coukos, G., Rubin, S.C., Kaiser, L.R. and June, C.H. (2001) Regulatory CD4(+)CD25(+) T cells in tumors from patients with early-stage non-small cell lung cancer and late-stage ovarian cancer. *Cancer Res*, 61, 4766–4772.
115. Yu, P., Rowley, D.A., Fu, Y.X. and Schreiber, H. (2006) The role of stroma in immune recognition and destruction of well-established solid tumors. *Curr Opin Immunol*, 18, 226–231.
116. Zlatnik, E. and Golotina, L. (2005) The use of roncoleukin for LAK therapy in ovarian cancer. *Vopr Onkol*, 51, 680–684.
117. Zou, W. (2006) Regulatory T cells, tumour immunity and immunotherapy. *Nat Rev Immunol*, 6, 295–307.
118. Carlsten M., Björkström N.K., Norell H., Bryceson Y., van Hall T., Baumann B.C., Hanson M., Schedvins K., Kiessling R., Ljunggren H.G., Malmberg K.J..(2007) DNAX accessory molecule-1 mediated recognition of freshly isolated ovarian carcinoma by resting natural killer cells. *Cancer Res*, 67,1317-1325.

Regulatory T Cells: A New Frontier in Cancer Immunotherapy

Brian G. Barnett, Jens Rüter, Ilona Kryczek, Michael J. Brumlik,
Pui Joan Cheng, Benjamin J. Daniel, George Coukos, Weiping Zou,
and Tyler J. Curiel

1 Introduction

Tumor-specific immune-mediate cancer therapy was documented in a mouse model about one and half century ago (1). Nonetheless, the success of immune-based cancer treatments in humans has remained quite modest despite advances in our understanding and technology. The current paradigm driving most immune strategies is that tumors express tumor-associated antigens (TAA), thereby making them the objects of immune attack. These TAA should then be captured by professional antigen-presenting cells, particularly dendritic cells, which in turn prime naïve T cells to become TAA-specific effector cells through T cell cosignaling molecules and other mediators. This paradigm predicts that the solution to improving the efficacy of tumor immunotherapy is to augment TAA expression, boost cosignaling, or increase the number of effector T cells or professional antigen-presenting cells. Experience shows, however, that with a few limited exceptions, such strategies do not yield durable clinical successes.

Recent work, including from our group, now demonstrates that tumors employ a wide variety of active mechanisms to thwart what could be an otherwise effective host antitumor immune response (2–4). These tumor-associated mechanisms include production of factors such as VEGF, TGF-β, or IL-10; induction of dysfunctional dendritic cells; or dysfunctional T cell cosignaling (5).

Much recent work implicates CD4$^+$CD25$^+$ regulatory T cells (Tregs) as an agent of this tumor-mediated anti-host defense (2, 6). CD4$^+$CD25$^+$ regulatory T cells normally mediate peripheral tolerance (7, 8). However, if they are abnormally elevated in numbers or function, they have the potential to perturb homeostatic immune functions or defeat a required immune response (6).

In 1999, it was demonstrated that depletion of CD4$^+$CD25$^+$ T cells in a mouse model for cancer using PC61 antibody improved immune-mediated tumor rejection (9). Soon thereafter, CD4$^+$CD25$^+$ T cell depletion was shown to boost endogenous TAA-specific immunity as well as the efficacy of active immunization or anti-CTLA-4 blockade (10, 11). CD4$^+$CD25$^+$ regulatory T cells are elevated in the peripheral blood of patients with a variety of cancers (2, 12–18). Nonetheless, most of the work performed until this time centered on mouse models of cancer.

We decided to study human Tregs and thus focused on ovarian cancer because we could obtain large quantities of immune cells from ascites and could grow autologous tumors from individual patients for studies of tumor-specific events (3, 19). We demonstrated that dendritic cells in the ovarian cancer microenvironment were dysfunctional and could induce T cells to produce IL-10 (3, 19), suggestive of Tregs. We then demonstrated that Tregs are accumulated in the human ovarian cancer microenvironment against a concentration gradient. These cells could inhibit tumor-specific cytotoxicity and cytokine production of tumor-specific CD8$^+$ T cells in vitro and in vivo in a chimeric SCID/NOD xenograft model. We further demonstrated an inverse correlation between tumor Treg content and patient survival even after adjusting for known factors predicting survival (2).

Taken together, data from mouse models and our findings in human ovarian cancer predicted that reducing the Treg content of patients with ovarian cancer might be of therapeutical use. We then identified denileukin diftitox (Ontak) as an agent that might be useful to deplete CD4$^+$CD25$^+$ Tregs in human cancer. Denileukin diftitox is a fusion protein of IL-2 and diphtheria toxin, which is licensed by the United States Food and Drug Administration to treat cutaneous T cell leukemia/lymphoma, a T cell malignancy with a CD4$^+$CD25$^+$ phenotype similar to regulatory T cells. It binds to cells through an IL-2 receptor and inhibits protein translation following internalization, leading to apoptosis (20). We undertook a phase 0/I clinical trial enrolling patients with any advanced stage epithelial carcinoma to test the hypothesis that denileukin diftitox depletes Tregs and thus boosts immunity. This phase 0/I design was to test immunologic end points only, but we subsequently obtained some clinical efficacy data as well, as discussed later.

2 Results

Six patients with advanced stage ovarian, breast, or lung cancer were enrolled in the phase 0/I study, the first three receiving a single intravenous dose of denileukin diftitox at 9 µg ml^{-1} and the next three a single intravenous dose of 12 µg ml^{-1}. A seventh patient with stage IV pancreatic cancer received five consecutive weekly doses of denileukin diftitox at 12 µg kg^{-1}. Blood was obtained just before and 3–7 days after the first single intravenous dose from all seven patients. Results for those receiving 9 or 12 µg kg^{-1} were similar and thus pooled for the day 3–7 analyses. All studies were approved by the Tulane IRB and patients provided written, informed consent.

Phenotypic Tregs (CD4$^+$CD25$^+$ T cells) were reduced in prevalence in 6 of the 7 patients and in absolute numbers in blood from all seven patients, 3–7 days after a single intravenous infusion of denileukin diftitox. This was not a simple panlymphopenia, as CD3$^+$CD8$^+$ T cells expressing interferon-γ increased significantly in number at the same time. CD4$^+$CD25$^+$ T cells expressing *foxp3* message and FOXP3 protein were reduced in number, and their capacity to inhibit T cell proliferation was significantly reduced in the week following denileukin diftitox administration. In three of the patients, we were able to test the capacity of sorted $CD4^+CD25^+$ T cells to suppress T cell proliferation before, in the first week after, and again 21–30 days after denileukin diftitox infusion. Suppression of T cell proliferation by sorted $CD4^+CD25^+$ T cells was significantly reduced 1 week after infusion and remained reduced at 21–30 days, the latest time points tested, in these three patients. These data are consistent with reduction in functional Treg activity for up to 30 days after a single denileukin diftitox infusion.

Patient 4 had stage IV epithelial ovarian carcinoma that had failed conventional chemotherapeutic and surgical approaches, and had no clearly beneficial standard treatment alternatives. She was enrolled in the stage 0/I trial at the $12\,\mu g\,kg^{-1}$ dose. Thirty days after this single dose, when her part in the phase 0/I study ended, she received six additional weekly cycles of denileukin diftitox at $12\,\mu g\,kg^{-1}$ with therapeutic intent in an IRB-approved study amendment. Following the first infusion in the phase 0/I study, her blood CA 125 fell significantly. After the first of six additional consecutive weekly infusions (which began 42 days after her initial treatment), her blood CA125 decreased to and remained within the normal range. A PET/CT fusion scan demonstrated resolution of bony, visceral, and lymphatic disease with a residual mass persisting in the left groin, which proved to be ovarian carcinoma upon biopsy. She failed local radiotherapy and then refused additional therapies and died of infection 13 months later.

Patient 7 with stage IV pancreatic cancer was treated with five consecutive weekly denileukin diftitox infusions at $12\,\mu g\,kg^{-1}$ in a separate protocol with therapeutic intent. After the third infusion, $CD3^+CD8^+$ interferon-γ^+ T cells, which had risen during the first two cycles, began to decline below baseline levels, and therapy was stopped after the fifth weekly dose as these cells continued to decline after each successive infusion. *Foxp3* message in $CD4^+CD25^+$ T cells also fell with each successive cycle, becoming undetectable after the fourth cycle. $CD3^+CD8^+$ interferon-γ^+ T cells in blood recovered to greater than 50% of baseline by 7 weeks after the final infusion. *Foxp3* message data during this recovery phase were discarded when RNA degradation in the samples was detected in quality control experiments, and no FOXP3 protein expression data was available at this time. The patient had stable disease for approximately 4 months, and then the disease progressed and she died 3 months after progression.

3 Conclusions

These preliminary data suggest the following points. A single infusion of denileukin diftitox at 9 or $12\,\mu g\,kg^{-1}$ decreases blood $CD4^+CD25^+$ T cell prevalence and numbers, and reduces *foxp3* message and FOXP3 protein expression in blood $CD4^+CD25^+$ T cells coincident with a reduction in the suppressive capacity of these cells in patients with ovarian, lung, breast, or pancreatic cancer. Thus, denileukin diftitox depletes regulatory T cell function from blood of such patients. Regulatory T cell function and *foxp3* message and FOXP3 protein expression remained significantly reduced for 3–4 weeks after the single infusion demonstrating a prolonged reduction in regulatory T cell function. A single infusion of denileukin diftitox at 9 or 12 μg kg^{-1} is well tolerated by patients with a variety of epithelial carcinomas, even with serum albumins as low as 2 gm dl^{-1}. Repeated weekly infusions at 12 μg kg^{-1} for up to five cycles are also well tolerated, but eventually deplete interferon-γ-secreting T cells that likely represent immune effector cells. Thus, the dosing interval for this strategy should be longer than weekly. Resolution of metastatic ovarian carcinoma was observed in conjunction with regulatory T cell depletion in the one patient so studied, suggesting potential for clinical efficacy, although the relation between any immune events and any clinical outcomes remains unknown at present. We hypothesize improved immunity through Treg reduction, augmenting immune-mediated tumor clearance as the mechanism which represents an area of active investigation for us.

These data suggest that Ontak may be useful to reduce CD4$^+$CD25$^+$ regulatory T cell function in blood of patients with certain epithelial carcinomas. Additional data (Barnett et al., manuscript in preparation) suggest that monthly denileukin diftitox infusions at 12 μg kg^{-1} may continue to suppress regulatory T cell function with little reduction of interferon-γ$^+$ T cells in blood. On this basis, we are conducting a phase II trial of monthly denileukin diftitox at 12 μg kg^{-1} to treat patients with stage III or IV epithelial ovarian carcinoma failing standard therapies. Study details and eligibility criteria can be found on the NCI PDQ Web site [http://www.cancer.gov/search/ViewClinicalTrials.aspx?cdrid=445063&version=patient&protocolsearchid=2853127]. Eight patients have been enrolled to date and there continues to be minimal toxicity.

Current data suggest that the IL-2 moiety of Ontak does not mediate its therapeutic effect and that an intact immune system is required for therapeutic efficacy (Barnett, Rüter, Cao et al., manuscript in preparation). Much additional work remains to be done to test whether reduced Treg numbers or function are a mechanism of action of increased interferon-γ production and to relate any clinical effects to Treg depletion.

Our results suggest that regulatory T cell depletion will be a useful immune strategy to treat certain epithelial carcinomas. Further questions include determining whether eliminated Tregs are antigen-specific and if so, how that influences immunity, and whether combining regulatory T cell depletion with active vaccination or additional treatments will be more efficacious than either treatment alone. Further investigations also include studying whether this strategy reduces regulatory T cell function at the tumor site and in draining lymph nodes, what the relationship between depletion in blood vs. local sites is, and whether these observations individually or collectively predict immunologic or clinical benefits. Additional agents suggested or tested to deplete regulatory T cells include cyclophosphamide and fludarabine among others (see our review in *Frontiers in Bioscience*, in press, for many additional details). We are also developing proprietary agents to deplete regulatory T cells selectively. Immunopathologic contributions of other regulatory cell populations such as CD8$^+$ regulatory T cells [21], immature myeloid cells [22], NKT cells [23], and B7-H4$^+$ myeloid cells [24] remain to be established and merit further investigations.

Acknowledgments This work was supported by a grant from the Ovarian Cancer Research Fund to BGB, and by FD003118, CA100425, and CA105207 to TJC.

References

1. Prehn, R. (1957) Immunity to methylcholanthrene-induced sarcomas. *J Natl Cancer Inst* 18, 769.
2. Curiel, T.J. et al. (2004) Specific recruitment of regulatory T cells in ovarian carcinoma fosters immune privilege and predicts reduced survival. *Nat Med* 10, 942–949.
3. Curiel, T.J. et al. (2003) Blockade of B7-H1 improves myeloid dendritic cell-mediated antitumor immunity. *Nat Med* 9, 562–567.
4. Zou, W. et al. (2001) Reciprocal regulation of plasmacytoid dendritic cells and monocytes during viral infection. *Eur J Immunol* 31, 3833–3839.
5. Zou, W. (2005) Immunosuppressive networks in the tumour environment and their therapeutic relevance. *Nat Rev Cancer* 5, 263–274.
6. Zou, W. (2006) Regulatory T cells, tumour immunity and immunotherapy. *Nat Rev* 6, 295–307.
7. Shevach, E.M. (2002) CD4+ CD25+ suppressor T cells: more questions than answers. *Nat Rev* 2, 389–400.
8. Francois Bach, J. (2003) Regulatory T cells under scrutiny. *Nat Rev* 3, 189–198.
9. Shimizu, J. et al. (1999) Induction of tumor immunity by removing CD25 + CD4+ T cells: a common basis between tumor immunity and autoimmunity. *J Immunol* 163, 5211–5218.
10. Sutmuller, R.P. et al. (2001) Synergism of cytotoxic T lymphocyte-associated antigen 4 blockade and depletion of CD25(+) regulatory T cells in antitumor therapy reveals alternative pathways for suppression of autoreactive cytotoxic T lymphocyte responses. *J Exp Med* 194, 823–832.
11. Steitz, J. et al. (2001) Depletion of CD25(+) CD4(+) T cells and treatment with tyrosinase-related protein 2-transduced dendritic cells enhance the interferon alpha-induced, CD8(+) T-cell-dependent immune defense of B16 melanoma. *Cancer Res* 61, 8643–8646.
12. Woo, E.Y. et al. (2001) Regulatory CD4(+)CD25(+) T cells in tumors from patients with early-stage non-small cell lung cancer and late-stage ovarian cancer. *Cancer Res* 61, 4766–4772.
13. Woo, E.Y. et al. (2002) Regulatory T cells from lung cancer patients directly inhibit autologous T cell proliferation. *J Immunol* 168, 4272–4276.
14. Javia, L.R., and Rosenberg, S.A. (2003) CD4 + CD25+ suppressor lymphocytes in the circulation of patients immunized against melanoma antigens. *J Immunother* 26, 85–93.
15. Somasundaram, R. et al. (2002) Inhibition of cytolytic T lymphocyte proliferation by autologous CD4+/CD25+ regulatory T cells in a colorectal carcinoma patient is mediated by transforming growth factor-beta. *Cancer Res* 62, 5267–5272.
16. Wolf, A.M. et al. (2003) Increase of regulatory T cells in the peripheral blood of cancer patients. *Clin Cancer Res* 9, 606–612.
17. Sasada, T. et al. (2003) CD4 + CD25+ regulatory T cells in patients with gastrointestinal malignancies: possible involvement of regulatory T cells in disease progression. *Cancer* 98, 1089–1099.
18. Liyanage, U.K. et al. (2002) Prevalence of regulatory T cells is increased in peripheral blood and tumor microenvironment of patients with pancreas or breast adenocarcinoma. *J Immunol* 169, 2756–2761.
19. Zou, W. et al. (2001) Stromal-derived factor-1 in human tumors recruits and alters the function of plasmacytoid precursor dendritic cells. *Nat Med* 7, 1339–1346.
20. Foss, F.M. (2000) DAB(389)IL-2 (ONTAK): a novel fusion toxin therapy for lymphoma. *Clin Lymphoma* 1, 110–116; discussion 117.

21. Wei, S. et al. (2005) Plasmacytoid dendritic cells induce CD8+ regulatory T cells in human ovarian carcinoma". *Cancer Res* 65, 5020–5026.
22. Almand, B. et al. (2001) Increased production of immature myeloid cells in cancer patients: a mechanism of immunosuppression in cancer. *J Immunol* 166, 678–689.
23. Taniguchi, M. et al. (2003) The regulatory role of Valpha14 NKT cells in innate and acquired immune response. *Annu Rev Immunol* 21, 483–513.
24. Kryczek, I. et al. (2006) B7-H4 expression identifies a novel suppressive macrophage population in human ovarian carcinoma. *J Exp Med* 203, 871–881.

Inhibitory B7 Family Members in Human Ovarian Carcinoma

Shuang Wei, Tyler Curiel, George Coukos, Rebecca Liu, and Weiping Zou

1 Introduction

Tumors express tumor-associated antigens (TAA), and thus should be the object of immune attack. Nonetheless, spontaneous clearance of established tumors is rare (1, 2). Much work has demonstrated that tumors have numerous strategies either to prevent presentation of TAA or to prevent TAA presentation in the context of T cell costimulatory molecules (3–12).

Thus, it was thought that a lack of TAA-specific immunity was largely a passive process; tumors simply did not present enough TAA, or antigen-presenting cells did not have sufficient stimulatory capacity. On this basis, attempts were made to bolster TAA-specific immunity by using optimal antigen-presenting cells (13–15), by growing TAA-specific effector T cells ex vivo followed by adoptive transfer (3–10, 16), by cytokine/chemokine immunotherapy (1, 17), or by peptide vaccination (18–21). These approaches were met with some success in mouse models of human tumors and showed some early clinical efficacy in human trials, although long-term efficacy remains to be established and logistical problems are considerable (3–12).

These studies and clinical trials established the concept that experimentally induced TAA-specific immunity is a rational and potentially efficacious, means to treat established human cancer. Nonetheless, more recent work, including from our laboratories, demonstrates that a lack of naturally induced TAA-specific immunity is not simply a passive process whereby adaptive immunity is shielded from detecting TAA (1–8, 10).

In fact, over the past few years, our research group took human ovarian cancer as a human tumor model and focused on defining tumor immune mechanisms. We have made important insights into cancer immunopathogenesis. Our prior research efforts clearly demonstrated that the tumor microenvironment is composed of dysfunctional immune cells that have been reprogrammed by active tumor-mediated processes to defeat tumor-specific immunity in a highly effective manner. Two major mediators of poor tumor immunity are dysfunctional APCs (including stromal macrophages) and Tregs. In this chapter, we will focus on inhibitory B7 family members, and discuss three fundamental areas: (1) tumor processes that render

APCs (including macrophages) and T cell dysfunction in the tumor microenvironment, (2) cellular and molecular mechanisms whereby dysfunctional APCs (including macrophage subpopulations) inhibit tumor immunity and induce Tregs in the tumor microenvironment, and (3) the immunologic and preclinical consequences of reprogramming tumor microenvironment.

2 B7-H4, Ovarian Cancer, and Other Human Cancers

B7-H4 (B7x, B7S1), a member of the B7 family of T cell costimulatory molecules was identified in 2003 (22–24). B7-H4 has about 25% homology in the extracellular portion with other B7-family members. The B7-H4 receptor has not been identified.

B7-H4 mRNA expression was found to be widely distributed in the peripheral tissues including kidney, liver, lung, spleen, thymus, and placenta. However, B7-H4 protein expression on the cells seems to be limited (22, 25). Interestingly, several groups have demonstrated that human ovarian cancers express high levels of B7-H4 protein (25–30). The expression is variable in different histological types of ovarian cancers. B7-H4 was detected in primary and metastatic serous, endometrioid and clear cell ovarian carcinoma. The expression of B7-H4 in mucinous ovarian carcinoma is limited (28, 30). More interestingly, low levels of B7-H4 protein were found in all sera from ovarian cancer patients. The levels of serum B7-H4 were significantly higher when compared with healthy controls or women with benign gynecologic diseases. The median B7-H4 concentration in endometrioid and serous histotypes was higher than in mucinous histotypes (28), consistent with results of immunohistochemical staining. The multivariate logistic regression analysis of B7-H4 suggests that B7-H4 is a promising new biomarker for ovarian carcinoma (29).

One study reported that overexpression of B7-H4 in a human ovarian cancer cell line with little endogenous B7-H4 expression increased tumor formation in SCID mice, suggesting that ovarian cancer B7-H4 is involved in tumorigenesis (30). Further work is warranted to confirm this observation and to define the potential action mode of B7-H4 in ovarian cancer pathology.

In addition to human ovarian cancer, high levels of B7-H4 were found in nonsmall cell lung cancer (31), ductal and lobular breast cancer (30, 32), and renal cell carcinoma (33). Vascular endothelial cells also express B7-H4 in renal cell carcinoma (33). Further, B7-H4 expression was associated with adverse clinical and pathologic features, including constitutional symptoms, tumor necrosis, and advanced tumor size, stage, and grade. Patients with tumors expressing B7-H4 were also three times more likely to die from renal cell carcinoma compared with patients lacking B7-H4. These data suggest that B7-H4 has the potential to be a useful prognostic marker for patients with renal cell carcinoma (33).

Although the functional activity, the regulatory mechanism, and the signal pathways of B7-H4 remain to be defined, the broad expression of B7-H4 in human tumor suggests a potential role for this protein in human tumor biology. We will further discuss the nature, role, and potential application of B7-H4 in the following section.

3 B7-H4 and Tumor Immunosuppression

It was shown that B7-H4 is a negative regulator of T cell responses in vitro by inhibiting T cell proliferation, cell-cycle progression, and cytokine production (22–24, 34). Antigen-specific T cell responses are impaired in mice treated with a B7-H4Ig fusion protein (22).

Antigen-presenting cells (APCs) are critical for initiating and maintaining tumor-associated antigen (TAA)-specific T cell immunity. Tumor-associated macrophages (TAMs) markedly outnumber other APCs, such as dendritic cells (DCs), and represent an abundant population of APCs in solid tumors (35–38). Strikingly, numerous studies have investigated the phenotypes and functions of DCs in tumor immunity (2, 3, 5–8, 39–42). Studies in mice have also revealed that TAMs promote tumor growth and metastasis by directly acting on tumor cells (35–38). Immunohistochemical assessment of the number and the distribution of TAMs in human tumors have yielded scant and often contradictory results regarding any potential role in tumor pathogenesis (43–45). We now report a novel TAM population in patients with ovarian carcinoma, namely B7-H4$^+$ macrophages. The expression and function of B7-H4 in TAM may provide a consensus of the role of TAMs in human tumor immunology.

We now document that B7-H4$^+$ macrophages significantly inhibit TAA-specific T cell proliferation, cytokine production, and cytotoxicity in vitro. These B7-H4$^+$ TAMs also inhibit TAA-specific immunity in vivo and foster tumor growth in chimeric SCID/NOD mice bearing autologous human tumors, despite the presence of potent TAA-specific effector T cells. The notion that TAM B7-H4 signals contribute to immunopathology is supported by several lines of evidence. First, B7-H4$^+$ TAMs are significantly more suppressive than B7-H4$^-$ TAMs. Second, blocking B7-H4 on tumor conditioned-macrophages disables their suppressive capacity. Third, forced B7-H4 expression renders normal macrophages suppression. Fourth, blocking B7-H1 and inhibiting iNOS and arginase have minor effects on B7-H4$^+$ macrophage-mediated T cell suppression. Although ovarian tumor cells express B7-H4, fixed or irradiated ovarian tumor cells do not induce T cell suppression. It suggests the role of B7-H4 may be different in tumor from in APCs. In addition to CD4$^+$ Tregs (46), these findings establish B7-H4$^+$ TAMs as a novel immune regulatory population in human ovarian cancer.

4 B7-H4 and Tumor Environmental Cytokines

Mouse B7-H4 ligation of T cells has an inhibitory effect on T cell activation (22, 24). We recently defined the regulatory mechanisms and the function of human B7-H4 in human immunology. We show for the first time that recombinant and tumor environmental IL-6 and IL-10 stimulate monocyte/macrophage B7-H4 expression, and that GM-CSF and IL-4 reduce B7-H4 expression (26, 47). We also observed similar regulatory mechanism for B7-H4 regulation on myeloid dendritic cells (26, 47). In human ovarian cancer, it appears that the dysfunctional tumor microenvironmental cytokine networks enable TAM B7-H4 expression. This conclusion is supported by the following evidence. We found high concentrations of IL-6 and IL-10 but not GM-CSF and IL-4 in the tumor microenvironment. IL-6 and IL-10 in the tumor environment strongly stimulate macrophage B7-H4 expression, whereas IL-4 and GM-CSF strongly suppress it. Our data and those of others taken together demonstrate a new immune evasion strategy, whereby tumors maximize local tolerizing conditions through suppressing dendritic cell differentiation while simultaneously inducing macrophage B7-H4 expression. These ends are mediated through maximal local accumulation of B7-H4 inducing cytokines, IL-6, and IL-10 in the virtual absence of B7-H4 reducing and dendritic cell differentiation cytokines, GM-CSF and IL-4. Strikingly, although ovarian tumor cells express B7-H4, it appears that tumor B7-H4 expression is exclusively intracellular and unable to induce T cell suppression. Further, cytokines IL-4, GM-CSF, IL-6, and IL-10 have no regulatory effects on tumor B7-H4 expression. These data suggest that tumor B7-H4 and APC B7-H4 may be functionally distinct and be differentially regulated (26, 47).

Tumor cells, tumor-associated macrophages, and regulatory T cells may be the source for IL-6 and IL-10 (40, 46). These data provide mechanisms for how tumor environmental IL-6 and IL-10 induce immune dysfunction. GM-CSF has been used to boost TAA-specific immunity in mouse cancer models (48, 49). A proposed mechanism for GM-CSF efficacy in these models is differentiation or attraction of dendritic cells that boost TAA-specific immunity. In light of our present work, it will be interesting and worthwhile to reexamine these GM-CSF studies to determine whether a GM-CSF-mediated reduction in APC B7-H4 expression accounts for efficacy.

5 B7-H4 and Regulatory T Cells

In human ovarian cancer, Treg cells and B7-H4+ TAMs are colocalized in the tumor environment (26, 46, 47). We demonstrate that Treg cells, but not conventional T cells, trigger high levels of IL-10 production by APCs, stimulate APC B7-H4 expression through IL-10, and render APC immunosuppressive. These data are in line with the observations that macrophages spontaneously produce IL-10 in ovarian tumor environment (40). Initial blockade of B7-H4 reduces the suppressive activity mediated by Treg cell-conditioned APCs. Further, APC, rather than Treg cell-derived IL-10, is responsible for APC B7-H4 induction. Therefore, Treg cells convey suppressive activity to APCs by stimulating B7-H4 expression through IL-10.

Our data mechanistically link IL-10, B7-H4, Treg cells, and APCs in the context of Treg biology. Our findings thus provide three pieces of novel information: (1) IL-10 is capable of inducing B7-H4 expression on human APCs; (2) Similar to murine B7-H4 fusion protein (22–24), human APC B7-H4 negatively regulates T cell responses; (3) Human Treg cells enable suppressor activity to APCs via triggering B7-H4 expression. Thus, as suppression partially relies on Treg-triggered, APC-dependent IL-10, our data reconcile the apparent contradiction in previous in vitro and in vivo studies regarding the role and source of IL-10 in Treg cell biological activity. These data provide a novel cellular and molecular mechanism for Treg cell-mediated immunosuppression at the level of APCs. This mode of suppression mediated by Treg cells may be particularly operative in patients with ovarian cancer.

6 B7-H1 Expression, Ovarian Cancer, and Other Human Cancers

B7-H1 is one of the B7 family molecules belonging to immunoglobulin (Ig) superfamily. Like other B7 family molecules, B7-H1 has one Ig V and one Ig C in its extracellular domain with a transmembrane and an intracellular domain. Program death one (PD-1) has been identified as the receptor for B7-H1 (50). It has been suggested that some of immunological functions of B7-H1 are mediated by PD-1, while others may be mediated by an additional unidentified receptor (34, 39, 51, 52).

B7-H1 mRNA expression is widely distributed in every tissues virtually, and cell surface B7-H1 is only found in macrophage-origin cells such as Kupffer cells in the liver and dust cells in the lung by immunohistochemistry analysis. Interestingly, B7-H1 protein is highly expressed in various human tumors including breast cancer (53), colon cancer (54), esophageal cancer (55), leukemia cells (56), lung cancer (54, 57), melanomas (54), oral squamous cell carcinoma (58), ovarian cancer (54), and renal cell carcinoma (59). The relationship between tumor-associated B7-H1 and clinical cancer progression has been studied in renal cell carcinoma (RCC) (60) and esophageal cancer (55). Patients with RCC harboring high intratumoral expression levels of B7-H1 exhibit aggressive tumors and are at a markedly increased risk of death from RCC (60) and esophageal cancer (55). Further, the expression was more pronounced in the advanced stage of tumor than in the early stage. Multivariate analysis indicated that B7-H1 was an independent prognostic factor for RCC (60) and esophageal cancer (55). Although the role of B7-H1 has been extensively studied in the context of tumor immunity (see discussion below), the biological activity of B7-H1 in tumor pathology has not been studied.

7 Regulation of B7-H1 Expression

Several cytokines are able to regulate B7-H1 expression. Although B7-H1 expression on APCs can be stimulated by TNF-α, IFN-γ, IL-10, and VEGF (39, 54, 61), IFN-γ is the most powerful stimulus among these cytokines. Further, tumor B7-H1 expression can be potently induced by IFN-γ (54) and type I IFN, but not by other cytokines (our unpublished data). Consistent with these observations, IFN-γ can induce high levels of B7-H1 expression on normal epithelial cells and vascular endothelial cells (62). The IFN-γ induced B7-H1 on vascular endothelial cells also mediate T cell suppression (62). These data suggest that strong Th1 responses may induce B7-H1 expression on APCs and epithelial cells and endothelial cells through IFN-γ, and in turn maintain the threshold of T cell activation and avoid tissue/organ damage. As B7-H1 can be induced on APCs and multiple human epithelial tumors, the upregulation of B7-H1 would defeat T cell mediated tumor immunity.

8 B7-H1 and Tumor Immunity

8.1 Tumor B7-H1 and Tumor Immunity

As we discussed earlier, multiple human tumor cells express high levels of B7-H1. It has been demonstrated that tumor B7-H1 reduces tumor immunity through two mechanisms. The first mechanism is that tumor-associated B7-H1 promotes apoptosis of tumor antigen-specific human T cells (54). In support of this, blocking B7-H1 reduces T cell apoptosis in vitro and in vivo (54). Further, CD8$^+$ T cell apoptosis is decreased in the liver in B7-H1$^{-/-}$ mice, and B7-H1 deficient mice were prone to experimentally induced hepatitis (63). These data reveal an inhibitory role of B7-H1 in T cell activation. The second mechanism is that tumor B7-H1 directly reduces CD8$^+$ T cell-mediated cytotoxicity. Expression of B7-H1 in mouse P815 mastocytoma cells results in a reduced killing activity of CTL (64). Similarly, B7-H1 transfected B16-F10 melanoma expressing H-2K^b binding peptide SIYRYYGL acquired resistance in vitro for cytotoxic lysis by 2C T cell receptor transgenic T cells (65). However, it remains to be defined how tumor B7-H1 mediates T cell apoptosis and suppression of CTL activity. The receptor PD-1 would be implicated in these processes.

8.2 B7-H1$^+$ APCs and Tumor Immunity

In addition to tumor cells, the expression and function of B7-H1 on APCs have been studied. For example, myeloid dendritic cells (MDCs) highly express B7-H1 in tumor draining lymph nodes and tumor environment in patients with ovarian cancer. These B7-H1 expressing MDCs engage T cells, leading to downregulation of MDC IL-12 and upregulation of MDC IL-10. As MDC IL-12 is critical for establishing tumor specific immunity and Th1-polarization (66), and as IL-10 inhibits tumor-specific immunity, including in ovarian carcinomas (40), MDC-associated B7-H1 signals could determine the nature of subsequent T cell activation. In support of this concept, we found that blockade of tumor MDC-associated B7-H1 decreases T cell

IL-10, increases T cell interferon-γ production and improves clearance of tumor in xenotransplanted mice, which is associated with tumor infiltration by interferon-γ secreting T cells (39). The study has been further confirmed and extended in different systems. For example, blockade of B7-H1 improves tumor regression in mouse models with CT26 colon cancer cells (67), B16 melanoma (67), oral squamous cell carcinoma (58), P815 tumor (64), squamous cell carcinomas of the head and neck (SCCHN) (68), and in human T cell leukemia model (56).

In summary, B7-H1 expressing APCs and tumor cells may mediate T cell suppression by inducing T cell apoptosis, reducing CTL cytoxicity, and inhibiting DC function. These mechanisms could operate either individually or cooperatively, depending on the tumor microenvironment. Thus, targeting the B7-H1 and PD-1 signal pathway provides a novel strategy to treat human cancer.

9 Tumor Immunological Therapy and B7-H1/PD-1 and B7-H4

We recently described several mechanisms in the human ovarian cancer microenvironment that actively defeat tumor immunity (2, 46, 69), including an immunopathologic role for regulatory T cells (46, 69) and inhibitory B7 family members. As many tumor-associated APCs and tumor cells, including the majority of ovarian carcinomas, express B7-H1 and B7-H4, tumor and associated APCs reduce T cell effector functions, induce T cell apoptosis or T cell cycle arrest through B7-H1/PD-1 or/and B7-H4 signal pathways. Blocking B7-H1/PD-1 and B7-H4 signals may be useful to treat certain cancers. It is possible to block these inhibitory pathways by designing genetic and pharmaceutical methods such as interference RNA or small molecules or humanized specific antibody. Blockade of inhibitory B7 family molecules may be combined with other immunotherapeutic interventions. For example, we observed that Treg cell depletion (2) and blocking B7-H1 synergistically stimulate tumor T cell immunity and significantly reduces tumor growth in mouse models. Administration of B7-H1 blocking mAb enhanced therapeutic effects of anti-CD137 agonistic mAb (70), which could augment tumor specific CTL activity (71). Given the broad expression and suppressive functions of these inhibitory molecules in tumor and tumor-associated APCs, these strategies will be promising approaches for broad application in cancer treatment.

Acknowledgments This work was supported by the Department of Defense (OC020173) and National Cancer Institute (CA092562, CA100227, CA099985).

References

1. Kaufman, H.L., and M.L. Disis. 2004. Immune system versus tumor: shifting the balance in favor of DCs and effective immunity. J Clin Invest 113:664–667.
2. Zou, W. 2005. Immunosuppressive networks in the tumour environment and their therapeutic relevance. Nat Rev Cancer 5:263–274.

3. Pardoll, D. 2003. Does the immune system see tumors as foreign or self? Annu Rev Immunol 21:807–839.
4. Khong, H.T., and N.P. Restifo. 2002. Natural selection of tumor variants in the generation of "tumor escape" phenotypes. Nat Immunol 3:999–1005.
5. Cerundolo, V., I.F. Hermans, and M. Salio. 2004. Dendritic cells: a journey from laboratory to clinic. Nat Immunol 5:7–10.
6. Gilboa, E. 2004. The promise of cancer vaccines. Nat Rev Cancer 4:401–411.
7. Munn, D.H., and A.L. Mellor. 2004. IDO and tolerance to tumors. Trends Mol Med 10:15–18.
8. Finn, O.J. 2003. Cancer vaccines: between the idea and the reality. Nat Rev Immunol 3:630–641.
9. Yu, P., Y. Lee, W. Liu, R.K. Chin, J. Wang, Y. Wang, A. Schietinger, M. Philip, H. Schreiber, and Y.X. Fu. 2004. Priming of naive T cells inside tumors leads to eradication of established tumors. Nat Immunol 5:141–149.
10. Schreiber, H., T.H. Wu, J. Nachman, and W.M. Kast. 2002. Immunodominance and tumor escape. Semin Cancer Biol 12:25–31.
11. Lotze, M.T., and M. Papamichail. 2004. A primer on cancer immunology and immunotherapy. Cancer Immunol Immunother 53:135–138.
12. Lotze, M.T., B. Hellerstedt, L. Stolinski, T. Tueting, C. Wilson, D. Kinzler, H. Vu, J.T. Rubin, W. Storkus, H. Tahara, E. Elder, and T. Whiteside. 1997. The role of interleukin-2, interleukin-12, and dendritic cells in cancer therapy. Cancer J Sci Am 3(Suppl 1):S109–S114.
13. Banchereau, J., and R.M. Steinman. 1998. Dendritic cells and the control of immunity. Nature 392:245–252.
14. Banchereau, J., F. Briere, C. Caux, J. Davoust, S. Lebecque, Y.J. Liu, B. Pulendran, and K. Palucka. 2000. Immunobiology of dendritic cells. Annu Rev Immunol 18:767–811.
15. Mullins, D.W., S.L. Sheasley, R.M. Ream, T.N. Bullock, Y.X. Fu, and V.H. Engelhard. 2003. Route of immunization with peptide-pulsed dendritic cells controls the distribution of memory and effector T cells in lymphoid tissues and determines the pattern of regional tumor control. J Exp Med 198:1023–1034.
16. Plautz, G.E., P.A. Cohen, and S. Shu. 2003. Considerations on clinical use of T cell immunotherapy for cancer. Arch Immunol Ther Exp (Warsz) 51:245–257.
17. Flanagan, K., and H.L. Kaufman. 2006. Chemokines in tumor immunotherapy. Front Biosci 11:1024–1030.
18. Slingluff, C.L., Jr., G.R. Petroni, G.V. Yamshchikov, D.L. Barnd, S. Eastham, H. Galavotti, J.W. Patterson, D.H. Deacon, S. Hibbitts, D. Teates, P.Y. Neese, W.W. Grosh, K.A. Chianese-Bullock, E.M. Woodson, C.J. Wiernasz, P. Merrill, J. Gibson, M. Ross, and V.H. Engelhard. 2003. Clinical and immunologic results of a randomized phase II trial of vaccination using four melanoma peptides either administered in granulocyte-macrophage colony-stimulating factor in adjuvant or pulsed on dendritic cells. J Clin Oncol 21:4016–4026.
19. Weber, J.S., and A. Aparicio. 2001. Novel immunologic approaches to the management of malignant melanoma. Curr Opin Oncol 13:124–128.
20. Disis, M.L., and S. Rivkin. 2003. Future directions in the management of ovarian cancer. Hematol Oncol Clin North Am 17:1075–1085.
21. Knutson, K.L., T.J. Curiel, L. Salazar, and M.L. Disis. 2003. Immunologic principles and immunotherapeutic approaches in ovarian cancer. Hematol Oncol Clin North Am 17:1051–1073.
22. Sica, G.L., I.H. Choi, G. Zhu, K. Tamada, S.D. Wang, H. Tamura, A.I. Chapoval, D.B. Flies, J. Bajorath, and L. Chen. 2003. B7-H4, a molecule of the B7 family, negatively regulates T cell immunity. Immunity 18:849–861.
23. Zang, X., P. Loke, J. Kim, K. Murphy, R. Waitz, and J.P. Allison. 2003. B7x: a widely expressed B7 family member that inhibits T cell activation. Proc Natl Acad Sci USA 100:10388–10392.
24. Prasad, D.V., S. Richards, X.M. Mai, and C. Dong. 2003. B7S1, a novel B7 family member that negatively regulates T cell activation. Immunity 18:863–873.

25. Choi, I.H., G. Zhu, G.L. Sica, S.E. Strome, J.C. Cheville, J.S. Lau, Y. Zhu, D.B. Flies, K. Tamada, and L. Chen. 2003. Genomic organization and expression analysis of B7-H4, an immune inhibitory molecule of the B7 family. J Immunol 171:4650–4654.
26. Kryczek, I., L. Zou, P. Rodriguez, G. Zhu, S. Wei, P. Mottram, M. Brumlik, P. Cheng, T. Curiel, L. Myers, A. Lackner, X. Alvarez, A. Ochoa, L. Chen, and W. Zou. 2006. B7-H4 expression identifies a novel suppressive macrophage population in human ovarian carcinoma. J Exp Med 203:871–881.
27. Bignotti, E., R.A. Tassi, S. Calza, A. Ravaggi, C. Romani, E. Rossi, M. Falchetti, F.E. Odicino, S. Pecorelli, and A.D. Santin. 2006. Differential gene expression profiles between tumor biopsies and short-term primary cultures of ovarian serous carcinomas: Identification of novel molecular biomarkers for early diagnosis and therapy. Gynecol Oncol 103:405–416.
28. Tringler, B., W. Liu, L. Corral, K.C. Torkko, T. Enomoto, S. Davidson, M.S. Lucia, D.E. Heinz, J. Papkoff, and K.R. Shroyer. 2006. B7-H4 overexpression in ovarian tumors. Gynecol Oncol 100:44–52.
29. Simon, I., S. Zhuo, L. Corral, E.P. Diamandis, M.J. Sarno, R.L. Wolfert, and N.W. Kim. 2006. B7-h4 is a novel membrane-bound protein and a candidate serum and tissue biomarker for ovarian cancer. Cancer Res 66:1570–1575.
30. Salceda, S., T. Tang, M. Kmet, A. Munteanu, M. Ghosh, R. Macina, W. Liu, G. Pilkington, and J. Papkoff. 2005. The immunomodulatory protein B7-H4 is overexpressed in breast and ovarian cancers and promotes epithelial cell transformation. Exp Cell Res 306:128–141.
31. Sun, Y., Y. Wang, J. Zhao, M. Gu, R. Giscombe, A.K. Lefvert, and X. Wang. 2006. B7-H3 and B7-H4 expression in non-small-cell lung cancer. Lung Cancer 53:143–151.
32. Tringler, B., S. Zhuo, G. Pilkington, K.C. Torkko, M. Singh, M.S. Lucia, D.E. Heinz, J. Papkoff, and K.R. Shroyer. 2005. B7-h4 is highly expressed in ductal and lobular breast cancer. Clin Cancer Res 11:1842–1848.
33. Krambeck, A.E., R.H. Thompson, H. Dong, C.M. Lohse, E.S. Park, S.M. Kuntz, B.C. Leibovich, M.L. Blute, J.C. Cheville, and E.D. Kwon. 2006. B7-H4 expression in renal cell carcinoma and tumor vasculature: Associations with cancer progression and survival. Proc Natl Acad Sci USA 103:10391–10396.
34. Chen, L. 2004. Co-inhibitory molecules of the B7-CD28 family in the control of T-cell immunity. Nat Rev Immunol 4:336–347.
35. Wyckoff, J., W. Wang, E.Y. Lin, Y. Wang, F. Pixley, E.R. Stanley, T. Graf, J.W. Pollard, J. Segall, and J. Condeelis. 2004. A paracrine loop between tumor cells and macrophages is required for tumor cell migration in mammary tumors. Cancer Res 64:7022–7029.
36. Pollard, J.W. 2004. Tumour-educated macrophages promote tumour progression and metastasis. Nat Rev Cancer 4:71–78.
37. Vakkila, J., and M.T. Lotze. 2004. Inflammation and necrosis promote tumour growth. Nat Rev Immunol 4:641–648.
38. Mantovani, A., S. Sozzani, M. Locati, P. Allavena, and A. Sica. 2002. Macrophage polarization: tumor-associated macrophages as a paradigm for polarized M2 mononuclear phagocytes. Trends Immunol 23:549–555.
39. Curiel, T.J., S. Wei, H. Dong, X. Alvarez, P. Cheng, P. Mottram, R. Krzysiek, K.L. Knutson, B. Daniel, M.C. Zimmermann, O. David, M. Burow, A. Gordon, N. Dhurandhar, L. Myers, R. Berggren, A. Hemminki, R.D. Alvarez, D. Emilie, D.T. Curiel, L. Chen, and W. Zou. 2003. Blockade of B7-H1 improves myeloid dendritic cell-mediated antitumor immunity. Nat Med 21:21.
40. Zou, W., V. Machelon, A. Coulomb-L'Hermin, J. Borvak, F. Nome, T. Isaeva, S. Wei, R. Krzysiek, I. Durand-Gasselin, A. Gordon, T. Pustilnik, D.T. Curiel, P. Galanaud, F. Capron, D. Emilie, and T.J. Curiel. 2001. Stromal-derived factor-1 in human tumors recruits and alters the function of plasmacytoid precursor dendritic cells. Nat Med 7:1339–1346.
41. Gabrilovich, D. 2004. Mechanisms and functional significance of tumour-induced dendritic-cell defects. Nat Rev Immunol 4:941–952.

42. O'Neill, D.W., S. Adams, and N. Bhardwaj. 2004. Manipulating dendritic cell biology for the active immunotherapy of cancer. Blood 104:2235–2246.
43. Bingle, L., N.J. Brown, and C.E. Lewis. 2002. The role of tumour-associated macrophages in tumour progression: implications for new anticancer therapies. J Pathol 196:254–265.
44. Zavadova, E., A. Loercher, S. Verstovsek, C.F. Verschraegen, M. Micksche, and R.S. Freedman. 1999. The role of macrophages in antitumor defense of patients with ovarian cancer. Hematol Oncol Clin North Am 13:135–144, ix.
45. Ohno, S., N. Suzuki, Y. Ohno, H. Inagawa, G. Soma, and M. Inoue. 2003. Tumor-associated macrophages: foe or accomplice of tumors? Anticancer Res 23:4395–4409.
46. Curiel, T.J., G. Coukos, L. Zou, X. Alvarez, P. Cheng, P. Mottram, M. Evdemon-Hogan, J.R. Conejo-Garcia, L. Zhang, M. Burow, Y. Zhu, S. Wei, I. Kryczek, B. Daniel, A. Gordon, L. Myers, A. Lackner, M.L. Disis, K.L. Knutson, L. Chen, and W. Zou. 2004. Specific recruitment of regulatory T cells in ovarian carcinoma fosters immune privilege and predicts reduced survival. Nat Med 10:942–949.
47. Kryczek, I., S. Wei, L. Zou, G. Zhu, P. Mottram, H. Xu, L. Chen, and W. Zou. 2006. Cutting edge: induction of B7-H4 on APCs through IL-10: novel suppressive mode for regulatory T cells. J Immunol 177:40–44.
48. van Elsas, A., R.P. Sutmuller, A.A. Hurwitz, J. Ziskin, J. Villasenor, J.P. Medema, W.W. Overwijk, N.P. Restifo, C.J. Melief, R. Offringa, and J.P. Allison. 2001. Elucidating the autoimmune and antitumor effector mechanisms of a treatment based on cytotoxic T lymphocyte antigen-4 blockade in combination with a B16 melanoma vaccine: comparison of prophylaxis and therapy. J Exp Med 194:481–489.
49. Levitsky, H.I., A. Lazenby, R.J. Hayashi, and D.M. Pardoll. 1994. In vivo priming of two distinct antitumor effector populations: the role of MHC class I expression. J Exp Med 179:1215–1224.
50. Freeman, G.J., A.J. Long, Y. Iwai, K. Bourque, T. Chernova, H. Nishimura, L.J. Fitz, N. Malenkovich, T. Okazaki, M.C. Byrne, H.F. Horton, L. Fouser, L. Carter, V. Ling, M.R. Bowman, B.M. Carreno, M. Collins, C.R. Wood, and T. Honjo. 2000. Engagement of the PD-1 immunoinhibitory receptor by a novel B7 family member leads to negative regulation of lymphocyte activation. J Exp Med 192:1027–1034.
51. Dong, H., G. Zhu, K. Tamada, and L. Chen. 1999. B7-H1, a third member of the B7 family, co-stimulates T-cell proliferation and interleukin-10 secretion. Nat Med 5:1365–1369.
52. Kanai, T., T. Totsuka, K. Uraushihara, S. Makita, T. Nakamura, K. Koganei, T. Fukushima, H. Akiba, H. Yagita, K. Okumura, U. Machida, H. Iwai, M. Azuma, L. Chen, and M. Watanabe. 2003. Blockade of B7-H1 suppresses the development of chronic intestinal inflammation. J Immunol 171:4156–4163.
53. Ghebeh, H., S. Mohammed, A. Al-Omair, A. Qattan, C. Lehe, G. Al-Qudaihi, N. Elkum, M. Alshabanah, S. Bin Amer, A. Tulbah, D. Ajarim, T. Al-Tweigeri, and S. Dermime. 2006. The B7-H1 (PD-L1) T lymphocyte-inhibitory molecule is expressed in breast cancer patients with infiltrating ductal carcinoma: correlation with important high-risk prognostic factors. Neoplasia 8:190–198.
54. Dong, H., S.E. Strome, D.R. Salomao, H. Tamura, F. Hirano, D.B. Flies, P.C. Roche, J. Lu, G. Zhu, K. Tamada, V.A. Lennon, E. Celis, and L. Chen. 2002. Tumor-associated B7-H1 promotes T-cell apoptosis: a potential mechanism of immune evasion. Nat Med 8:793–800.
55. Ohigashi, Y., M. Sho, Y. Yamada, Y. Tsurui, K. Hamada, N. Ikeda, T. Mizuno, R. Yoriki, H. Kashizuka, K. Yane, F. Tsushima, N. Otsuki, H. Yagita, M. Azuma, and Y. Nakajima. 2005. Clinical significance of programmed death-1 ligand-1 and programmed death-1 ligand-2 expression in human esophageal cancer. Clin Cancer Res 11:2947–2953.
56. Salih, H.R., S. Wintterle, M. Krusch, A. Kroner, Y.H. Huang, L. Chen, and H. Wiendl. 2006. The role of leukemia-derived B7-H1 (PD-L1) in tumor-T-cell interactions in humans. Exp Hematol 34:888–894.
57. Konishi, J., K. Yamazaki, M. Azuma, I. Kinoshita, H. Dosaka-Akita, and M. Nishimura. 2004. B7-H1 expression on non-small cell lung cancer cells and its relationship with tumor-infiltrating lymphocytes and their PD-1 expression. Clin Cancer Res 10:5094–5100.

58. Tsushima, F., K. Tanaka, N. Otsuki, P. Youngnak, H. Iwai, K. Omura, and M. Azuma. 2006. Predominant expression of B7-H1 and its immunoregulatory roles in oral squamous cell carcinoma. Oral Oncol 42:268–274.
59. Thompson, R.H., M.D. Gillett, J.C. Cheville, C.M. Lohse, H. Dong, W.S. Webster, L. Chen, H. Zincke, M.L. Blute, B.C. Leibovich, and E.D. Kwon. 2005. Costimulatory molecule B7-H1 in primary and metastatic clear cell renal cell carcinoma. Cancer 104:2084–2091.
60. Thompson, R.H., M.D. Gillett, J.C. Cheville, C.M. Lohse, H. Dong, W.S. Webster, K.G. Krejci, J.R. Lobo, S. Sengupta, L. Chen, H. Zincke, M.L. Blute, S.E. Strome, B.C. Leibovich, and E.D. Kwon. 2004. Costimulatory B7-H1 in renal cell carcinoma patients: Indicator of tumor aggressiveness and potential therapeutic target. Proc Natl Acad Sci USA 101:17174–17179.
61. Brown, J.A., D.M. Dorfman, F.R. Ma, E.L. Sullivan, O. Munoz, C.R. Wood, E.A. Greenfield, and G.J. Freeman. 2003. Blockade of programmed death-1 ligands on dendritic cells enhances T cell activation and cytokine production. J Immunol 170:1257–1266.
62. Mazanet, M.M., and C.C. Hughes. 2002. B7-H1 is expressed by human endothelial cells and suppresses T cell cytokine synthesis. J Immunol 169:3581–3588.
63. Dong, H., G. Zhu, K. Tamada, D.B. Flies, J.M. van Deursen, and L. Chen. 2004. B7-H1 determines accumulation and deletion of intrahepatic CD8(+) T lymphocytes. Immunity 20:327–336.
64. Iwai, Y., M. Ishida, Y. Tanaka, T. Okazaki, T. Honjo, and N. Minato. 2002. Involvement of PD-L1 on tumor cells in the escape from host immune system and tumor immunotherapy by PD-L1 blockade. Proc Natl Acad Sci USA 99:12293–12297.
65. Blank, C., I. Brown, A.C. Peterson, M. Spiotto, Y. Iwai, T. Honjo, and T.F. Gajewski. 2004. PD-L1/B7H-1 inhibits the effector phase of tumor rejection by T cell receptor (TCR) transgenic CD8+ T cells. Cancer Res 64:1140–1145.
66. Trinchieri, G., and P. Scott. 1999. Interleukin-12: basic principles and clinical applications. Curr Top Microbiol Immunol 238:57–78.
67. Iwai, Y., S. Terawaki, and T. Honjo. 2005. PD-1 blockade inhibits hematogenous spread of poorly immunogenic tumor cells by enhanced recruitment of effector T cells. Int Immunol 17:133–144.
68. Strome, S.E., H. Dong, H. Tamura, S.G. Voss, D.B. Flies, K. Tamada, D. Salomao, J. Cheville, F. Hirano, W. Lin, J.L. Kasperbauer, K.V. Ballman, and L. Chen. 2003. B7-H1 blockade augments adoptive T-cell immunotherapy for squamous cell carcinoma. Cancer Res 63:6501–6505.
69. Zou, W. 2006. Regulatory T cells, tumour immunity and immunotherapy. Nat Rev Immunol 6:295–307.
70. Hirano, F., K. Kaneko, H. Tamura, H. Dong, S. Wang, M. Ichikawa, C. Rietz, D.B. Flies, J.S. Lau, G. Zhu, K. Tamada, and L. Chen. 2005. Blockade of B7-H1 and PD-1 by monoclonal antibodies potentiates cancer therapeutic immunity. Cancer Res 65:1089–1096.
71. Melero, I., W.W. Shuford, S.A. Newby, A. Aruffo, J.A. Ledbetter, K.E. Hellstrom, R.S. Mittler, and L. Chen. 1997. Monoclonal antibodies against the 4-1BB T-cell activation molecule eradicate established tumors. Nat Med 3:682–685.

Role of Vascular Leukocytes in Ovarian Cancer Neovascularization

Klara Balint, Jose R. Conejo-Garcia, Ron Buckanovich, and George Coukos

1 Tumor Angiogenesis and Vasculogenesis

Tumor angiogenesis is a process that allows primary tumors to grow beyond the approximate size of 1–2 mm^3. It has been shown that if cancer cells are placed in an avascular site like rabbit cornea, and the capillaries are physically prevented from reaching the implant or were inhibited from undergoing angiogenesis, tumor growth is dramatically impaired, restricting the tumor size to 0.4 mm (15). In the absence of adequate vasculature, tumor cells become apoptotic or necrotic (4). It is now well-accepted that antiangiogenic therapy, originally proposed by Judah Folkman (14), is a promising strategy against cancer. Recent trials with bevacizumab, an antibody against vascular endothelial growth factor (VEGF) used alone or in combination with chemotherapy, showed that systemic antiangiogenic therapy may indeed have a measurable impact on cancer progression and patient survival (31).

During embryogenesis, the formation and remodeling of new blood vessels occurs in two different ways: (1) angiogenesis-new vessels sprout and mature from preexisting vasculature, and (2) vasculogenesis-the new vessels are born from progenitor cells. In adults, new vessels are produced physiologically only via angiogenesis. For example in females, reproductive organs are formed during the follicular and menstrual cycles. Adult neovascularization occurs largely under pathologic situations, such as wound healing and tumor growth. The vascularization in malignant tumors happens through several different mechanisms, which are not mutually exclusive and very often occur concurrently (10, 17).

The mechanisms of tumor angiogenesis can be summarized as follows: endothelial sprouting (induction of new capillaries from preexisting host vessels); vessel cooption (tumors arise or metastasize to a preexisting, well-vascularized organ); intussusceptive microvascular growth (fast and economic vessel network formation by insertion of connective tissue columns or pillars, resulting in partitioning the vessel lumen); glomeruloid angiogenesis (several closely associated microvessels surrounded by a thickened basement membrane and pericytes, best known in high-grade glial malignancies); vasculogenic mimicry (aggressive melanoma cells form vessel-like network in three-dimensional culture, also found in breast, prostate, ovarian, chorio- and lung-carcinomas, sarcomas, and phaeochromocytomas); and the *postnatal vasculogenesis*. The resulting tumor vessels differ from the normal vasculature. They are morphologically dilated, leaky, disorganized, immature, and unstable, and are also distinct in their molecular signature.

2 Endothelial and Hematopoetic Stem Cells in Neovascularization

Vasculogenesis has long been thought to occur only during embryogenesis, but recent studies have found that bone marrow-derived endothelial progenitor cells (EPCs) as well as hematopoietic stem cells (HSCs) were able to home to sites of neovascularization and differentiate into endothelial cells (29). Endothelial progenitor cells and hematopoietic stem cells originate from the same cell called hemangioblast. EPCs and HCSc are in close interaction as the site of hematopoiesis shifts from the yolk sac or aorta-gonad-mesonefros region to the fetal liver, and then ultimately to the spleen, as well as to the adult bone marrow. EPCs have the capacity to proliferate, migrate, and differentiate into endothelial lineage cells. Immature EPCs and primitive HSCs, which reside in the bone marrow's stem cell niche, share similar cell surface markers: $CD133^+$ $CD34^+$ $VEGFR2^+$ $CD117^+$ VE-cad^- Tie-2^+ and c-kit^+ (11). Circulating EPCs become $CD133^-$ KDR^+ $CD34^+$ $CD146^+$ $CD14^-$ and start expressing endothelial markers such as VE-cadherin or E-selectin. Accumulating evidence indicates that EPCs can be specifically recruited from the bone marrow and can be attracted to a site of neovascularization in response to physiological and pathological stimuli. Several studies have demonstrated that bone marrow-derived EPCs functionally contribute to vasculogenesis during tumor growth as well as wound healing, limb ischemia, postmyocardial infarction, and endothelialization of vascular grafts (21, 25, 27). EPCs have been detected at increased frequency in the circulation of cancer patients and also in lymphoma-bearing mice, where the tumor volume and production of VEGF were found to correlate with EPC mobilization. Recruitment of EPCs is a dynamic process that needs sequential activation of molecular switches and release of active cytokines. Several mobilizing factors have been identified: VEGF, angiopoetin-1, stroma-derived factor 1 (SDF-1)/CCL22, placental growth factor (PlGF), estrogen, and erythropoietin (EPO) (11, 18, 19, 20). Treatment of tumor-bearing mice with vascular disrupting agents also leads to acute mobilization of EPCs (33).

The absolute requirement of EPCs and $VEGF$-$R1^+$ proangiogenic hematopoietic cells for human tumor angiogenesis has not been established with clarity to date. Estimates of the contribution of EPCs to tumor vasculature in untreated tumors range from as much as 10–50% to 5% or less (33). Differences likely depend on species (human vs. mouse) but also in experimental conditions and type of tumor examined. Although the level of EPCs is typically low in untreated tumors, their level can suddenly rise in response to acute stress (33) – similarly to pathological cardiovascular events (myocardial infarct).

Hematopoietic stem cells, the best-characterized somatic stem cells, derive from the adult bone marrow and are able to self-renew and provide a lifelong production of all blood-cell lineages. Hematopoietic stem cells are $CD133^+$ $CD34^+$ KDR^+ $CD117^+$ VE-cadherin⁺ Tie-2⁺, Sca⁺, and c-kit⁺, while residing in their stem cell niche. Transplantation of hematopoietic stem cells has been shown to reconstitute hematopoiesis, while progenitors provide only short-term reconstitution of lineage specific precursors. HSCs traffic regularly in and out of the bone marrow. In mice, approximately 100–400 cells are present in the circulation at any given time (1). HSCs are thought to possess the ability for plasticity. It has been demonstrated that HSCs cultured under proangiogenic condition develop an endothelial phenotype. In a mouse model, when HSCs expressing green fluorescent protein (GFP) were transplanted into ischemic retina, HSCs could clonally differentiate into all hematopoietic lineages, as well as into endothelial cells leading to the revascularization of the retina (16).

3 Cells of Myeloid Origin in Angiogenesis

Several studies have demonstrated recently that besides myeloid progenitor cells (2), differentiated monoctyes and monocyte-like cells can also be recruited to sites of neovascularization and can incorporate into vessels (28). Tumor-associated macrophages can blunt antitumor immunity and stimulate angiogenesis, cell migration, invasion, and metastasis (5, 26). Macrophages, which represent a major component of leukocyte infiltrate of solid tumors, were also shown to promote the progression of cancer (24, 26), and were found to be even a regulator of the angiogenic switch in a mouse breast cancer model (23). A specific subset of proangiogenic Tie-2 expressing monocytes was found to be specifically recruited to spontaneous and orthotopic tumors, and was required for tumor neovascularization in the mouse (9). These Tie-2⁺ monocytes were also recently described in human tumors (34). Antigen-presenting cells have remarkable plasticity and divergent functions: It has been reported that CD14⁺ monocytes cultured under angiogenic conditions coexpress endothelial lineage markers (CD31, von Willebrand factor, VE-cadherin, VEGFR-1, endothelial nitric oxide synthase), as well as macrophagocytic lineage markers (CD45, low level of CD14), and stain positive for the macrophage receptor of ox-LDL. The same cells were able to acquire endothelial functions by forming tubular-like structures in Matrigel and also exhibit other functions of endothelial cells such as LDL uptake and lectin binding (12, 13, 22, 32, 36). Rohde et al. (30) performed subtractive "colony-forming units of endothelial progenitor cells" (CFU-EP) analysis on peripheral mononuclear cells obtained from 19 healthy donors and found that depletion of CD14⁺ cells abrogates colony formation, thus progenitor cells are part of the CD14⁺ cell population.

4 Vascular Leukocytes

4.1 Vascular Leukocytes: A Novel Subset

Our laboratory has recently identified a novel leukocyte cell subset within ovarian carcinoma. This population simultaneously expresses endothelial cell markers such as VE-cadherin, CD31, CD34, CD146, and dendritic cell markers like CD45 and CD11c (6). These cells, termed as vascular leukocytes (VLCs), are highly frequent in ovarian cancer. Characterization of the cell surface markers of sorted CD45$^+$ VE-cadherin$^+$ cells from ovarian carcinoma samples showed high levels of MHC-II, CD86, CD11c, intermediate levels of CD8α and CCR6, as well as low levels of CD14. The functional analysis of this highly purified population has revealed that these cells are able to mimic endothelial cells, as they were able to form tubular structures in vivo and incorporate into tumor vessels in an animal model.

CD45$^+$ VE-cadherin$^+$ human sorted vascular leukocytes were cultured on fibronectin-coated plates for 2 weeks and underwent phenotypic changes: development of intercellular junctions and cytoplasmic interdigitations similar to endothelial cells. To confirm the human vascular leukocytes' angiogenic potential in vivo we transplanted freshly sorted CD45$^+$ VE-cadherin$^+$ human VLCs labeled with CFSE in Matrigel (containing an inhibitor against natural killer cell (NK)-mediated rejection) into the flanks of immunodeficient SCID mice. At day 14, when the Matrigel plugs were harvested, sectioned, and analyzed microscopically, we detected tomato-lectin perfusable capillaries assembled by CFSE-labeled cells in the Matrigel transplants. As the cells were CFSE^{bright}, we believe that the cells did not proliferate significantly (7). To summarize, VLCs have the capacity to build functional blood vessels in vivo.

4.2 Animal Model to Study Vascular Leukocytes in Ovarian Cancer

To analyze the behavior of vascular leukocytes, we utilized the ID-8/VEGF syngenic mouse ovarian carcinoma model (35). ID-8 tumor cells with or without the β-defensin-29 (*Defb29*) gene and with high or low expression of VEGF-A were injected subcutaneously or intraperitoneally into C57BL/6 mice (6). We found that the expression of Defb29 significantly increased tumor growth and ascites formation, and accelerated death, but only for tumors expressing high level of VEGF-A. The micro vessel density in ID-8/VEGF/Defb29 tumors was significantly higher than in control ID8/Defb-29 tumors, which expressed low levels of VEGF. As β-defensins, which are antimicrobial inflammatory peptides, chemoattract dendritic cells, we investigated the spatial distribution of CD11c$^+$ cells in ID-8/VEGF/Defb29 tumors. We found that a significant proportion of these cells (up to 40%) were localized to capillary-like structures identified by CD31 and CD11c double immunostaining. Because we attracted antigen-presenting cells into the tumor, we expected to see accelerated phagocytosis and tumor antigen presentation, as well as tumor immune rejection. Instead, we observed that the attracted dendritic cells under the influence of high level of VEGF-A dramatically increased tumor growth and reduced the survival of the animals (7). When CD11c$^+$ cells derived from these tumor ascites were isolated and cotransplanted with ID8/VEGF cells into the flank of C57BL/6 mice, their presence promoted tumor growth and vessel formation compared with the tumor cells injected alone.

To test whether CD11c$^+$ cells are capable of vessel formation, a highly purified population of these cells derived from tumor ascites (labeled with CFSE) was mixed with ID8/VEGF supernatant containing Matrigel and injected subcutaneously into C57BL/6 mice. After 14 days, the Matrigel plugs were harvested and analyzed. Most of the transplanted cells appeared to be incorporated into dextran perfusable mature capillaries mixed with host vascular cells. The cells forming these capillaries strongly expressed CD11c, CD45, DEC-205, and CD8α indicating their dendritic lineage.

To explore the underlying mechanism of increased tumor growth in ID-8/VEGF/Defb29 mice, we tested whether the enrichments of tumors with CD11c$^+$ cells could repeat the effects of Defb29. For this purpose, we injected an equal number of ID-8/VEGF or ID8 cells alone or mixed with bone marrow-derived CD34$^-$CD11c$^+$ cells at 20:1 ratio subcutaneously. We found that the presence of CD11c$^+$ dendritic cells increased the level of vascularization only in the ID8 tumors (compared with ID8 cell injected alone). However, if VEGF was highly expressed (ID8/VEGF tumors), the addition of dendritic cells increased the tumor growth ten-fold but did not effect tumor growth in tumors expressing normal levels of VEGF-A. CD11c$^+$ dendritic cells were able to reproduce the effects of Defb29, namely increasing growth of tumors producing high level of VEGF-A.

The next step was to test whether Defb29 was a chemoattractant for CD11c$^+$ cells. Bone marrow-derived murine immature dendritic cells were treated with either ID-8/VEGF or ID-8/VEGF/Defb29 conditioned supernatants or murine recombinant macrophage inflammatory protein-3-alpha (MIP-3α, a known dendritic cell chemotactic cytokine) or nothing. ID-8/Defb29 or ID-8/VEGF/Defb29 conditioned supernatants chemoattracted immature DCs similarly to MIP-3α. Defb29 and MIP-3α attenuated each other's chemotactic effect for dendritic cells, which suggested that the effect of Defb29 was mediated via CCR6 (6). Addition of CCR6-specific antibody significantly reduced the migration of CD11$^+$ dendritic cells toward undiluted ID-8/Defb29 and ID-8/VEGF/Defb29 conditioned media in vitro. In an in vivo experiment, when we injected CD11c$^+$ cells intraperitoneally into C57BL/6 mice, which were bearing an ID-8/VEGF as well as a contra lateral subcutaneous ID-8/VEGF/Defb29 tumor, we detected a significantly higher migration rate of dendritic cells toward the Defb29 expressing tumor. Pretreatment of CD11c$^+$ cells with a CCR6 specific antibody, but not with an isotype control, largely impaired their chemotaxis to ID-8/VEGF/Defb29 tumors. Tumors recruited CD11c$^+$ cells via Defb29 and CCR6.

To verify the role of the CCR6 receptor in dendritic cell-dependent tumor growth, we transplanted ID-8/VEGF/Defb29 tumors together with a CCR6 specific rat antibody or with an antibody isotype control, in addition with a secondary antibody against rat conjugated to ribosome-inactivating saporin. The treatment of CCR6 expressing dendritic cells with a CCR6-specific immunotoxin complex reduced the number of tumor-infiltrated dendritic cells by 95%, as well as reduced tumor growth compared with controls treated with rat isotype control/immunotoxin. This confirms that targeting of CCR6⁺ cells results in impaired tumor growth.

As vascular leukocytes represent the majority of CD45⁺ tumor-infiltrating leukocytes in ovarian cancer, the depletion of this cell population may provide a good strategy against ovarian cancer. Bak et al. has recently demonstrated that targeted depletion of tumor-associated myeloid cells via intraperitoneally injected carrageenan (oligosaccharide, blocker of phagocytosis) prevented CB6Fl, ID8 tumor-bearing mice from rapid tumor progression (3). Furthermore, they have identified the scavenger receptor-A as a VLC-specific cell surface marker, which could present a promising target for antitumor treatment. If mice transplanted with ID8-C3 ovarian tumor cells were treated with weekly injections of saporin toxin conjugated to anti-SR-A antibody (to eliminate vascular leukocytes), depletion of VLCs could block ovarian tumor progression.

4.3   Endothelial-Like Differentiation of Dendritic Cells In Vitro

To further study the ontogenesis of VLCs and understand our findings, we cultured murine bone marrow cells in the presence of granulocyte-macrophage colony-stimulating factor cells to obtain CD11c⁺ CD45⁺ CD34⁻ dendritic cells. Cells exhibited the classical dendritic cell phenotype and functions: dendritic shape, phagocytosis, and the ability to induce antigen specific T cell response (i.e., antigen-specific proliferation and IL-2 production). After we further cultured these cells in tumor-conditioned media (of ID-8/VEGF cells), dendritic cells acquired spindle-shaped phenotype and upregulated CD31. At week 3, more than 80% of cells were CD31 and vWF positive; the cells assembled into cord-like structures, and more than 90% were able to uptake fluorescent acetylated LDL (8). Electron microscopy revealed that these DCs contained Weibel-Palade bodies and had developed intercellular junctions, meaning that their morphological features are very close to the endothelial cell (6).

Next, we tested the role of VEGF-A in the endothelialization process of dendritic cells in vitro. We used neutralizing antibodies against VEGFR-1, VEGFR-2, VEGFR-3, and CCR6 on CD11c⁺ cells cultured in ID8/VEGF/Defb29 tumor-conditioned media. Blocking VEGFR-1 or VEGFR-2 inhibited the downregulation of CD45 and the upregulation of CD34 cell surface marker, whereas blocking VEGFR-3 and CCR6 had no effect. VEGF-A, but not β-defensin 29, caused the endothelialization of the dendritic cells.

5   Summary

Vascular leukocytes are a unique population of CD45⁺ VE-cadherin⁺ cells with diverse functions. VLCs are capable of antigen presentation, as well as formation of endothelial-like structures in vitro and in vivo. VLCs are largely present among

CD45⁺ cells infiltrating human ovarian carcinomas and are highly represented in the β-defensin-VEGF-ID8 syngenic mouse model. Vascular leukocytes are a new and promising novel therapeutic target for anti-angiogenic therapy. Their unique mechanism merits thorough and extensive exploration in the future.

References

1. Adams, G. B. and D. T. Scadden (2006). "The hematopoietic stem cell in its place." Nat Immunol 7(4): 333–7.
2. Bailey, A. S., H. Willenbring, et al. (2006). "Myeloid lineage progenitors give rise to vascular endothelium." Proc Natl Acad Sci USA 103(35): 13156–61.
3. Bak, S. P., J. J. Walters, et al. (2007). "Scavenger receptor-A-targeted leukocyte depletion inhibits peritoneal ovarian tumor progression." Cancer Res 67(10): 4783–9.
4. Brem, S. (1976). "The role of vascular proliferation in the growth of brain tumors." Clin Neurosurg 23: 440–53.
5. Condeelis, J. and J. W. Pollard (2006). "Macrophages: obligate partners for tumor cell migration, invasion, and metastasis." Cell 124(2): 263–6.
6. Conejo-Garcia, J. R., F. Benencia, et al. (2004). "Tumor-infiltrating dendritic cell precursors recruited by a beta-defensin contribute to vasculogenesis under the influence of Vegf-A." Nat Med 10(9): 950–8.
7. Conejo-Garcia, J. R., R. J. Buckanovich, et al. (2005). "Vascular leukocytes contribute to tumor vascularization." Blood 105(2): 679–81.
8. Coukos, G., F. Benencia, et al. (2005). "The role of dendritic cell precursors in tumour vasculogenesis." Br J Cancer 92(7): 1182–7.
9. De Palma, M., M. A. Venneri, et al. (2005). "Tie2 identifies a hematopoietic lineage of proangiogenic monocytes required for tumor vessel formation and a mesenchymal population of pericyte progenitors." Cancer Cell 8(3): 211–26.
10. Dome, B., M. J. Hendrix, et al. (2007). "Alternative vascularization mechanisms in cancer: Pathology and therapeutic implications." Am J Pathol 170(1): 1–15.
11. Eguchi, M., H. Masuda, et al. (2007). "Endothelial progenitor cells for postnatal vasculogenesis." Clin Exp Nephrol 11(1): 18–25.
12. Fernandez Pujol, B., F. C. Lucibello, et al. (2000). "Endothelial-like cells derived from human CD14 positive monocytes." Differentiation 65(5): 287–300.
13. Fernandez Pujol, B., F. C. Lucibello, et al. (2001). "Dendritic cells derived from peripheral monocytes express endothelial markers and in the presence of angiogenic growth factors differentiate into endothelial-like cells." Eur J Cell Biol 80(1): 99–110.
14. Folkman, J. (1971). "Tumor angiogenesis: therapeutic implications." N Engl J Med 285(21): 1182–6.
15. Gimbrone, M. A., Jr., S. B. Leapman, et al. (1972). "Tumor dormancy in vivo by prevention of neovascularization." J Exp Med 136(2): 261–76.
16. Grant, M. B., W. S. May, et al. (2002). "Adult hematopoietic stem cells provide functional hemangioblast activity during retinal neovascularization." Nat Med 8(6): 607–12.
17. Hanahan, D. and J. Folkman (1996). "Patterns and emerging mechanisms of the angiogenic switch during tumorigenesis." Cell 86(3): 353–64.
18. Hattori, K., B. Heissig, et al. (2003). "The regulation of hematopoietic stem cell and progenitor mobilization by chemokine SDF-1." Leuk Lymphoma 44(4): 575–82.
19. Hattori, K., B. Heissig, et al. (2002). "Placental growth factor reconstitutes hematopoiesis by recruiting VEGFR1(+) stem cells from bone-marrow microenvironment." Nat Med 8(8): 841–9.
20. Heeschen, C., A. Aicher, et al. (2003). "Erythropoietin is a potent physiologic stimulus for endothelial progenitor cell mobilization." Blood 102(4): 1340–6.

21. Igreja, C., M. Courinha, et al. (2007). "Characterization and clinical relevance of circulating and biopsy-derived endothelial progenitor cells in lymphoma patients." Haematologica 92(4): 469–77.
22. Kuwana, M., Y. Okazaki, et al. (2006). "Endothelial differentiation potential of human monocyte-derived multipotential cells." Stem Cells 24(12): 2733–43.
23. Lin, E. Y., J. F. Li, et al. (2006). "Macrophages regulate the angiogenic switch in a mouse model of breast cancer." Cancer Res 66(23): 11238–46.
24. Lin, E. Y., A. V. Nguyen, et al. (2001). "Colony-stimulating factor 1 promotes progression of mammary tumors to malignancy." J Exp Med 193(6): 727–40.
25. Lyden, D., K. Hattori, et al. (2001). "Impaired recruitment of bone-marrow-derived endothelial and hematopoietic precursor cells blocks tumor angiogenesis and growth." Nat Med 7(11): 1194–201.
26. Pollard, J. W. (2004). "Tumour-educated macrophages promote tumour progression and metastasis." Nat Rev Cancer 4(1): 71–8.
27. Rafii, S., D. Lyden, et al. (2002). "Vascular and haematopoietic stem cells: novel targets for anti-angiogenesis therapy?" Nat Rev Cancer 2(11): 826–35.
28. Rehman, J., J. Li, et al. (2003). "Peripheral blood "endothelial progenitor cells" are derived from monocyte/macrophages and secrete angiogenic growth factors." Circulation 107(8): 1164–9.
29. Reyes, M., A. Dudek, et al. (2002). "Origin of endothelial progenitors in human postnatal bone marrow." J Clin Invest 109(3): 337–46.
30. Rohde, E., C. Bartmann, et al. (2007). "Immune Cells Mimic the Morphology of Endothelial Progenitor Colonies in Vitro." Stem Cells 25(7): 1746–52.
31. Ruegg, C., J. Y. Meuwly, et al. (2003). "The quest for surrogate markers of angiogenesis: a paradigm for translational research in tumor angiogenesis and anti-angiogenesis trials." Curr Mol Med 3(8): 673–91.
32. Schmeisser, A., C. D. Garlichs, et al. (2001). "Monocytes coexpress endothelial and macrophagocytic lineage markers and form cord-like structures in Matrigel under angiogenic conditions." Cardiovasc Res 49(3): 671–80.
33. Shaked, Y., A. Ciarrocchi, et al. (2006). "Therapy-induced acute recruitment of circulating endothelial progenitor cells to tumors." Science 313(5794): 1785–7.
34. Venneri, M. A., M. D. Palma, et al. (2007). "Identification of proangiogenic TIE2-expressing monocytes (TEMs) in human peripheral blood and cancer." Blood 109(12): 5276–85.
35. Zhang, L., N. Yang, et al. (2002). "Generation of a syngeneic mouse model to study the effects of vascular endothelial growth factor in ovarian carcinoma." Am J Pathol 161(6): 2295–309.
36. Zhang, R., H. Yang, et al. (2005). "Acceleration of endothelial-like cell differentiation from CD14+ monocytes in vitro." Exp Hematol 33(12): 1554–63.

Heparin-Binding Epidermal Growth Factor-Like Growth Factor as a New Target Molecule for Cancer Therapy

Shingo Miyamoto, Hiroshi Yagi, Fusanori Yotsumoto, Tatsuhiko Kawarabayashi, and Eisuke Mekada

1 Introduction

ErbB receptors belong to the tyrosine kinase family and consist of four ErbB members including EGFR/ErbB1, ErbB2, ErbB3, and ErbB4 [37, 40, 55] (Fig. 1). Activation of ErbB receptors is controlled by the spatiotemporally regulated expression and liberation of their ligands, which are members of the EGF family of growth factors. Ligand binding induces formation of homo- or heterodimeric complexes and activation of the intrinsic kinase domain, resulting in phosphorylation of specific tyrosine residues that serve as docking sites for adaptor molecules, which in turn leads to activation of intracellular signaling pathways. The EGF family of ligands is divided into four groups: (1) EGF, transforming growth factor-α (TGF-α), and amphiregulin, which bind to EGFR; (2) heparin-binding EGF-like growth factor (HB-EGF), epiregulin, epigen, and betacellulin, which bind to both EGFR and ErbB4; (3) neuregulin (NRG)1 and NRG2, which bind to ErbB3 and ErbB4; and (4) NRG3 and NRG4, which bind only to ErbB4, but not to ErbB3 (Fig. 1). ErbB receptors and their cognate ligands play fundamental roles in development, proliferation, and differentiation.

Downward et al. [7] reported that EGFR (epidermal growth factor receptor) was the cellular homolog of avian erythroblastosis virus v-erbB oncogene. Accumulating evidences have revealed that alterations in the EGFR signaling pathway contribute to malignant transformation. Malignant transformation as a consequence of EGFR dysregulation can occur in human cancers by different mechanisms including receptor overexpression, activation of mutations, alterations in dimerization processes, activation of the autocrine loop of growth factors, limited or enhanced endocytosis of activated receptors, deficiency in specific phosphatases inactivating phosphorylated EGFR tyrosine residues, and limited turnover [42]. EGFR gene overexpression without gene amplification and EGFR activation, by EGFR ligands in an autocrine loop, are two of the main frequent mechanisms implicated in cancer development and progression [42].

Overexpressions of EGFR and ErbB2 were shown to induce malignant transformation in NIH-3T3 cells [6]. The ErbB family, and in particular EGFR, has been found to be altered in a variety of human cancers [34, 39]. Signaling through

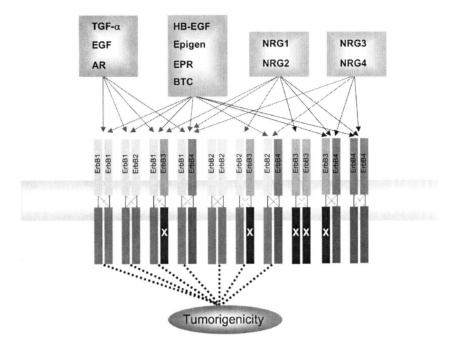

Fig. 1 Binding specificites of members of the ErbB receptor family to their cognate ligands. ErbB receptor homo- and heterodimer combinations activated by ErbB ligands. *EGF* epidermal growth factor; *TGF-α* transforming growth factor-α; *HB-EGF* heparin-binding EGF-like growth factor; *AR* amphiregulin; *BTC* betacellulin; *EPR* epiregulin; *NRG* neuregulin. ErbB3 is deficient in kinase activity (X). Blue bar, yellow bar, green bar, and purple bar indicate ErbB 1, 2, 3, and 4, respectively. The red bar indicates transforming activity in NIH3T3 cells for ErbB receptors. The red arrow indicates binding of ErbB ligands to ErbB receptors with a transforming activity for NIH3T3 cells

EGFR is intricately involved in human cancer, and therefore serves as a target for cancer therapy. Several strategies exist to target EGFR including monoclonal antibodies (mAbs) directed towards the extracellular domain of EGFR such as Cetuximab (Erbitux), and low molecular weight tyrosine kinase inhibitors for EGFR that interfere with receptor signaling (TKIs) such as Gefitinib (Iressa) and Erlotinib (Tarceva). Cetuximab is the most extensively studied and clinically approved chimeric mAb designed to specifically inhibit EGFR [2]. In addition to Cetuximab, several mAbs have already entered Phase I clinical trials, e.g., Panitumumab, Matuzumab, and h-R3 are the closest to clinical development [3]. Some studies indicated that Gefitinib affected many of the same intracellular signaling pathways inhibited by anti-EGFR mAb therapy. The identification of somatic activating mutations associated with Gefitinib hypersensitivity has established a new paradigm for mutated EGFR signaling in cancer and Gefitinib sensitivity [43, 44]. Further clinical studies are required to identify the most effective antibody- or small-molecule-based treatments for particular tumor types and for particular patients.

Autocrine loops, in which both the receptor and the ligand are produced by the same tumor cells, may be important contributors to growth autonomy of cancer cells [45]. However, In contrast to their receptors, ligands comprising the EGF family of growth factors have not yet been investigated as targets for cancer therapy. This is possibly due to the redundancy of ErbB ligands for each receptor, and the fact that inhibition of receptor function is more effective than inhibition of multiple ligands for cancer therapy. However, recent studies have indicated that expression levels of each individual EGFR ligand vary in different cancers, and a particular ligand is specifically expressed in some human cancers [26, 49]. These evidences will help us develop therapeutic strategies to target EGFR ligands in some human cancers. In the present chapter, we would like to highlight the features of HB-EGF among EGFR ligands as a candidate target for cancer therapy.

2 The Physiological Role of HB-EGF

HB-EGF is initially synthesized as a membrane-bound precursor (pro-HB-EGF) [12]. The soluble form of HB-EGF (sHB-EGF) is released from the cell membrane by ectodomain shedding of pro-HB-EGF, in a manner similar to that of other EGFR ligands [41]. A number of physiological and pharmacological stimuli including G protein-coupled receptor (GPCR) ligands such as lysophosphatidic acid (LPA) induce the ectodomain shedding of pro-HB-EGF [9]. Ectodomain shedding of pro-HB-EGF is critical for growth factor activity, and dysregulated release of sHB-EGF results in lethal severe hyperplasia in mice [53]. Interestingly, the transmembrane form of HB-EGF (pro-HB-EGF) also acts in a juxtacrine manner to signaling neighboring cells [16].

The transmembrane form of HB-EGF forms complexes with several molecules. In epithelial cells, pro-HB-EGF interacts with CD9, integrin α3β1, and heparan-sulfate proteloglycan (HSPG) [29]. CD9 modulates juxtacrine activity of pro-HB-EGF; integrin α3β1 and HSPG may also be implicated in biological functions mediated by HB-EGF such as adhesion and signaling [13]. A yeast two-hybrid screening identified BAG1 and PLZF (promyelocytic leukemia zinc finger) as the proteins that bind to the cytoplasmic domain of pro-HB-EGF [21, 30]. BAG-1, which has been demonstrated to bind to Bcl-1 and several other signaling molecules, is capable of suppressing apoptosis [47]. PLZF has been recognized as a transcriptional repressor and a negative regulator of the cell cycle [20]. Through the multifunction of HB-EGF, HB-EGF participates in a variety of physiological and pathological processes including wound healing, blast implantation, atherosclerosis, and tumor formation [17, 25, 31, 32, 35].

3 Expression of HB-EGF in Human Ovarian Cancer

Ovarian cancer is the most frequent cause of death among all gynecologic cancers; in the last 30 years, current therapies have not improved cure rates. High mortality is predominantly caused by occult progression of the tumor into the peritoneal cavity with the initial diagnosis usually being made at an advanced stage. Tumor growth is characterized by local spreading into the peritoneal cavity following the circulatory pathway of the peritoneal fluid produced by peritoneal epithelial and cancer cells. Accumulated evidence from many studies revealed that ascites from patients with ovarian cancer were a rich source of growth factor activity for ovarian cancer cells, termed ovarian cancer activating factors (OCAFs) [23]. Dissemination of cancer cells activated by OCAFs results in exaggerated increase in peritoneal fluid, leading to tumor spreading of ovarian cancer. To gain an insight into the role of HB-EGF as OCAF in ovarian cancer, we previously examined cell proliferation-promoting activities and levels of EGFR ligands in peritoneal fluids obtained from patients with ovarian cancer [54]. Proliferating-promoting activities in peritoneal fluids obtained from patients with ovarian cancer were much higher than those in peritoneal fluids from patients with benign ovarian cysts and normal ovaries, and activity was suppressed only by antibodies against EGFR and HB-EGF (Fig. 2a). In addition, cell survival activity mediated by peritoneal fluid obtained from patients with ovarian cancer was significantly elevated, compared to those from patients with benign ovarian cysts and normal ovaries (Fig. 2b). This cell survival activity was also prohibited by antibodies against HB-EGF (Fig. 2b). Significant differences were observed in levels of HB-EGF and TGF-α or amphiregulin in patients with ovarian cancer [26]. These results indicated that HB-EGF in peritoneal fluid of ovarian cancer was sufficient for cancer cells to survive and proliferate, suggesting that HB-EGF in peritoneal fluid played a key role in tumor spreading of ovarian cancer.

Moreover, to reconfirm the clinical significance of HB-EGF in human ovarian cancer, we investigated expressions of EGFR ligands and the ADAM family, which induces the ectodomain shedding of EGFR ligands, using real-time PCR, immunohistochemistry, and in situ hybridization [48]. Large differences in expression were found between HB-EGF and other EGFR ligands and between ADAM17 and other

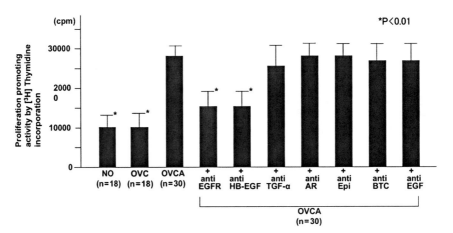

Fig. 2 (continued)

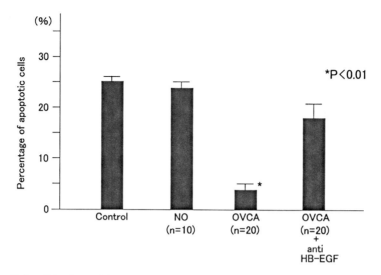

Fig. 2 Cell proliferation activity and cell survival activity mediated by peritoneal fluid in patients with ovarian cancer. (a) [^{3}H]thymidine incorporation in SKOV3 cells incubated with patients' peritoneal fluids from normal ovaries (NO, $n = 18$), ovarian cysts (OVC, $n = 18$), and ovarian cancer (OVCA, $n = 30$); and alterations in [^{3}H]thymidine incorporation of an OVCA patient's peritoneal fluid by anti-EGFR ligand antibodies or anti-EGFR antibody. Bars indicate mean values and standard errors. P-values represent comparison between different levels in patients in the absence of inhibitory antibodies. (b) Alteration in percentages of apoptotic cells in SKOV3 cells after incubation with peritoneal fluid from a normal ovary or ovarian cancer. Control indicates the percentage of apoptotic cells under serum-free conditions. Bars indicate mean values and standard errors. P-values represent comparisons between different levels in patients and level from a normal ovary (Reproduced from [54], with permission)

ADAM family members (Fig. 3a). In addition, HB-EGF expression was significantly increased in advanced ovarian cancer compared to that in normal ovaries, and was significantly associated with clinical outcomes (Fig. 3b). ADAM17 expression was significantly enhanced in both early and advanced ovarian cancers compared to that in normal ovaries. Immunohistochemistry showed that HB-EGF protein was abundantly seen in interstitial tissues, but not in cancer cells while diffuse staining for HB-EGF mRNA was found only in cancer cells, but not in interstitial tissues using in situ hybridization. These results suggested that HB-EGF protein was only produced by cancer cells, and by interstitial tissues, and the proteolytic form of HB-EGF accumulated in the extracellular matrix with heparin sulfate in the interstitial tissues surrounding cancer cells. Taken together, these clinical studies suggested that HB-EGF might contribute to tumor progression of ovarian cancer, and HB-EGF was a putative target molecule for ovarian cancer therapy. To identify HB-EGF as a novel target for ovarian cancer therapy, we have to prove that HB-EGF is intensely involved in peritoneal dissemination as well as in chemoresistance of ovarian cancer.

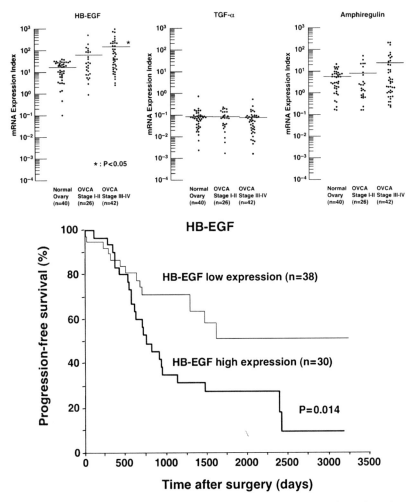

Fig. 3 (a) Differences in expressions of EGFR ligands between normal ovaries and ovarian cancer (OVCA). mRNA expression indices of HB-EGF, TGF-α, and amphiregulin in patients with normal ovaries ($n = 40$), early ovarian cancer (stages I–II, $n = 26$), and advanced ovarian cancer (stages III–IV, $n = 42$). Lines indicate mean values of mRNA expression indices for each group. *P*-values represent comparison of levels of mRNA expression indices in patients and those in patients with normal ovaries. (b) Clinical significance of HB-EGF expression in ovarian cancer. Progression-free survival of patients with ovarian cancer in relation to tumor HB-EGF expression status (Reproduced from [48], with permission)

4 Peritoneal Dissemination Mediated by HB-EGF in Ovarian Cancer

Acquisition of malignant phenotype in peritoneal dissemination of ovarian cancer is involved in four key steps: (1) survival of cancer cells detached from the primary tumor in peritoneal fluid, (2) adhesion of cancer cells to the peritoneum, (3) motility and invasion of cancer cells into the peritoneum, (4) tumor formation through angiogenesis induced by cancer cells (Fig. 4). From previous studies, HB-EGF was shown to be sufficient for cancer cells to survive in the peritoneal fluid of ovarian cancer patients [54].

To examine the involvement of HB-EGF in cell adhesion, a spreading assay was performed on fibronectin, collagen type I, and collagen type III, which are components of extracellular matrices in the abdominal peritoneum, using ovarian cancer cell lines RMG1 and SKOV3 cells as models of ovarian cancer. Transfection of small interfering RNAs (siRNAs) against HB-EGF or EGFR but not against TGF-α or amphiregulin into RMG1 cells resulted in a significant decrease in cell adhesion properties on extracellular matrices. Suppression of HB-EGF expression in RMG1 cells also inhibited activations of EGFR and FAK, and expression of integrin β1. In addition to these results, presence of the soluble form of HB-EGF enhanced cell adhesion properties of SKOV3 and RMG1 cells on extracellular matrices. These results suggested that HB-EGF was responsible for cell adhesion properties on extracellular matrices in abdominal peritoneum.

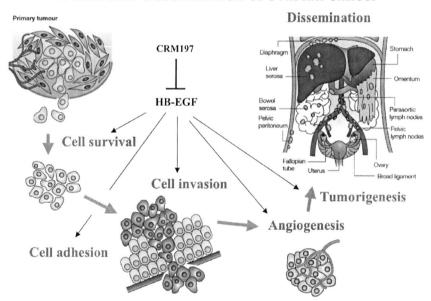

Fig. 4 Steps in peritoneal dissemination of ovarian cancer. Peritoneal dissemination mainly consists of four steps: (1) survival of cancer cells in peritoneal fluid detached from the primary cancer lesion, (2) adhesion of cancer cells to extracellular matrices of the peritoneum covering the peritoneal cavity, (3) invasion of cancer cells into extracellular matrices of the peritoneum, (4) angiogenesis and tumorigenicity mediated by cancer cells at disseminating sites. Our studies revealed that HB-EGF was intricately implicated in each step of peritoneal dissemination, and behavior of cancer cells at each step was suppressed by CRM197

In the invasion assay, the bottom of the chamber was coated with Matrigel, and then, RMG1 and SKOV3 cells were introduced into the wells and cultured. Migrated cells were quantified by counting numbers of cells. The numbers of migrated cells transfected with siRNA against EGFR or HB-EGF significantly decreased, compared to those transfected with siRNA for TGF-α or amphiregulin, and compared to untransfected cells. Addition of the soluble form of HB-EGF to SKOV3 cells also significantly increased cell invasion properties. Transfection of siRNA against HB-EGF or EGFR, but not against TGF-α or amphiregulin into ovarian cancer cells resulted in a significant decrease in expressions of VEGF and interleukin-8. Finally, constitutive suppression of HB-EGF, which was mediated by transfection of the HB-EGF siRNA vector, markedly prohibited tumor growth in xenografted mice. These results indicated that HB-EGF contributed to the aggressive behavior of a tumor such as invasiveness, angiogenesis, and tumorigenicity. According to these results, HB-EGF seems to be implicated in each key step of peritoneal dissemination. To reconfirm promotion of peritoneal dissemination of ovarian cancer mediated by HB-EGF, tumor volume in peritoneal cavity was analyzed using RMG1 cells and transfected RMG1 cells with siRNAs against HB-EGF, TGF-α, or amphiregulin. RMG1 cells, which highly express HB-EGF, formed definite peritoneal dissemination in mice by intraperitoneal inoculation. RMG1 cells transfected with siRNA against HB-EGF failed to form disseminated tumors in the peritoneal cavity, while RMG1 cells transfected with siRNA against TGF-α or amphiregulin resulted in similar tumor volumes in peritoneal dissemination compared to parental cells. SKOV3 cells harboring relatively low expression of HB-EGF showed no peritoneal dissemination in mice by intraperitoneal inoculation. Consistently, after transfection with a constructed plasmid of human pro-HB-EGF cDNA, transfected SKOV3 cells highly expressed HB-EGF, and resulted in significantly big tumor volumes in the peritoneal cavity of mice by intraperitoneal inoculation. According to our observations, HB-EGF plays a pivotal role in peritoneal dissemination including cell survival, cell adhesion, angiogenesis, and tumorigenicity.

5 Association between HB-EGF Expression and Chemoresistance in Ovarian Cancer

In our clinical study, enhanced expression of HB-EGF was significantly associated with clinical outcomes and chemoresistance in ovarian cancer. As shown previously, Taxol inhibits tumor formation in SKOV3 cells growing subcutaneously in nude mice, but only weakly inhibits tumor formation after transfection with SKOV3 and overexpression of HB-EGF (SK-HB) [26, 54]. Taxol partially, but dose-dependently, suppresses in vitro proliferation of SKOV3 cells, while no inhibitory effect was observed in SK-HB cells even at the highest concentration of Taxol. When comparing SKOV3 cells with SK-HB cells, Taxol induces an increase in the number of apoptotic cells and activation of

JNK and p38 in SKOV3 cells compared to SK-HB cells, whereas Akt activation is clearly found in SK-HB cells, but not in SKOV3 cells. Accordingly, enhanced expression and/or presence of HB-EGF modulates Taxol-induced anti-apoptotic effects such as ERK- and Akt-signaling pathways, or pro-apoptotic effects such as JNK- and p38-signaling pathways, leading to acquisition of chemoresistant properties in cells.

6 Roles of HB-EGF in Other Human Cancers

Recently, there has been growing evidence of increased HB-EGF expression in tumors compared to normal tissues including pancreatic, liver, esophageal, colon, gastric, ovarian, bladder cancer, melanoma, and glioblastoma [25]. The relationship between HB-EGF expression and human cancer has been minutely investigated in human ovarian cancer. In bladder cancer, HB-EGF is highly expressed, at least 10–100 times more than the expression levels of other EGFR ligands [49]. It has been demonstrated that only HB-EGF is an abundantly expressed molecule among EGFR ligands, and HB-EGF, but not other EGFR ligands, possibly contributes to tumor growth signaling via EGFR activation in both ovarian and bladder cancers. Several laboratories reported that HB-EGF mRNA was overexpressed in pancreatic cancer compared to normal tissues [19]. HB-EGF mRNA expression also correlated with clinical prognosis in patients with gastric cancer [28]. HB-EGF gene expression, as measured by in situ hybridization and immunostaining, was elevated in 100% (17/17) of human hepatocellular carcinoma biopsies compared to surrounding liver tissues, which were only faintly positively stained in normal hepatocytes [14]. On the other hand, in colon and pancreatic cancers, HB-EGF expression is associated with early stages as described by [15]. According to these reports, it is plausible that HB-EGF expression, which is predominantly found in a variety of human cancers, is associated with the aggressive behavior of a tumor.

We also studied expressions of EGFR ligands in a variety of human cancer cell lines using real-time PCR. In ovarian, bladder, gastric, endocervical, and endometrial cancer cell lines, HB-EGF was the primarily expressed EGFR ligand whereas expression levels of other EGFR ligands appeared to vary. In melanoma and glioblastoma cell lines, HB-EGF was also recognized as the most prevalent EGFR ligand, although TGF-α was as highly expressed as HB-EGF in some of these cell lines. In many human cancer cells, HB-EGF was the predominantly expressed EGFR ligand.

In human gastric cancer, HB-EGF was identified as one candidate DDP-resistance-related gene [46]. Chemotherapy induces elevated expression of HB-EGF, which is largely dependent on chemotherapy-resistant genes including activator protein-1 and NF-κB activation, suggesting that chemotherapy-induced HB-EGF activation represents a critical mechanism of inducible chemotherapy resistance [51]. Thus, HB-EGF is a key molecule in resistance to cancer agents.

7 CRM197 as Anticancer Agent

Cross-reacting material 197 (CRM197) is a nontoxic mutant of diphtheria toxin that shares the same immunological properties of the native molecule, and binds to human HB-EGF to block its mitogenic activity by prohibiting binding to EGFR [24]. Since CRM197 does not inhibit the mitogenic activity of other EGFR ligands, CRM197 has been used as a specific inhibitor of HB-EGF. Our previous observations demonstrated that CRM197 attenuated cell survival properties including the ones in the peritoneal fluid of ovarian cancer. In RMG1 cells, CRM197 blocked cell adhesion mediated by integrin on extracellular matrices, accompanied with inhibition of FAK and EGFR activation. The number of migrated cells in ovarian cancer was significantly reduced in the presence of CRM197. In addition, expressions of VEGF and IL-8 were also suppressed in the presence of CRM197.

To investigate antitumor effects of CRM197 in xenografted mice, RMG1, SKOV3, and OVMG1 cells were subcutaneously injected into nude mice, and tumor sizes were measured weekly at the injection site [26]. After confirmation of a definite tumor on subcutaneous tissues, CRM197 or control saline was injected intraperitoneally weekly for 10 weeks. Each tumor formation in SKOV3, RMG1, and OVMG1 cells was completely suppressed by CRM197 treatment. To further evaluate whether CRM197 inhibits peritoneal dissemination in ovarian cancer, RMG1 cells were injected into the peritoneal cavity, and tumor volume formed in the abdominal cavity was estimated after 6 weeks. Administration of CRM197 or control saline was given weekly for 6 weeks from the following day after intraperitoneal inoculation. Volume of the tumor that formed in the abdominal cavity was small after CRM197 treatment, while tumor volume was estimated as 4.2 ± 1.2 g (mean ± standard deviation) in control mice.

To further examine the combined antitumor effects of CRM197 and Taxol in xenografted mice, SKOV3 and OVMG1 cells were subcutaneously injected into nude mice, and tumor sizes were measured weekly at the injection site. When CRM197 was administered alone, tumor growth in mice was suppressed in a dose-dependent manner in both cell types (partial suppression required 5 mg kg^{-1} CRM197; complete suppression required 50 mg kg^{-1} CRM197). Taxol (10 mg kg^{-1}) alone did not significantly inhibit tumor formation in SKOV3 cells or OVMG1 cells. However, co-administration of 10 mg kg^{-1} Taxol and 5 mg kg^{-1} CRM197 completely blocked tumor formation in both SKOV3 cells and OVMG1 cells. These synergistic in vivo antitumor effects occurred by combined treatment with CRM197 with Taxol. Moreover, CRM197 displayed antitumor effects in mice xenografted with human cancer cells including gastric, bladder, prostate, breast, endometrial cancer, melanoma, and glioblastoma.

In spite of its contradictory nature, Diphtheria toxin deserves careful consideration as a potential therapeutic agent in human cancer. Buzzi [4] reported that a clinical trial investigating the use of small amounts of Diphtheria toxin for patients with advanced human cancer revealed that 24 of 50 patients indicated partial or complete response after this treatment. In his study, Buzzi revealed that a small amount of Diphtheria toxin was not dangerous, and resulted in no early antibody response according to the phenomenon referred to as low-dose tolerance, although reversible side effects were detected in 17 patients. Since it remained unclear whether antitumor activity of Diphtheria toxin depended on toxicity of the molecule only or on its strong inflammatory immunological properties, another clinical trial for patients with advanced human cancer was further performed by Buzzi et al., using CRM197 [5]. Twenty-five outpatients with advanced cancer, who were refractory to standard therapies or had refused conventional therapies, were treated with CRM197 injected subcutaneously in the abdominal wall. Two, one, and six patients indicated complete response, partial response, and stable disease, respectively. Toxicities were minimal since only one patient developed irritating skin reactions at the injection sites and a flu-like syndrome with fever. Taken together, these results suggested that CRM197-mediated inhibition of HB-EGF contributed to the loss of in vitro and in vivo tumor formation in cancer, and HB-EGF was a potential target for cancer therapy. We are developing clinical trials for CRM197 for ovarian cancer patients. The use of CRM197 will allow us to improve clinical outcomes in patients with cancer.

8 Future Directions

Many clinical trials for cancer therapy have been performed to develop tools against EGFR. However, most of these clinical trials have not always been successful. In principle, EGFR antagonists interfere with the activation of several intracellular pathways that control cell proliferation, survival, apoptosis, invasion, and metastasis. The acquired resistance to EGFR antagonists can occur as a result of several different molecular mechanisms: autocrine/paracrine production of ligands, receptor mutations, constitutive activation of downstream pathways, and activation of alternative pathways [27]. There are two receptor mutations of EGFR in human cancers. The first one which involves EGFRvIII (variant III), which is generated from deletion of exons 2–7 of the EGFR gene, is overexpressed in glioblastoma multiforms [52]. This mutant EGFR does not bind to EGFR ligands, and is more tumorigenic than the wild-type receptor. However, EGFRvIII induces expression of HB-EGF as well as other genes, and inhibition of HB-EGF activity with neutralizing antibodies reduces cell proliferation induced by expression of EGFRvIII, suggesting that the EGFRvIII-HB-EGF-wild-type EGFR autocrine loop plays an important role in signal transduction by EGFRvIII in glioma cells [36]. The second mutation of EGFR is exemplified in lung cancer, where three different mutations, all located in exons 18–21, have been identified: missense mutation, deletion, and in-frame insertion [43, 44]. These mutant EGFRs are hyperreactive to EGFR ligands compared to wild-type EGFR, and selectively activate the Akt and STAT pathways. Mutations of K-ras, which is a signaling molecule in the EGFR downstream pathway, have been detected in most patients with pancreatic cancer, approximately 50% of patients with colon cancer, and 20–30% of patients with other cancers [10]. Ras point mutations have been implicated in the accelerated signaling of tumor growth, not only through the downstream signaling pathways of

Ras/Raf/ERK, but also through the upstream signaling pathways of EGFR mediated by increased expressions of EGFR ligands [11, 50]. In addition, increased expressions of EGFR ligands have also been reported to occur in ErbB-2-transformed human mammary epithelial cells [33]. Therefore, in some epithelial cancers harboring Ras point mutations or enhanced expression of ErbB-2, autocrine loops of EGFR/EGFR ligands may play pivotal roles in cancer progression. In breast and prostate cancer cells, acquired resistance to Genitinib or Trasuzumab has been shown to be associated with increased signaling via the insulin-like growth factor I receptor (IGF-IT) pathway [18, 22]. The crosstalk between IGF-IR and EGFR occurs via an autocrine mechanism involving matrix metalloprotease-dependent release of HB-EGF, and accounts for the majority of IGF-I-stimulated Shc phosphorylation and activation of the ERK cascade in COS-7 cells [38]. On the basis of these evidences, abundant increase in a particular EGFR ligand should contribute to resistance to EGFR targeting therapy. In the near future, development of a targeting agent against each EGFR ligand, like EGFR antagonists, should improve clinical outcomes of cancer patients.

References

1. Athale CA, Deisboeck TS. The effects of EGF-receptor density on multiscale tumor growth patterns. J Theor Biol 2006, 238:771–779
2. Azemar M, Schmidt M, Arlt F, Kennel P, Brandt B, Papadimitriou A, Groner B, Wels W. Recombinant antibody toxins specific for ErbB2 and EGF receptor inhibit the in vitro growth of human head and neck cancer cells and cause rapid tumor regression in vitro. Int J Cancer 2000, 86:269–275
3. Buter J, Giaccone G. EGFR inhibitors in lung cancer. Oncology (willston Park) 2005, 19:1707–1711
4. Buzzi S. Diphtheria toxin treatment of human advanced cancer. Cancer Res 1982, 42:2054–2058
5. Buzzi S, Rubboli D, Buzzi G, Buzzi AM, Morisi C, Pironi F. CRM197 (nontoxic diphtheria toxin): effects on advanced cancer patients. Cancer Immunol Immunother 2004, 53:1041–1048
6. Di Fiore PP, Pierce JH, Fleming TP, Hazan R, Ullrich A, King CR, Schlessinger J, Aaronson SA. Overexpression of the human EGF receptor confers an EGF-dependent transformed phenotype to NIH 3T3 cells. Cell 1987, 51:1063–1070
7. Downward J, Yarden Y, Mayer E, Scrace G, Totty N, Stockwell P, Ullich A, Schlessinger J, Waterfield MD. Close similarity of epidermal growth factor receptor and v-erb-B oncogene protein sequences. Nature 1984, 307:521–527
8. El-Obeid A, Hesselager G, Westermark B, Nister M. TGF-alpha-driven tumor growth is inhibited by an EGF receptor tyrosine kinase inhibitor. Biochem Biophys Res Commun 2002, 290:349–358
9. Fischer OM, Hart S, Gschwind A, Ullrich A. EGFR signal transactivation in cancer cells. Biochem Soc Trans 2003, 31(Pt 6):1203–1208
10. Friday BB, Adjei AA. K-ras as a target for cancer therapy. Biochem Biophys Acta (2005), 1756:127–144
11. Gangarosa LM, Sizemore N, Graves-Deal R, Oldham SM, Der CJ, Coffey RJ. A raf-independent epidermal growth factor receptor autocrine loop is necessary for ras transformation of rat intestinal epithelial cells. J Biol Chem (1997) 272:18926–18931

12. Higashiyama S, Abraham JA, Miller J, Fiddes JC, Klagsbrun MA. Heparin-binding growth factor secreted by macrophage-like cells that is related to EGF. Science 1991, 251:936–939
13. Higashiyama S, Iwamoto R, Goishi K, Raab G, Taniguchi N, Klagsbrun M, Mekada E. The membrane protein CD9/DRAP 27 potentiates the juxtacrine growth factor activity of the membrane-anchored heparin-binding EGF-like growth factor. J Cell Biol 1995, 128:929–938
14. Ito Y, Takeda T, Higashiyama S, Sakon M, Wakasa KI, Tsujimoto M, Monden M, Matsuura N. Expression of heparin binding epidermal growth factor-like growth factor in hepatocellular carcinoma: an immunohistochemical study. Oncol Rep 2001, 8(4):903–907
15. Ito Y, Higashiyama S, Takeda T, Yamamoto Y, Wakasa KI, Matsuura N. Expression of heparin-binding epidermal growth factor-like growth factor in pancreatic adenocarcinoma. Int J Pancreatol 2001, 29:47–52
16. Iwamoto R, Higashiyama S, Mitamura T, Taniguchi N, Klagsbrun M, Mekada E. Heparin-binding EGF-like growth factor, which acts as the diphtheria toxin receptor, forms a complex with membrane protein DRAP27/CD9, which up-regulates functional receptors and diphtheria toxin sensitivity. EMBO J 1994, 13:2322–2330
17. Iwamoto R, Mekada E. Heparin-binding EGF-like growth factor: a juxtacrine growth factor. Cytokine Growth Factor Rev 2000, 11(4):335–344
18. Jones HE, Goddard L, Gee JM, Hiscox S, Rubini M, Barrow D, Knowlden JM, Williams S, Wakeling AE, Nicholson RI. Insulin-like growth factor-I receptor signaling and acquired resistance to gefitinib (ZD1839); Iressa) in human breast and prostate cancer cells. Endocr Relat Cancer 2004, 11:793–814
19. Kobrin MS, Funatomi H, Friess H, Buchler MW, Stathis P, Korc M. Induction and expression of heparin-binding EGF-like growth factor in human pancreatic cancer. Biochem Biophys Res Commun 1994, 202(3):1705–1709
20. Krug U, Ganser A, Koeffler HP. Tumor suppressor genes in normal and malignant hematopoiesis. Oncogene 2002, 21:3475–3495
21. Lin J, Hutchinson L, Gaston SM, Raab G, Freeman MR. BAG-1 is a novel cytoplasmic binding partner of the membrane form of heparin-binding EGF-like growth factor: a unique role for proHB-EGF in cell survival regulation. J Biol Chem 2001, 276:30127–30132
22. Lu Y, Zi X, Zhao Y, Mascarrenhas D, Pollak M. Insulin-like growth factor-I receptor signaling and resistance to trastuzumab (Herceptin). J Natl Cancer Inst 2001, 93:1852–1857
23. Mills GB, May C, McGill M, Roifman CM, Mellers A. A putative new growth factor in ascetic fluid from ovarian cancer patients: identification, characterization, and mechanism of action. Cancer Res 1988, 48:1066–1071
24. Mitamura T, Higashiyama S, Taniguchi N, Klgasbrun M, Mekada E. Diphtheria toxin binds to the epidermal growth factor (EGF)-like domain of human heparin-binding EGF-like growth factor/diphtheria toxin receptor and inhibits specifically its mitogenic activity. J Biol Chem 1995, 270:1015–1019
25. Miyamoto S, Yagi H, Yotsumoto F, Kawarabayashi T, Mekada E. Heparin-binding epidermal growth factor-like growth factor as a novel targeting molecule for cancer therapy. Cancer Sci 2006, 97(5):341–347
26. Miyamoto S, Hirata M, Yamazaki A, Kageyama T, Hasuwa H, Mizushima H, Tanaka Y, Yagi H, Sonoda K, Kai M, Kanoh H, Nakano H, Mekada E. Heparin-binding EGF-like growth factor is a promising target for ovarian cancer therapy. Cancer Res 2004, 64:5720–5727
27. Morgillo F, Lee HY. Resistance to epidermal growth factor receptor-targeted therapy. Drug Resist Updat 2005, 8:298–310
28. Naef M, Yokoyama M, Friess H, Buchler MW, Korc M. Co-expression of heparin-binding EGF-like growth factor and related peptides in human gastric carcinoma. Int J Cancer 1996, 66(3):315–321
29. Nakamura K, Iwamoto R, Mekada E. Membrane-anchored heparin-binding EGF-like growth factor (HB-EGF) and diphtheria toxin receptor-associated protein (DRAP27)/CD9 form a complex with integrin alpha 3 beta 1 at cell–cell contact sites. J Cell Biol 1995, 129:1691–1705

30. Nanba D, Mammoto A, Hashimoto K, Higashiyama S. Proteolytic release of the carboxy-terminal fragment of proHB-EGF causes nuclear export of PLZF. J Cell Biol 2003, 163:489–502
31. Nanba D, Higashiyama S. Dual intracellular signaling by proteolytic cleavage of membrane-anchored heparin-binding EGF-like growth factor. Cytokine Growth Factor Rev 2004, 15(1):13–19
32. Nishi E, Klagsbrun M. Heparin-binding epidermal growth factor-like growth factor (HB-EGF) is a mediator of multiple physiological and pathological pathways. Growth Factors 2004, 22(4):253–260
33. Normanno N, Selvam MP, Saeki T, Qi C, Johnson GR, Kim N, Ciardiello F, Shoyab M, Plowman GD, Todaro GJ, Salmon DS. Amphiregulin as an autocrine growth factor for c-Ha-ras- and c-erbB-2-transformed human mammary epithelial cells, Proc Natl Acad Sci USA (1994), 91:2790–2794
34. Normanno N, Bianco C, De Luca A, Maiello MR, Salomon DS. Target-based agents against ErbB receptors and their ligands: a novel approach to cancer treatment. Endocr Relat Cancer 2003, 10(1):1–21
35. Raab G, Klagsbrun M Heparin-binding EGF-like growth factor. Biochimica et Biophysica Acta 1997, 1333(3):F179–F199
36. Ramnarain DB, Park S, Lee DY, Hatnpaa KJ, Scoggin SO, Out H, Libermann TA, Raisanen JM, Ashfaq R, Wong ET, Wu J, Elliott R, Habib AA. Differential gene expression analysis reveals generation of an autocrine loop by a mutant epidermal growth factor receptor in glioma cells. Cancer Res 2006, 66:867–874
37. Riese DJ, II, Stern DF. Specificity within the EGF family/ErbB receptor family signaling network. Bioessays 1998, 20(1):41–48
38. Roudabush FL, Pierce KL, Maudsley S, Khan KD, Luttrell LM. Transactivation of the EGF receptor mediates IGF-I-stimulated shc phosphorylation and ERK1/2 activation in COS-7 cells. J Biol Chem 2000, 275:22583–22589
39. Salomon DS, Brandt R, Ciardiello F, Normanno N. Epidermal growth factor-related peptides and their receptors in human malignancies. Crit Rev Oncol Hematol 1995, 19(3):183–232
40. Schlessinger J. Common and distinct elements in cellular signaling via EGF and FGF receptors. Science 2004, 306(5701):1506–1507
41. Seals DF, Courtneidge SA. The ADAMs family of metalloproteases: multidomain proteins with multiple functions. Genes Dev 2003, 17(1):7–30
42. Sebastian S, Settleman J, Reshkin SJ, Azzariti A, Bellizzi A, Paradiso A. The complexity of targeting EGFR signaling in cancer: from expression and turnover. Biochem Biophys Acta 2006, 1766:120–139
43. Settleman J. Inhibition of mutant EGF receptors by gefitinib: targeting an achilles' heel of lung cancer. Cell Cycle 2004, 3:1496–1497
44. Sordella R, Bell DW, Haber DA, Settleman J. Gefitinib-sensitizing EGFR mutations in lung cancer activate anti-apoptotic pathways. Science 2004, 305:1163–1167
45. Sporn MB, Todaro GJ. Autocrine secretion and malignant transformation of cells. N Engl J Med 1980, 303:878–880
46. Suganuma K, Kubota T, Saikawa Y, Abe S, Otani Y, Furukawa T, Kumai K, Hasegawa H, Watanabe M, Kitajima M, Nakayama H, Okabe H. Possible chemoresistance-related genes for gastric cancer detected by cDNA microarray. Cancer Sci 2003, 94:355–359
47. Takayama S, Sato T, Krajewski S, Kochel K, Irie S, Millan JA, Reed JC. Cloning and functional analysis of BAG-1: a novel Bcl-2-binding protein with anti-cell death activity. Cell 1995, 80:279–284
48. Tanaka Y, Miyamoto S, Suzuki SO, Oki E, Yagi H, Sonoda K, Yamazaki A, Mizushima H, Maehara Y, Mekada E, Nakano H. Clinical significance of heparin-binding EGF (epidermal growth factor)-like growth factor and ADAM (a disintegrin and metalloprotease) 17 expression in human ovarian cancer. Clin Cancer Res 2005, 11:4783–4792
49. Thogersen VB, Sorensen BS, Poulsen SS, Orntoft TF, Wolf H, Nexo E. A subclass of HER1 ligands are prognostic markers for survival in bladder cancer patients. Cancer Res 2001, 61:6227–6233

50. Tzahar E, Waterman H, Chen X, Levkowitz G, Karunagaran D, Lavi S, Ratzkin BJ, Yagen Y. A hierarchial network of interreceptor interactions determines signal transduction by neu differentiation factor/neuregulin and epidermal growth factor. Mol Cell Biol 1996, 16:5276–5287
51. Wang F, Liu R, Lee SW, Sloss CM, Couget J, Cusack JC. Heparin-binding EGF-like growth factor is an early response gene to chemotherapy and contributes to chemotherapy resistance. Oncogene, in press
52. Wong AJ, Ruppert JM, Binger SH, Grzeschik CH, Humphrey PA, Bigner DS, Vogelstein B. Structural alterations of the epidermal growth factor receptor gene in human gliomas. Proc Natl Acad Sci USA 1992, 89:2965–2969
53. Yamazaki S, Iwamoto R, Saeki K, Asakura M, Takashima S, Yamazaki A, Kimura R, Mizushima H, Moribe H, Higashiyama S, Endoh M, Kaneda Y, Takagi S, Itami S, Takeda N, Yamada G, Mekada E. Mice with defects in HB-EGF ectodomain shedding show severe developmental abnormalities. J Cell Biol 2003, 163:469–475
54. Yagi H, Miyamoto S, Tanaka Y, Sonoda K, Kobayashi H, Kishikawa T, Iwamoto R, Mekada E, Nakano H. Clinical significance of heparin-binding epidermal growth factor-like growth factor in peritoneal fluid of ovarian cancer. Brit J Cancer 2005, 92:1737–1745
55. Yarden Y, Sliwkowski MX. Untangling the ErbB signalling network. Nat Rev Mol Cell Biol. 2001, 2(2):127–137

Index

A
ABCC3 gene, 30
ABCC6 gene, 30
Activator protein-1, 289
Active immunization, 245
Acute myelogenous leukemia (AML), 237
Acute myeloid leukemia, 241, 242
ADAM17 expression, 284–285
Adenocarcinomas, of lung, 19
Adnexal carcinoma, 79, 81
ADNP gene, 27
Adoptive NK-cell therapy, 241–243
Adoptive T cell therapy, 240–241
A549 (lung cancer cell lines), 72
Amphiregulin, 284, 288
Anaplastic cells, 173
Androgens, 58, 60, 123
Angiogenesis, 141
Angiogenesis inhibitor, 156
Anti-apoptotic proteins, 160
Anti-apoptotic signaling pathways, 156
Antibody-dependent cellular cytotoxicity (ADCC), 236
Antibody-directed enzyme prodrug therapy (ADEPT), 155
Anti-CD137 agonistic mAb, 267
Anti-CTLA-4 blockade, 255
Antigen-mAb complex, 237
Antigen-presenting cells, 223, 235, 263
Antineoplastic drugs, 170
Antioxidants, use in prevention of ovarian cancer, 122
Antitumor immune response, 223–224
ANXA4 gene, 28
Aotus nancymae, 222
Apaf-1 dysfunction
  mechanisms of, 202
  in ovarian cancer, 201–202
Apc^{D716} mice, 106

APC gene, 38, 43
APCs. *See* Antigen-presenting cells
Apoptosis
  and fat metabolism, 69
  pathways and defects of, 198–202, 242, 256
Apoptotic cascade, 198
Apoptotic evasion, 159–160
Arbeitsgemeinschaft Gynäkologische Onkologie (AGO) Group
  clinical trials in ovarian cancer, 135
ARF gene, 61
Argonaute 2, 70
ARHI (Ras homologue member 1) gene, 37, 39
ARLTS1 gene, 38
Array-based comparative genomic hybridization (aCGH), 73
ASK gene, 27
Aspirin, 106
ATP-binding cassette (ABC), 157
Autoantibodies, to mesothelin, 17–18
Autocrine loops, 283
Autologous bone marrow transplantation (ABMT), 154
Avian erythroblastosis virus v-erbB oncogene, 281
5-Aza-2′-deoxycytidine, 72

B
BACH1 gene, 61
BARD1 gene, 61
B cell chronic lymphocytic leukemia (CLL), 71
BCL6, 72
BCL-2 gene, 101
Bcl-2-related proteins, 200
Bead-based assays, 10–11

Bead-based flow cytometric method, 70
Beta-tubulin, 160
Bevacizumab, 141, 237
B16-F10 melanoma, 266
B7-H4 (B7x, B7S1) molecule
    as negative regulator of T cell, 263
    and regulatory T cells, 264–265
    in TAM, 263
    and tumor environmental cytokines, 263–264
    in various carcinoma, 262
B7-H1 expression, 265
    regulation of, 265–266
    and tumor immunity, 266–267
B7-H4$^+$ macrophages, 263
B7-H4$^+$ myeloid cells, 258
B7-H1/PD-1 and B7-H4 signals blocking, 267
B7-H4 serum biomarker, 6
Bladder cancer, 71
BRCA1 (breast cancer susceptibility gene 1), 53, 55, 57–59, 61, 79–80, 82, 102, 106, 108, 121
    cancer predisposition in, 94
    as classical tumor suppresser, 89–90
    epigenetic regulation of, 36–37, 41–43
    expression and menstrual cycle regulation, 90–92
    granulosa cell-specific inactivation, 92–93
    implications for ovarian epithelial tumors, 95
    inactivation on estrus cycle, 93–94
    induced ovarian oncogenesis, 89–96
    mutation carriers, 89, 93–95
BRCA2 (breast cancer susceptibility gene 2), 42–43, 53, 55, 57–59, 61, 79–80, 121
BRCA genes polymorphisms, 58–59. *See also* *BRCA1* (breast cancer susceptibility gene 1); *BRCA2* (breast cancer susceptibility gene 2)
BRCA germline mutation, and fallopian tube carcinoma, 80
Brca1$^{-/-}$ mouse embryos, 102
Breast cancer, 60, 70–73, 79, 89, 99, 191
Breastfeeding, and ovarian cancer, 56
Buthionine sulphoximine (BSO), 158

C
CA125 antigens, 236, 257
CA19-9 biomarker, 15
Caenorhabditis elegans, 103
CA125II RIA assay, 11
CA125II serum biomarker, 11

Cancer
    and miRNAs
        deregulation in, 70–71
        DNA copy number alterations in, 72–73
        mutations identified in, 72–73
        regulation by epigenetic alterations in, 71–72
    predisposition in BRCA1 mutation carriers, 94
Cancer biomarkers, 188
Cancer vaccines, 236–238
    dendritic cell-based cancer vaccines, 239–240
    peptide vaccines, 238–239
Carboplatin, 156
Carboplatin plus paclitaxel
    in clinical trials in ovarian cancer, 132, 134–136, 141
CARDs. *See* Caspase-associated recruitment domains
CA125 serum biomarker, 5–8, 10–12
    developing decision rules for, 6–7
    for ovarian carcinoma, 15–19
Caspase-9, 198
Caspase-associated recruitment domains, 200
Caspases enzyme, 198
Catalase, 121
CCNB1 gene, 26
CCNB2 gene, 27
CCNDBP1 gene, 26
CCNE1 gene, 27
CD4+ and CD8+ T lymphocytes, 235
CDC2 gene, 26
CDC20 gene, 27
CDC42 gene, 30
CDK inhibitor *p21* gene, 101
CDKN1A gene, 27
CD40 ligand (CD40L), 224
CD56 receptor, 241
Cell penetrating peptides (CPP), 236
Cell proliferation and death, 69
Cells with immunosuppressive properties, 243–244
Cellular proteolysis, 198
CENP-A gene, 27
Cervix cancer, 99, 122
Cetuximab, 237, 282
Chemoresistance, development of, 197
Chemotherapeutic agents, 153, 156
Chemotherapeutic drugs, 155
Chemotherapy regimens, in clinical trials in ovarian cancer, 133–136
Chemotherapy-response profile (CRP), 161

Index

Chromatin immunoprecipitation (ChIP) analysis, 39
Chromatin modifications, influence on genes, 38–39
Chromatin remodeling, 35
Cisplatin, 42
Cisplatin plus cyclophosphamide, in clinical trials in ovarian cancer, 132–133, 137
Cisplatin plus paclitaxel, in clinical trials in ovarian cancer, 133–135
Claudin 3 biomarker, 15
Claudin-4 (CLDN4) gene, 39
Clear cell EOC, 99–100
Clear cell ovarian cancer, 27–28, 31
Clinical drug resistance
  causes of, 153
  mechanisms for, 153
  potential targets in modulating, 157
  to wide variety of cytotoxic agents, 158
Clinical epigenetic markers
  early diagnostic markers, 42–43
  prognostic markers, 40–41
  of therapeutic responsiveness, 41–42
Clinical trials in ovarian cancer
  carboplatin plus paclitaxel, 132, 134–136, 141
  chemotherapy regimens, 133–136
  cisplatin plus cyclophosphamide, 132–133, 137
  cisplatin plus paclitaxel, 133–135
  early stage, 131–133
  by GOG USA, 131–141
  IP chemotherapy, 140–141
  IP phosphorus-32, 132
  maintenance and consolidation therapy, 139–140
  molecular-targeted therapy in, 141
  oral melphalan, 132
  second-look laparotomy (SLL), 137–139
  surgical issues in, 136–137
Colon cancer, 70, 106
Colony-forming units of endothelial progenitor cells (CFU-EP), 275
Colorectum cancer, 122
Composite marker (CM), 11
Consolidation therapy, in clinical trials in ovarian cancer, 139–140
Cortactin gene, 30
Corynebacterium diphtheriae, 210
COX2 inhibitors, 244
CpG dinucleotides. *See* CpG islands
CpG islands, 35, 40–42
  hypermethylation, 36
Cre-*loxP* technology, 105

Cross-reacting material 197 (CRM197), 290–291
CTLs. *See* Cytotoxic T lymphocytes
Cutaneous breast cancer, 191
Cyclin D1/CDK4, 103
Cyclin D-dependant kinases, 102–103
Cyclooxygenase-1 (COX-1), 106–107
Cyclooxygenase-2 (COX-2), 106–107
Cyclooxygenase (COX) protein, 62, 106
Cyclophosphamide, 132–133, 137, 244, 258
CYP17, 58
CYP3A4, 62
CYP2C9, 62
Cystadenocarcinomas, 29
Cytokeratin 8 (CK8) immunostaining, 105
Cytokines, 236
Cytosine deaminase (CD), 211
Cytotoxic agents, 154
Cytotoxic T lymphocytes, 224

D

DAB2 gene, 121
DAPK gene, 43
DC-LAMP. *See* Dendritic cell lysosome associated membrane protein
Death by checkpoint hypothesis, 102
Dendritic cell lysosome associated membrane protein, 224
Dendritic cells (DCs), 236, 263
Denileukin diftitox (Ontak), 256–257
DF3 biomarker, 15
DGCR8 protein, 70
Diphtheria toxin, 290, 291
DIRAS3 gene, 121
DLEC1 (deleted in lung and esophageal cancer 1) gene, 37, 39
DNA
  copy number, alterations in cancer, 72–73
  hypomethylation, 36
  methylation and ovarian cancer, 35–36, 38–40, 42–44, 71
  process for rectification and repairing of, 158–159
  repair genes polymorphisms, 61
DNAJ *(MCJ)* gene, 42
Drosha, 70
Drug delivery
  process for improving, 155
  tumour-cell-specific, 155
Drug detoxification mechanisms, 157
Drug efflux proteins, 157
Drug resistance. *See* Clinical drug resistance
Dysfunctional cellular signaling network, 188

E

Early Detection Research Network (EDRN) sites, 10
Early diagnostic markers, 42–43
E-cadherin, 83
ECT2 gene, 30
E2F3 gene, 26
EGF family of ligands, 281
EGFR and ErbB2 overexpressions, 281
EIF4G1 gene, 27
Embryologic Müllerian duct, 79, 85
Endometrial cancers, 60
    histological analysis, 28
Endometrioid EOC, 99–100
Endometrioid tumors, 23
Endometriosis, 60, 124
Endothelial progenitor cells (EPCs), 274
Epidermal growth factor receptor (EGFR), 281
Epigenetically regulated genes
    chromatin modifications influence, 38–39
    hypermethylated and silenced, 36–38
    hypomethylated, 39–40
    in ovarian cancer, 36–40
Epigenetic markers, of ovarian cancer, 35–44
Epithelial ovarian cancers (EOCs), 29, 37, 54, 99, 119. *See also* Ovarian cancer
    classification, 99
    etiology, 99–100
    genetics, 100–106
        mutations in p53 pathway, 100–102, 109
        mutations in RB pathway, 102–103, 109
        p53 and RB function in, 103–106, 109
    histotypes, 23
    mouse models for, 106–108
        drug design, 106–107
        imaging techniques development, 107–108
    ovulation role in, 123
    treatment
        cationic polymers, role of, 212–213
        CD/5-FC and *HSV-TK/GCV* gene, 211–212
        diphtheria toxin, role of, 210–211
        late-stage disease and, 214
        poly(β–amino ester)s, role of, 212–213
Epithelial ovarian carcinoma (EOC)
    immunosuppressive properties of, 224–225
    treatment of, 222–223
Epoxide hydrolase, 58
ErbB receptors, 281
ERCC1 genes, 159
Erlotinib, 282
Erythropoietin (EPO), 274
Estrogens, 60, 91, 123
Estrus cycle, BRCA1 inactivation consequences on, 93–94
European Organization for Research and Treatment of Cancer (EORTC), 137, 147
Exfoliation theory, 84–85
Exportin-5, 70
Extracellular matrix (ECM), 156

F

Fallopian tube carcinoma, 53, 79, 82, 89
    BRCA germline mutation and, 80
Fallopian tube inner surface epithelium (TSE), 79, 84–85
Fallopian tubes cancer. *See* Fallopian tube carcinoma
Familial adenomatous polyposis (FAP), 106
FANCD2 gene, 61
FANCF gene, 42
FAS gene, 101
Fat metabolism and apoptosis, 69
FDA-approved cytokines, 237
Female adnexal carcinoma, 79, 81
FEN1 gene, 27
FIGO at diagnosis stages, in ovarian cancer and survival, 3–4
Flp recombinase, 210
Fludarabine, 244, 258
5-Flurocytosine (5-FC), 211
Forkhead box (FOXO) transcription factors, 190
FOXP3 protein, 256, 257
Functional proteomics, 185

G

GADD45 gene, 61
Ganciclovir (GCV), 211
GATA4 gene silencing, 39
GATA6 gene silencing, 39
GDF15 gene, 27
Gefitinib, 282
Gene chip array technology, 187
Gene-directed enzyme prodrug therapy (GDEPT), 155
Genes expression, of tumor grades in ovarian cancer, 26–27
Genes polymorphisms
    BRCA genes, 58–59
    candidate gene approaches, 62–63
    DNA repair genes, 61

Index   301

inflammation pathways genes, 61–62
progesterone receptor genes, 59–61
role in susceptibility of ovarian cancer, 53–63
whole genome studies, 63
Genome hypomethylation, 36
Germ-line mutation, 73
Glioblastoma cancer, 70
Glucocorticoids (GCs), 242
Glutathione perioxidase, 121
Gluthatione-s-transferase-pi (GST), 157
GM-CSF (Sargramostim), 237
Gonadotropins, 58, 123
G protein-coupled receptor (GPCR), 283
Granulosa cells, 91–95
    BRCA1 gene inactivation in, 92–93
Green fluorescent protein (GFP), 275
Growth factors, EGF family, 281
GSTP1 gene, 38, 41, 43
Guanine, 121
Gynecologic cancer. *See* Ovarian cancer
Gynecologic Cancer Intergroup (GCIG), clinical trials in ovarian cancer, 136
Gynecologic Oncology Group (GOG), USA
    clinical trials in ovarian cancer, 131–141
        carboplatin plus paclitaxel, 132, 134–136, 141
        chemotherapy regimens, 133–136
        cisplatin plus cyclophosphamide, 132–133, 137
        cisplatin plus paclitaxel, 133–135
        early stage, 131–133
        IP chemotherapy, 140–141
        IP phosphorus-32, 132
        maintenance and consolidation therapy, 139–140
        molecular-targeted therapy in, 141
        oral melphalan, 132
        second-look laparotomy (SLL), 137–139
        surgical issues in, 136–137

H
Haploidentical hematopoietic stem cell (HSC), 241
HATs. *See* Histone acetyltransferases
HDACs. *See* Histone deacetylases
HDM2 gene, 101, 103
HE4. *See* Human epididymis protein 4
Heat shock proteins (HSPs), 225
Hematopoiesis, 69
Hematopoietic stem cells (HSCs), 274

Hens
    intraperitoneal carcinomas in, 120
    serous tubal carcinomas in, 82–83
Heparan-sulfate proteloglycan (HSPG), 283
Heparin-binding EGF-like growth factor (HB-EGF)
    expression and chemoresistance in ovarian cancer, 288–289
    in human ovarian cancer, 283–286
    in other human cancers, 289
    peritoneal dissemination in ovarian cancer, 286–288
    physiological role, 283
HER-2/neu expression, 236, 244
Herpes simplex virus type 1 thymidine kinase, 211
Herpesvirus (HSV), 221
HE4 serum biomarker, 6–7, 10–11, 15
    for other tumors detection by assaying, 19
    for ovarian carcinoma, 18–19
HIC1 gene, 38
Hierarchical clustering, 188
Histone acetyltransferases, 203
Histone deacetylases, 71–72, 203
Histone H3, 38–39
Histone H4, 38–39
Histone protein acetylation and methylation, 35
HK6 biomarker, 15
HK11 serum biomarker, 6
HLA class I antigens, 243
HNPCC genes, 55, 57, 61
Hodgkin disease, 242
Hodgkin's lymphoma, 190
Homologous recombination (HR) pathway genes, 159
HSV-TK. *See* Herpes simplex virus type 1 thymidine kinase
hTR gene, 38
Human cancer. *See* Cancer
Human epididymis protein 4, 6, 211, 214
Human genome, miRNAs in, 69–70
Human ovarian carcinomas, defects of *p16* and *Rb* in, 103–104
Hybridoma technology, 236
Hypermethylated genes, 36–38
Hypomethylated genes, 39–40

I
IFN-α2b (Intron A) cytokine, 237
IGFBP-3 methylation, 40
IL-6, 62
IL-10, 62
IL-β, 62

IL-2 (Proleukin) cytokine, 237
IL-1RA, 62
Immortalized ovarian surface epithelium (IOSE) cell lines, 24–25
Immunotherapeutic strategy for cancer, 245–246
Immunotherapy of cancer, antibody based, 236
Incessant ovulation hypothesis, 100
Inclusion cyst epithelium (ICE), 83–84, 119–120
Infected cell protein (ICP), 221
Inflammation pathways genes polymorphisms, 61–62
Inhibitor of apoptosis (IAP) proteins, 200
INK4 protein $p16^{INK4a}$ (p16), 103
Insulin-like growth factor binding protein 3 (IGFBP-3), 38
Interferon-alpha (IFN-α), 225
Interferon-beta (IFN-β), 225
Interferon-γ, 256–258
International HapMap Project (HapMap), 62
Intraperitoneal antineoplastic therapy, 170
Intraperitoneal (IP) chemotherapy
   clinical trails
      phase I, 146–147, 171
      phase II, 146–147, 171
      phase III, 147–149
   effect of increasing dose exposure, 154
   feasibility of employing, 150
   optimizing schedules for, 155
   theory of, 145–146
   therapy for treatment of ovarian cancer, 170–171
   toxicity and complications associated with, 149
Intraperitoneal (IP) chemotherapy, in ovarian cancer, 140–141
Intraperitoneal tumors, 172
Intratumoral interstitial fluid pressure, 156
Intra-tumoral drug penetration, 156
Invasive epithelial ovarian cancer. *See* Ovarian cancer
Invasive ovarian carcinoma, 38
Invasive tumors, 29–30
   LMP tumors and, 26
IP cisplatin chemotherapy, 147
IP phosphorus-32, in clinical trials in ovarian cancer, 132
IQGAP2 gene, 30

K
Killer cell immunoglobulin-like receptor (KIR), 241
KIR-incompatible NK cells, 241

L
LAK-based immunotherapy, 241
LAQ824 (HDAC inhibitor), 71–72
Leukemia, 70–71
Leukocytes, 119
Lipopolysacharide (LPS), 225
Liposomal doxorubicin, 159
Liver cancer, 70
Loss of heterozygosity (LOH)
   at *BRCA1* locus, 37
   in ovarian cancer, 36
Low malignant potential (LMP) tumors, 23, 29–31, 38
   and invasive tumors, 26–27
Luminex® assays, 10
Lung cancer, 70, 122
   let-7 miRNA family in, 71
Lymphokine-activated killer cells (LAK), 241
Lymphoma cancer, 70
Lysophosphatidic acid (LPA), 160, 283

M
MAb569, 15–17
Macrophage inflammatory protein-beta (MIP-b), 225
Maintenance therapy, in clinical trials in ovarian cancer, 139–140
Major histocompatibility class II (MHC-II), 224
Mammalian target of rapamycin (mTOR) kinase signaling, 160
Maspin *(SERPINB5)* gene, 39
Matuzumab, 282
Maximum tolerated dose (MTD), 154
MCJ methylation, 42
MCM4 gene, 27
MCM5 gene, 27
MCM7 gene, 27
MDM2 gene, 61
Melanoma, 73
Melphalan, 132
Menstrual cycle regulation, and *BRCA1* gene expression, 90–92
Mesothelin (MSLN) serum biomarker, 6–8, 10
   autoantibodies to, 17–18
Mesothelin protein, 211
Mesothelioma, 16
Metaplasia theory, 83
MethyLight technique, 40
MGMT gene, 38, 41
MHC-peptide antigen complexes, 235

Index 303

MicroRNAs (miRNAs)
  in cancer
    deregulation, 70–71
    diagnosis and treatment, 74
    DNA copy number alterations, 72–73
    mutations identified in, 73–74
    regulation by epigenetic alterations, 71–72
  in human cancer, 69–74
  in human genome, 69–70
Microsatellite instability (MSI), 158
MINT25 gene, 38
Mir-16-1, 72–73
Mir-22, 74
Mir-127, 72
Mir-146, 74
Mir-221, 74
Mir-15a, 72–73
MISIIR. *See* Mullerian Inhibiting Substance II Receptor
MISIIR/Tag transgenic mouse model, 215
Mismatch repair (MMR) pathway proteins, 158
Mixed epithelial neoplasms EOC, 99
MLH1 gene, 38, 42
MLH1 promoter methylation, 158
MMR genes, 158
Molecular markers, 184
Molecular-targeted therapy, in clinical trials in ovarian cancer, 141
Monoclonal antibodies (mAbs), 15–16, 236, 282
MOSEC. *See* Mouse ovarian surface epithelial cells
Mouse granulosa cells, BRCA1 gene inactivation in, 92–93
Mouse ovarian surface epithelial cells, 215–216
  isolation and culture of, 171–172
  similarities with human ovarian cancer, 173
MSLN gene, 211, 214
MUC1 biomarker, 15
Mucinous adenocarcinomas, 30
Mucinous cystadenomas, 29–30
Mucinous EOC, 99–100
Mucinous tumors, 23, 29–30
Mullerian ducts, 92
Mullerian Inhibiting Substance II Receptor, 105, 214
Mullerian tumors, 159
Multidimensional scaling (MDS), of ovarian specimens, 25
Multi-drug resistance protein 1 (MRP1), 157

Multiphoton microscopy (MPM), 108
Mutations, of miRNAs identified in cancer, 73–74
Myeloid dendritic cells (MDCs), 266
MYO18B gene, 37

N
Nanotax®, 174. *See also* Paclitaxel
  cancer progression following treatment with, 177–179
  intraperitoneal delivery, effects of, 176–177
  intravenous delivery, effects of, 175–176
  survival of mice bearing ovarian cancer, effects on, 175
Natural killer (NK) cells, 235, 241–243
NCI-H157 (lung cancer cell lines), 72
Neuregulin (NRG), 281
NF-κB activation, 289
NIH-3T3 cells, 281
NK cells, 17
Non-immunosuppressive cyclosporine analogue, 157. *See also* PSC-833
Nonsteroidal antiinflammatory drugs (NSAIDs), 106
Normal ovarian surface epithelium (NOSE) cell lines, 24–25
Nucleotide excision repair (NER) system, 158

O
Oncolytic HSV
  efficiency of, 222
  immunological effects of, 225
  strains, 221–222
OPCML gene, 37
OPN biomarker, 15
Oral contraceptives (OCs) use, and ovarian cancer, 56, 58, 90–91, 100, 119, 124
Oral melphalan, in clinical trials in ovarian cancer, 132
Ovarian cancer, 197. *See also* Sporadic ovarian cancer
  AGO Group clinical trials in, 135
  animal model systems of, 169
  anti-mesothelin antibodies in, 17–18
  antioxidants use in prevention, 122
  Apaf-1 dysfunction in, 201–202
  apoptosis intrinsic pathway defects, 198–200
  biological treatment
    antitumor immune response, 223–224

Ovarian cancer (cont.)
    EOC, immunosuppressive properties
        of, 224–225
    oncolytic therapy, 222–223
    tumor vaccination, in situ, 226–228
biomarkers
    candidates, 5–9
    detection panel, 9–13
    need for detection, 15
breastfeeding and, 56
candidate gene approaches, 62–63
classification, 23–31
clear cell tumors, 27–28, 31
clinical characteristics, 23
clinical epigenetic markers, 40–43
    early diagnostic markers, 42–43
    prognostic markers, 40–41
    of therapeutic responsiveness, 41–42
clinical trials, 131–141
defects of $p16$ and Rb in, 103–104
development of drug resistance in cells of,
    157
diagnosis stages and survival, 3–4
DNA methylation and, 35–36, 38–40,
    42–44, 71
early detection prior to symptoms, 9–13
epigenetically regulated genes, 36–40
    chromatin modifications influence,
        38–39
    hypermethylated and silenced, 36–38
    hypomethylated, 39–40
epigenetic markers of, 35–44
genes expression profiling tumor
    grades, 26–27
    histotypes, 27–31
genetics, 100–106, 109
    mutations in p53 pathway, 100–102,
        109
    mutations in RB pathway, 102–103,
        109
    $p53$ and RB function in, 103–106, 109
    polymorphisms in susceptibility of,
        53–63
genomic analysis, 23–31
GOG-USA clinical trials in, 131–141
healthy controls and, 8
HE4 biomarker for, 18–19
histological analysis, 28
IP chemotherapy, 140–141
IP drug administration in, 145
loss of heterozygosity in, 36
microsimulation screening model, 6
miRNAs expression deregulated in, 70–71
molecular-targeted therapy in, 141

mouse models for, 214–216
drug design, 106–107
imaging techniques development,
    107–108
normal ovarian control, 24–25
OCAC studies, 53–55, 58–63
    $BRCA$ genes polymorphisms, 58–59
    candidate gene approaches, 62–63
    DNA repair genes polymorphisms, 61
    inflammation pathways genes
        polymorphisms, 61–62
    progesterone receptor genes
        polymorphisms, 59–61
    whole genome studies, 63
oral contraceptives use and, 56, 58, 90–91,
    100, 119, 124
paclitaxel for treatment of, 169–170
precursor cell, 91
pregnancy and, 56, 58, 90–91, 100, 119
process for maximising drug exposure in,
    154
screening and mortality, 3–5, 13
second-look laparotomy (SLL),
    137–139
serum biomarker, 3, 5–13
SMRP marker for, 15–19
surgical issues in, 136–137
survival rates, 3–4
susceptibility polymorphisms
    clinical utility, 55–56
    epidemiology, 56
    genetic and epidemiological risk
        factors, 58
    genetic susceptibility, 56–57
theory of incessant ovulation and
    development of, 171
treatment of, 153, 155, 162
VLCs in, 275–278
women at risk, 79–81
Ovarian cancer activating factors (OCAFs),
    284
Ovarian Cancer Association Consortium
    (OCAC) studies, 53–55, 58–63
    $BRCA$ genes polymorphisms, 58–59
    candidate gene approaches, 62–63
    DNA repair genes polymorphisms, 61
    inflammation pathways genes
        polymorphisms, 61–62
    progesterone receptor genes
        polymorphisms, 59–61
    whole genome studies, 63
Ovarian Cancer Specialized Program of
    Research Excellence (SPORE)
    sites, 10

Ovarian carcinogenesis. *See also* Ovarian cancer
   animal models, 82–83
   cell of origin, 83–84
   ovulation correlations, 119–122
      epidemiological evidence, 120–121
   sporadic ovarian cancer, 81–82
   women at risk, 79–81
Ovarian carcinoma. *See* Ovarian cancer
Ovarian epithelial tumors, 92
   cell of origin of, 95
Ovarian oncogenesis, BRCA1-induced, 89–96
Ovarian specimens, MDS of, 25
Ovarian surface epithelial cells, 171
Ovarian surface epithelium (OSE), 79, 81, 83, 85, 99–100, 102, 105, 107, 109, 121–123
   brushings, 24–25, 29–30, 37
Ovarian tumorigenesis, and BRCA1 gene, 91
Oviduct epithelium, 79, 85
Ovulation-cancer correlations, epidemiological evidence for, 120–121
Ovulatory genotoxicity, carcinogenic implication of, 121–122

P

Paclitaxel, 42, 132, 134–136, 141. *See also* Nanotax®
   formulation of nanoparticulate, 174–175
   side effects of, 170
   for treatment of cancers, 169–170
Panitumumab, 282
Papillary serous tumors, 23
Papillary thyroid carcinoma (PTC), 74
Paracrine-autocrine modulators, 123
Parametric empirical bayes (PEB) decision rule, 6
Pasha, 70
p16 (CDKN2A) gene, 37, 43
PCNA gene, 27
PDC4 gene, 26
Peripheral blood mononuclear cells (PBMC), 240
Peripheral blood stem cell transplantation (PBSCT), 154
Peritoneal cancer, 53
Permeability glycoprotein (PGP), 157
p14 gene, 37, 43
p73 gene, 38
p53 gene mutation, 58, 160
   frequency in histological subtypes of EOC, 101
   and ovarian cancer, 100–102, 109

Pharmacokinetic treatment, 153
4-Phenylbutyric acid, 72
Phosphatidylinositol-dependent kinase 1 (PDK1), 189
Phosphatidylinositol 3-kinase (PI3K) pathway, 189
Phosphoinositide-3-kinase (PTEN/PI3K) pathway, 160
PIG gene, 61
Pituitary adenomas, 72
Placental growth factor (PlGF), 274
Platinum-refractory ovarian cancer, 156, 158
Platinum-resistant disease, 156
$p53^{loxP/loxP}$ mice, 105
$p53^{loxP/loxP}$ $Rb^{loxP/loxP}$ mice, 105
PML gene, 27
PMS2 gene, 61
Poly(ADP-ribose) polymerase (PARP), 159
Poly(β–amino ester)s, 212–213. *See also* Epithelial ovarian cancers (EOCs)
Polycystic ovarian syndrome, 124
Poly(ethylene imine) (PEI), 212
Poor tumor immunity mediators, 261
PPM1A gene, 27
Precipitation with compressed antisolvents (PCA), 170, 174
Pregnancy and ovarian cancer, 56, 58, 90–91, 100, 119
Primary transcripts (pri-miRNA), 69–70
Progesterone, 60, 91, 121–122
Progesterone receptor, 58
   genes polymorphisms, 59–61
Progestins, 120, 123
PROGINS, 59
Prognostic markers, 40–41
Progression-free survival (PFS), 147
Prohibitin, 58
Prolactin serum biomarker, 6
Promyelocytic leukemia zinc finger (PLZF), 283
Prophylactic oophorectomy, 92
Prophylactic salpingoophorectomy, 53
Prostate cancer, 70, 122
Prostate, Lung, Colon, and Ovary (PLCO) trials, 9–13
Proteosomes, 141
PSC-833, 157
PTGS1, 62
PTTG gene, 27

R

RASSF1A tumor suppressor gene, 37–38, 43
$Rb^{loxP/loxP}$ mice, 105

Real-time RT-PCR-based TaqMan miRNA assay, 70
Regulatory T cells (Tregs), 255, 256, 258
Relative light units (RLU), 216
Retinoblastoma 1 (RB) gene mutation
  loss of heterozygosity of, 103
  and ovarian cancer, 102–103, 109
Reverse phase protein lysate array (RPPA), 186
  utility of, 188–189
RFC4 gene, 27
Ribonucleotide reductase (RR), 221
Risk of ovarian cancer algorithm (ROCA) method, 6
RNA-induced silencing complex (RISC), 70
miRNA microarray, 70
RNA-primed array-based Klenow enzyme (RAKE) assay, 70
RNase III endonuclease Dicer, 70
miRNA serial analysis of gene expression (miRAGE), 70

S

Satellite 2 (Sat2) DNA hypermethylation, 39–40
SCID. *See* Severe combined immune deficiency
Secondary Müllerian system, 79
Second-look laparotomy (SLL), in clinical trials in ovarian cancer, 137–139
Serous adenocarcinomas, 102
Serous carcinomas, *p53* gene mutation frequency in, 101–102
Serous EOC, 99
Serous ovarian adenocarcinomas, 79
Serous tubal carcinomas, 82–84
  exfoliation theory, 84–85
  metaplasia theory, 83
Sertoli cells, 92
Serum, SMRP marker for diagnostic assays of, 15–17
Severe combined immune deficiency, 222
SFRP1 gene, 37
Silenced genes, 36–38
Single nucleotides (SNPs), 57, 59, 62–63
SKBr3 (human breast cancer cell line), 71
SKOV3 cells, 285, 287–289
SOCS1 gene, 38
SOCS2 gene, 38
Soluble mesothelin-related proteins (SMRP) marker
  for diagnostic assays of serum and urine, 15–19
  for other tumors detection by assaying, 19

SPINT2 serum biomarker, 6
Sporadic ovarian cancer, 37, 81–82.
  See also Ovarian cancer
Squamous cell EOC, 99
Stem cell division, 69
STK6 gene, 27
STMN1 gene, 27
Stroma-derived factor 1 (SDF-1), 274
Stromal fibroblasts, 156
Suicide gene therapy, 210–212
Sulindac, 106
Supercritical fluid (SCF), 174
Superoxide dismutase, 121
Surgical issues, in clinical trials in ovarian cancer, 136–137
SV40 large T antigen (SV40 Tag) technique, 24, 105
Syngeneic mouse model, of ovarian cancer, 169, 171
Synuclein-γ *(SNCG)* gene, 39
Systemic drug delivery, 156

T

TAA-specific immunity, 255, 261, 263
Taxane-refractory ovarian cancer, 161
Taxane-resistant cells, 157
Taxol®, 175
T cell apoptosis, 267
T cell receptor zeta chain (TCR-ξ), 224
TCF2 gene, 38
Telomerase immortalization technique (TIOSE), 24–25
Temozolomide, 244
Therapeutic responsiveness markers, 41–42
Thrombocytopenia, carboplatin-associated, 155
T24 (human bladder cancer cell line), 71
Thymidine kinase gene *(TK)*, 221
Thyroid cancer, 70
TMS1 (target of methylation-induced silencing) gene, 38
TNF-α, 62
α–Tocopherol, 122
D-α–Tocopherol, 122
TP53 gene, 121
TRAIL receptor *DR4*, 37
Transforming growth factor beta (TGF-β), 224
Transitional cell EOC, 99
Transvaginal sonography (TVS), 5–6, 10, 12
Trastuzumab therapy, 236, 245
Treatment of cancer, new therapeutic modalities, 246

Tregs cells
  depletion, 267
  in malignancy, 244
Trichostatin A (TSA), 202
Tumor-associated antigens (TAA), 235, 255, 261
Tumor-associated macrophages (TAMs), 263
Tumor cell
  alterations of drug-specific targets, 160–161
  cytotoxicity of, 159
  heterogeneous, 156
Tumorigenesis, histone deacetylation in, 203–204
Tumor infiltrating lymphocytes (TIL), 240
Tumors
  angiogenesis and vasculogenesis, 273–274
  biopsy, 186
  genes expression profiling
    grades, 26–27
    histotypes, 27–31
  low malignant potential, 23, 26–27
  nodules, 146
  suppressor gene, 89–90
  vaccination for, 226–228
Tumor-specific cytotoxicity, 256
Tyrosine kinase inhibitors, 282

U
UBE2D1 gene, 27
UGT1A1 gene, 28

UL39 gene, 221
Urine, SMRP marker for diagnostic assays of, 15–17

V
Vascular-endothelial growth factor (VEGF), 156, 173, 224, 273
Vascular leukocytes (VLCs), in ovarian cancer, 275–278
VEGF biomarker, 15
Vitamin D receptor pathways, 60
Vitamin E, 122

W
WFDC2 (HE4) gene, 18. *See also* HE4 serum biomarker
Whole ovary (WO) samples, 24–25
Women's Health Initiative (WHI), 11–12
WT1 (Wilms tumor suppressor 1) gene, 38

X
XPD gene, 61
XRCC1 gene, 61

Z
70-kDa Zeta-associated protein (ZAP-70), 71